ORIGIN AND EARLY EVOLUTION OF ANGIOSPERMS

Edited by **CHARLES B. BECK**

ORIGIN AND EARLY EVOLUTION OF ANGIOSPERMS

COLUMBIA UNIVERSITY PRESS
New York and London
1976

Library of Congress Cataloging in Publication Data
Main entry under title:

Origin and early evolution of angiosperms.

"Most of the papers . . . presented . . . at the First International Congress of Systematic and Evolutionary Biology in Boulder."
Includes bibliographies and index.
1. Angiosperms, Fossil. 2. Angiosperms. 3. Plants —Evolution. I. Beck, Charles B. II. International Congress of Systematic and Evolutionary Biology, 1st, Boulder, Colo., 1973.
QE980.074 561'.2 75-15939
ISBN 0-231-03857-7

Printed in the United States of America

Preface

THE PROBLEM of the origin of angiosperms is a long way from solution, but progress in recent years has been exciting and substantial. This progress has been the result of the collective efforts of the best minds utilizing new approaches and methodologies, considering problems in new ways, and rising above the restrictive prejudices and sometimes false assumptions of the past.

Most of the papers that comprise this book were presented initially in a symposium at the First International Congress of Systematic and Evolutionary Biology in Boulder, Colorado during August, 1973. Included among the contributors are many of the most distinguished scientists whose interests include the origin and early evolution of angiosperms. Their opinions, new hypotheses, and research results as recorded herein provide a view of the diversity of approaches, the controversy, and many of the major ideas in the field today. More importantly, there emerges some unanimity of opinion on several major issues and one feels strongly, not only for this reason but also because of the critical, analytical approaches of many of the workers, that important new ideas and information of the type presented in these papers will be forthcoming at an increasingly rapid rate in the future.

I wish to acknowledge with appreciation the fine cooperation of the contributors, the services provided by the office staff of the Department of Botany, University of Michigan, the assistance of Judith Sayenga in preparing the index, and especially the excellent editorial work of Maria Caliandro of Columbia University Press.

Ann Arbor
June 1974

CHARLES B. BECK

Contents

ORIGIN AND EARLY EVOLUTION OF ANGIOSPERMS

Origin and Early Evolution of Angiosperms: A Perspective

CHARLES B. BECK, ***Department of Botany University of Michigan, Ann Arbor***

IN A SPEECH in 1960 to the British Association on the origin of angiosperms, Tom Harris set the stage by the following rather pessimistic statement: "I ask you to look back, not on a proud record of the success of famous men, but on an unbroken record of failure." Indeed, the mystery of the origin and early evolution of the angiosperms is as pervasive and as fascinating today as it was when Darwin emphasized the problem in 1879. Progress has been made, however, and some of the most significant accomplishments have come since 1960. Still we have no definitive answers because we are forced to base our conclusions largely on circumstantial evidence and they must usually of necessity be highly speculative and interpretive.

In considering the origin of a major group of organisms, time, place, and ancestral source are all important. The last is, of course, ultimately the most important, but a knowledge of time and place of origin will be very helpful in discovering the source. Indeed, it may well be that a general misconception about the time of origin of the angiosperms is the primary reason that progress toward the solution of their origin has been so slow. The predominant older view that angiosperms originated in the late Paleozoic implied that major diversification of the group took place prior to the Cretaceous. Consequently, many workers of the recent past have forced Cretaceous angiosperm fossils into modern families with the result, in the view of James Doyle (personal communication, 1973), that "progress in the field has been set back at least 20 years."

In my view, progress toward a better understanding of the time of origin of angiosperms represents perhaps the most significant contribution in recent years toward a solution of the larger problem of angiosperm origin.

The older predominant view of a possible late Paleozoic origin of angiosperms (Axelrod, 1952; Camp, 1947; Němejc, 1956; Thomas, 1957; Eames, 1959, 1961) seems to have largely given way to acceptance of at least a Mesozoic origin (Scott, Barghoorn, and Leopold, 1960; Takhtajan, 1969; Axelrod, 1970) or, most significantly it seems to me, even a Lower Cretaceous origin (Hughes, 1961a; Doyle, 1969).

Hughes (1961a) hypothesized that angiosperms evolved during Barremian and Aptian times, and diversified and gradually occupied all vegetated areas of the earth from Albian through Cenomanian times. If the pro-angiosperms and most primitive angiosperms were woody trees, as has been widely supposed, it is reasonable to assume a slow early evolution of angiosperms; it is perhaps largely on this basis that many morphologists have postulated a late Paleozoic or early Mesozoic origin (see Axelrod, 1952). Stebbins (1965), however, has postulated that the primitive angiosperms were small, shrubby, possibly insect-pollinated plants that lived in hilly or mountainous regions. Because little of their pollen would have been shed and because, as Axelrod (1952, 1961) has emphasized, plants living in such regions would be unlikely to be preserved as fossils, their absence in the geologic column would be understandable. Stebbins further believes that herbaceous angiosperms evolved very early; in fact, soon after the origin of the group. As Muller (1970) points out, however, this view is not supported by current palynological evidence. If Stebbins's analysis is correct, it is quite conceivable that the early evolution and initial major diversification of angiosperms could have taken place primarily in the Cretaceous. This view is strongly supported by solid evidence, presented by Doyle (1969) and Muller (1970), of progressive specialization in pollen-grain morphology between Lower and Upper Cretaceous. It is supported also by Wolfe's determination (1972) that leaves of Lower Cretaceous age from the Patuxent Formation of Virginia, compared by earlier workers to *Celastrus, Eucalyptus, Ficus, Protea, Quercus, Salix,* and *Sapindus,* are characterized predominantly by magnolialean characters. This is consistent with palynological evidence, presented by Muller (1970), who concludes that the evolution of Magnoliales probably preceded that of other groups of angiosperms in the Cretaceous.

It seems possible, therefore, that we may not have to look beyond the Cretaceous to find very primitive angiosperms or possibly even the immediate precursors of angiosperms. Indeed, past failure to look for them in the Cretaceous may well be a primary reason that the origin of angiosperms is still a mystery. While finding the available evidence for a possible Cretaceous origin compelling, I think it only prudent to point out that others, particularly Axelrod (1970), support their conclusions of an earlier origin with some reasonable and persuasive arguments. Furthermore, amino-acid sequence differences in cytochrome *c* have been interpreted by Ramshaw et al.

(1972) to suggest the origin of angiosperms between 400 and 500 million years ago, that is, in the Silurian or Ordovician. Their data suggest to them that even moderately highly specialized groups such as *Sesame, Abutilon,* mungbean, and cauliflower originated "well in excess of 200×10^6 years ago, i.e., well before the beginning of the Cretaceous." Such interpretations, while interesting, are difficult at present to accept because they are at variance with the great volume of morphological and paleobotanical evidence.

Another important question is the place of origin of the angiosperms. It is widely believed that the angiosperms originated in the tropics, a view supported by the distribution of angiosperm families as determined from both the extant and fossil records. Because of the predominance of primitive ranalean forms in the southwestern Pacific and southeastern Asia, Bailey (1949), Axelrod (1952, 1960), Takhtajan (1969), and others have suggested that this was probably the area of origin and early diversification of angiosperms.

Interestingly, the pollen studies of Muller (1970) and Doyle (1969) show no significant differences in the trends of morphological specialization, or in the time of appearance or relative proportion of types, between sediments of present tropical and temperate zones.

We must remember, of course, that the equatorial zone may well have shifted during geological time, and it is not certain that the present tropics represent exactly the location of the Cretaceous tropics. Furthermore, the climate over all of the earth during Cretaceous time was apparently less varied than it is today. Consequently, the palynological studies of Doyle and Muller, cited above, do not necessarily negate the possibility, indeed the probability, of a tropical origin of angiosperms.

Doyle (1969) effectively summarizes the situation when he says that "the pollen record provides no conclusive evidence on the hypothesis that the angiosperms appeared and diversified first in the tropics." If the angiosperms did originate somewhere in or near the Cretaceous equatorial zone, the problem of determining the actual site or sites of origin may be complicated by past continental movements and, further, by the possibility of a polyphyletic origin of angiosperms.

Most morphologists consider angiosperms to be monophyletic on the basis that all are characterized by a large set of unique or highly distinctive features. The very strong case for monophylesis has been summarized recently by Takhtajan (1969). Several morphologists nevertheless have argued for a polyphyletic origin. Indeed, this was a fairly popular viewpoint about 30 years ago when it was advocated by Hagerup, Emberger, Fagerlind, and Lam, to list only the more prominent exponents. (For a summary of these workers' viewpoints, see Just [1948].)

A polyphyletic origin of angiosperms is being advocated currently by Meeuse (1961, 1967), who believes that angiosperms originated from at least four different ancestral groups. For example, he has suggested that monocotyledons evolved from the Pentoxylales by way of *Pandanus*. He believes that monocots with secondary growth, such as *Dracaena, Cordyline,* and *Nolina,* exhibit a primitive type of stem structure similar to that of *Pentoxylon,* and from which that of *Pandanus* and other monocots has been derived. In contrast, Stebbins (1965) has argued forcefully that secondary cambial activity in monocots such as *Dracaena* is a derived rather than a primitive condition, a viewpoint expressed earlier by Bailey and Sinnott (1914).

We come now to perhaps the most difficult and certainly the most important problem: the ancestral source of the angiosperms. An important source of difficulty in solving this problem is the tendency of some workers to dismiss as unimportant, or as only gymnospermous, fossils that are not indisputably of angiosperm origin. Whatever the source or sources of the angiosperms, be they gymnosperms (which seems most probable) or pteridophytes, it must be clear to all that the evolution of this major group was a gradual process. Consequently, it is only logical to expect that the pro-angiosperms should exhibit a range of characteristics intermediate between those of their precursors and typical angiosperms. It is therefore highly counterproductive to conclude that every Lower Cretaceous or pre-Cretaceous fossil exhibiting some characteristics shared by gymnosperms and angiosperms is nonangiospermous. Indeed, these are the fossils that should be studied most carefully. This view gains weight from the fact that "nonangiospermous" features such as incompletely enclosed ovules, vesselless wood, dichotomous venation, and monosulcate pollen are characteristic of some living angiosperms.

Whether a particular fossil might be called a primitive angiosperm depends, of course, on how far one is willing to stretch the concept of angiospermy. *Genomosperma* (Long, 1960) is widely accepted as a very primitive gymnosperm ovule, but a strong argument could be made for calling it a specialized pteridophyte fructification. Such a negative approach would, however, do nothing to further our understanding of the origin and evolution of gymnosperms. Surely we must take a positive approach to fossils such as *Furcula* (Harris, 1932), *Sanmiguelia* (Brown, 1956), and others including, importantly, angiospermlike pollen.

Takhtajan (1969) suggests that pollen of the 'earliest angiosperms' was probably monosulcate and 'of the magnoleaceous or cycadaceous type,' a view supported by the studies of Doyle (1969), Muller (1970), and Pacltová (1971). Pacltová emphasizes that identification of very primitive angiosperms by palynological methods is difficult "because the primitive monocolpate pollens of angiosperms are difficult to distinguish safely from those of gymnosperms [such as cycadeoids, cycads, *Welwitschia,* and *Ginkgo*]." But she suggests further that it is the critical study of pollen of this very type, especially that found *in situ,* that may ultimately provide evidence of the earliest angiosperms. It is clear that study of pollen *in situ* may also be required to prove the gymnosperm origin of such pollen.

Eucommiidites (Erdtman, 1948), extending from Lower Jurassic to Barremian, was earlier thought to be angiospermlike and possibly the pollen of a primitive angiosperm. But both Hughes (1961b), who reinterpreted its morphology, and Brenner (1967) have found it in the micropylar canals of Lower Cretaceous gymnosperm seeds, and very recently Van Konijnenburg–Van Cittert (1971) has discovered it *in situ* in a cycadlike fructification resembling *Androstrobus.*

Clearly, therefore, *Eucommiidites* does not represent an angiosperm. This fact in no way precludes the possibility that some of the monosulcate pollen grains of the Jurassic and early Cretaceous might, indeed, represent primitive angiosperms or pro-angiosperms.

We are all familiar with the early view that the Caytoniaceae, extending from Upper Triassic to Cretaceous and characterized by netted-veined leaves and ovules almost completely enclosed in carpellike cupules, might be angiosperm precursors. Bierhorst (1970) has recently advised that we should not overlook these and the probably closely related peltasperms and corystosperms (see Thomas, 1955) in the search for angiosperm ancestors. He notes that a folded peltasperm disk bearing a ring of ovules would be remarkably similar to a winteraceous carpel.

One thing is clear: we shall not be able to determine the ancestral affinities of the angiosperms or even to recognize the significance of certain angiospermlike fossils until we have a series of fructifications connecting the primitive angiosperms with their nonangiospermous precursors. The absence of any known series of such intermediates imposes severe restrictions on morphologists interested in the ancestral source of angiosperms and leads to speculation and interpretation of homologies and relationships on the basis of the most meager circumstantial evidence. Much of the significant discussion has logically centered around the morphological nature of the ovule and the carpel and the homologies of these organs with those of possible ancestral groups.

The angiosperm ovule is generally interpreted as possessing two integuments, the gymnosperm ovule (except that of the chlamydospermales) as possessing only one. Excluding the chlamydosperms as unlikely ancestors because of their high degree of specialization, most recent workers have compared angiosperm and pteridosperm ovules. The problem in attempts to homologize angiosperm and pteridosperm ovules has been the origin of the second integument of the angiosperm ovule.

Long (1966) and Smith (1964) have considered this problem, and each has come to a different, highly speculative conclusion. Long believes that the second integument of the angiosperm is an outgrowth from either the chalaza or first integument. Smith homologizes it with the pteridosperm cupule.

Camp and Hubbard (1963) and Meeuse (1964) "solve" the problem by interpreting the pteridosperm ovule as having two integuments, thus suggesting that pteridosperm and angiosperm ovules are more similar than has been previously thought.

Eames (1961), supported by Puri (1967), does not believe that angiosperm ovules are homologous with those of any gymnosperms and suggests their possible origin from eusporangiate ferns.

The morphological nature and origin of the carpel are equally controversial. We are all familiar with the concept of the carpel as a modified, enrolled, or conduplicate leaf. Recently, Long (1966) has presented a detailed case in support of the carpel as a modified pteridosperm cupule, a possibility suggested also by Thomas (1934) and Andrews (1963). Meeuse (1964) also considers angiosperm ovules to be cupule-borne and interprets the aril of uniovulate arillate forms as homologous with the cupule. He considers the pistils to be aril derivatives even in genera that are not, in traditional morphological terms, arillate. Meeuse believes, consequently, that "a ranalean carpel is not a leaf homologue or a foliar sporophyll . . . and if the heuristically convenient term, 'carpel' is to be retained for such a structure it must be re-defined." It is especially interesting to note that Sporne (1969), from a study of statistical correlations between the crassinucellate condition and other characters of extant dicotyledons, has concluded tentatively that "the ancestral type of ovule in angiosperms was crassinucellate with three envelopes (two integuments and an aril)," a viewpoint corresponding at least in part with those of Camp and Hubbard (1963) and Meeuse (1964) but derived in an entirely different way.

Perhaps the most unusual speculative interpretation of the angiosperm flower is that of Melville (1960, 1962, 1963); he interprets the ovary as a compound structure of leaves, from which the wall is derived, and an equal number of epiphyllous branches bearing ovules. This hypothesis is based on the presumed nature of glossopteridean fructifications, such as *Scutum*, which consist of a leaflike structure that has reticulate venation and bears upon one surface a cluster of apparently spherical structures. Plumstead (1956) assumes such fructifications to be either ovuliferous, polliniferous, or even bisporangiate, although as Walton (1963) notes, "the existence in them of ovules or pollen sacs has not been demonstrated." Walton states further that he does not believe Melville is justified in adopting Plumstead's interpretation of the *Glossopteris* fructifications.

In support of his hypothesis that the angiosperms originated from the Permian Glossopteridales, Melville emphasizes the occurrence of *Glossopteris*-like venation patterns in some angiosperms. Alvin and Chaloner (1970), however, have called attention to the presence of similar venation patterns in several other presumably distantly related groups and conclude that "we clearly cannot use venation alone as a basis for reading angiosperm ancestry back to the Glossopteridae." That patterns of organization in leaf venation may be important, however, as indicators of levels of evolutionary specialization has been strongly suggested by the studies of Hickey and Doyle (1972) and Doyle and Hickey (1972).

We do not yet know when the angiosperms originated, although we may be getting close to an answer. We do not know exactly where they originated. We are even less certain of the ancestral group from which they originated. Nevertheless, I do not entirely agree with Tom Harris's pessimistic statement of 1960. Certainly, there have been many failures—indeed, some very recent ones—but I am optimistic about future successes, and I believe we shall ultimately recognize a series of successes of the past. An exceptionally outstanding group of scientists, many of whom are contributors to this volume, are bringing their talents to bear on the problems of angiosperm origin and early evolution. They are using sophisticated techniques and approaches and accumulated information not available to earlier workers.

Part of the success that I foresee will be the result simply of tedious, detailed, accurate work in collecting, identifying, and describing new plant fossils. Part will come from a better understanding of the most primitive living angiosperms. Part will result from the accurate interpretation of evolutionary trends, and very importantly the correlation of trends in different characters. Studies of pollen (including electron microscope studies of wall structure), detailed morphological analysis of leaves and other macrofossils of presumed angiosperm affinity from the Cretaceous and Jurassic, a reinvestigation of fossils of possible primitive or pro-angiosperms, especially fructifications of Jurassic and Cretaceous pteridosperms, continuing critical analyses of paleophytogeography, and considerations of the influence on angiosperm evolution of plant-animal interactions—and finally, of course, the continuing search for new fossil localities: all of these will play important roles in our unceasing efforts to solve the problems of the origin and early evolution of angiosperms.

Many of these problems are dealt with in detail in the chapters that follow, which together provide a comprehensive view of the diversity of approaches being used, the prevalent attitudes and viewpoints, past progress, and current progress in this important area.

References

Alvin, K., and Chaloner, W. G. 1970. Parallel evolution in leaf venation: An alternative view of angiosperm origins. *Nature* 226: 662–63.

Andrews, H. N. 1963. Early seed plants. *Science* 142: 925–31.

Axelrod, D. I. 1952. A theory of angiosperm evolution. *Evolution* 6: 29–60.

Axelrod, D. I. 1960. The evolution of flowering plants. In *Evolution after Darwin*, ed. S. Tax, vol. 1, pp. 227–305. Univ. of Chicago Press, Chicago.

Axelrod, D. I. 1961. How old are the angiosperms? *Amer. J. Sci.* 259: 447–59.

Axelrod, D. I. 1970. Mesozoic paleogeography and early angiosperm history. *Bot. Rev.* 36: 277–319.

Bailey, I. W. 1949. Origin of the angiosperms: Need for a broadened outlook. *J. Arnold Arbor.* 30: 64–70.

Bailey, I. W., and Sinnott, E. W. 1914. The origin and dispersal of herbaceous angiosperms. *Ann. Bot. (London)* 28: 547–600.

Bierhorst, D. W. 1970. *Morphology of Vascular Plants.* Macmillan, New York.

Brenner, G. J. 1967. Early angiosperm pollen differentiation in the Albian to Cenomanian deposits of Delaware (U.S.A.). *Rev. Palaeobot. Palynol.* 1: 219–27.

Brown, R. W. 1956. *Palm-like Plants from the Dolores Formation (Triassic) in Southwestern Colorado.* U.S. Geological Survey Professional Papers 274H: 205–9.

Camp, W. H. 1947. Distribution patterns in modern plants and problems of ancient dispersals. *Ecol. Monogr.* 17: 159–83.

Camp, W. H., and Hubbard, M. M. 1963. On the origins of the ovule and cupule in lyginopterid pteridosperms. *Amer. J. Bot.* 50: 235–43.

Doyle, J. A. 1969. Cretaceous angiosperm pollen of the Atlantic coastal plain and its evolutionary significance. *J. Arnold Arbor.* 50: 1–35.

Doyle, J. A., and Hickey, L. J. 1972. Coordinated evolution in Potomac Group angiosperm pollen and leaves. *Amer. J. Bot.* 59: 660. (Abstr.)

Eames, A. J. 1959. The morphological basis for a Paleozoic origin of the angiosperms. *Recent Adv. Bot.* 1: 721–25.

Eames, A. J. 1961. *Morphology of the Angiosperms*. McGraw, New York.
Erdtman, G. 1948. Did dicotyledonous plants exist in early Jurassic time? *Geol. Fören. Stockholm Förh.* 70: 265–71.
Harris, T. M. 1932. The fossil flora of Scoresby Sound, Greenland. Part 2. Description of seed plants *Incertae sedis* together with a discussion of certain cycadophyte cuticles. *Medd. Grønland* 85: 1–112.
Harris, T. M. 1960. The origin of Angiosperms. *Advancem. Sci.* 17: 207–213.
Hickey, L. J., and Doyle, J. A. 1972. Fossil evidence on evolution of angiosperm leaf venation. *Amer. J. Bot.* 59: 661. (Abstr.)
Hughes, N. F. 1961a. Fossil evidence and angiosperm ancestry. *Sci. Progr.* 49: 84–102.
Hughes, N. F. 1961b. Further interpretation of *Eucommiidites* Erdtman 1948. *Palaeontology* 4: 292–99.
Just, T. 1948. Gymnosperms and the origin of angiosperms. *Bot. Gaz.* 110: 91–103.
Long, A. G. 1960. On the structure of *Calymmatotheca kidstoni* Calder (emended) and *Genomosperma latens* gen. et sp. nov. from the Calciferous Sandstone Series of Berwickshire. *Trans. Roy. Soc. Edinburgh* 64: 29–44.
Long, A. G. 1966. Some Lower Carboniferous fructifications from Berwickshire, together with a theoretical account of the evolution of ovules, cupules and carpels. *Trans. Roy. Soc. Edinburgh* 66: 345–75.
Meeuse, A. D. J. 1961. The Pentoxylales and the origin of the monocotyledons. *Proc. Koninkl. Ned. Akad. Wetenschap. ser. C*, 64: 543–59.
Meeuse, A. D. J. 1964. The bitegmic spermatophytic ovule and the cupule—a re-appraisal of the so-called pseudomonomerous ovary. *Acta Bot. Neerl.* 13: 97–112.
Meeuse, A. D. J. 1967. Again: The growth habit of the early angiosperms. *Acta Bot. Neerl.* 16: 33–41.
Melville, R. 1960. A new theory of the angiosperm flower. *Nature* 188: 14–18.
Melville, R. 1962. A new theory of the angiosperm flower I. *Kew Bull.* 16: 1–50.
Melville, R. 1963. A new theory of the angiosperm flower II. The androecium. *Kew Bull.* 17: 1–63.
Muller, J. 1970. Palynological evidence on early differentiation of angiosperms. *Biol. Rev. Cambridge Phil. Soc.* 45: 417–50.
Němejc, F. 1956. On the problem of the origin and phylogenetic development of angiosperms. *Sb. Národ. Musea Praze, ser. B*, 12: 65–143.
Pacltová, B. 1971. Palynological study of Angiospermae from the Peruc Formation (?Albian–Lower Cenomanian) of Bohemia. *Sb. Geol. Ved, Ser. P*, 13: 105–41.
Plumstead, E. 1956. Bisexual fructifications on *Glossopteris* leaves from South Africa. *Palaeontographica* 100B: 1–25.
Puri, V. 1967. The origin and evolution of angiosperms. *J. Indian Bot. Soc.* 56: 1–14.
Ramshaw, J. A. M., Richardson, D. L., Meatyard, B. T., Brown, R. H., Ri-

chardson, M., Thompson, E. W., and Boulter, D. 1972. The time of origin of the flowering plants determined by using amino acid sequence data of cytochrome c. *New Phytol.* 71: 773–79.

Scott, R. A., Barghoorn, E. S., and Leopold, E. B. 1960. How old are the angiosperms? *Amer. J. Sci.* 258A: 284–99.

Smith, D. L 1964. The evolution of the ovule. *Biol. Rev. Cambridge Phil. Soc.* 39: 137–59.

Sporne, K. R. 1969. The ovule as an indicator of evolutionary status in angiosperms. *New Phytol.* 68: 555–66.

Stebbins, G. L. 1965. The probable growth habit of the earliest flowering plants. *Ann. Missouri Bot. Garden* 52: 457–68.

Takhtajan, A. 1969. *Flowering Plants: Origin and Dispersal.* Oliver, Edinburgh.

Thomas, H. H. 1934. The nature and origin of the stigma. *New Phytol.* 33: 173–98.

Thomas, H. H. 1955. Mesozoic pteridosperms. *Phytomorphology* 15: 177–85.

Thomas, H. H. 1957. Plant morphology and the evolution of the flowering plants. *Proc. Linn. Soc. London* 168: 125–33.

Van Konijnenburg–Van Cittert, J. H. A. 1971. In situ gymnosperm pollen from the middle Jurassic of Yorkshire. *Acta Bot. Neerl.* 20: 1–96.

Walton, J. 1963. The pteridosperms. *Proc. Bot. Soc. Edinburgh* 39: 449–59.

Wolfe, J. A. 1972. Phyletic significance of lower Cretaceous dicotyledonous leaves from the Patuxent Formation, Virginia. *Amer. J. Bot.* 59: 664. (Abstr.)

Cretaceous Paleobotanic Problems

N. F. HUGHES, ***Department of Geology***
Sedgwick Museum, Cambridge, England

THE ORIGIN and early evolution of angiosperms will eventually be fully elucidated by study of the now extensive possibilities of the fossil record of paleopalynology and small-fragment fossils. Most of the theory hitherto built around this problem is irrelevant or misleading and has actually inhibited progress; this particularly applies to pre-Cretaceous records of supposed angiosperms, to the theory of upland origin and development, and to widely accepted taxonomy and classification of both Cretaceous gymnosperms and angiosperms. Time, patience, and data-handling skill will all be required in studying the land-plant and insect life of early Cretaceous rocks, in order to arrive at a solution by the classic method of continuous work on the correct set of data.

With the advent of paleopalynology, the origin and early evolution of the angiosperms has become a geological rather than a botanical problem. During decades of paleontological neglect, several relatively unsupported biological theories have filled the literature without clarifying the main issue, and many workers have concluded that the problem is insoluble for lack of evidence. Coupled with this failure, and almost certainly caused by it, is the failure to classify satisfactorily the living angiosperms. This must be by far the largest group of living organisms in which it is still necessary to specify on each occasion the classification arrangement being followed.

The immediate effect of the paleopalynologic advance is to confirm that the only likely solution to the problem will be through examination of all available fossils from early Cretaceous (Berriasian to Cenomanian) rocks. The quantity of pollen presents a data-handling challenge, which is extended by the inclusion of small-fragment fossils and by the neutral reexamination of all the megafossils that were optimistically but unhelpfully named in the last century. Petrifaction fossils are, so far, conspicuously lacking in the Cretaceous; although there must be many unsearched successions in the world, particularly those including pyroclastics, which are more likely to yield such fossils. Improvements to the stratigraphic reference scale and to methods of correlation in time are essential to proper handling of the nonmarine rock sequences involved. Many investigations sink into speculation because of poor stratigraphic control, whose importance is not always appreciated by paleobotanists. A new approach to expression of the significance of observations is needed; too much time has been wasted on discussion of observations which the original author should have been obliged to repeat or upgrade before publication (Hughes, 1971).

Angiosperm Evolution in Theory and Practice

It is now more than a decade since two papers (Scott, Barghoorn, and Leopold, 1960; Hughes, 1961a) arising out of the 1959 International Botanical Conference (Montreal) suggested in slightly different ways that none of the published records genuinely indicated the existence of pre-Cretaceous angiosperms. One apparent result of this deliberate challenge has been that remarkably few claims of new discoveries have been published since 1960. None of the older records has been successfully rehabilitated in this period, and almost none of the new records stands free of serious doubt. The contention that angiosperms were not present anywhere in the world before Barremian time appears, therefore, to be greatly strengthened.

This approach to greater certainty appears, however, to have stimulated some evolutionists to even more elaborate attempts to explain the failings which they believe to be inherent in the fossil record. The main product of these attempts is a theory of "early angiosperm existence undetected in upland areas." Such a theory could be considered a harmless diversion to be appraised in the future, but unfortunately the field of actual (post-Albian) angiosperm history is also still dominated by theorists; the joint effect throughout has been to devalue the study of the fossils, which can still be entirely dismissed from consideration in an otherwise serious work (e.g., Davis and Heywood, 1966).

The current need is for unrelenting study of the fossil record of Cretaceous and Cenozoic time to demonstrate the nature of the evolution which is generally believed to have occurred. The proper time to explain discrepancies in this evolution is still some distance away.

Pre-Barremian Angiospermlike Fossils

Of the fossils discovered before 1960 and previously discussed (Hughes, 1961a), several cases have been further investigated in which numerous specimens were available and repeated observations were possible.

The leaf material of the late Triassic *Sanmiguelia* has now been examined by many paleobotanists. Doyle (1973) considers that it fails to qualify as a palm by several critical details of venation. It clearly remains an interesting fossil, of which better-preserved material with cuticle is urgently needed, but it is unlikely to be angiospermous.

The leaf *Furcula* from the late Triassic rocks of Greenland and the U.S.S.R. now appears most likely to belong to the fairly large number of Mesozoic gymnosperms known loosely as pteridosperms. No other organs have yet been attributed to the plant.

Eucommiidites has become a much better known fossil pollen grain and has now been found in three separate occurrences in the pollen chamber at the base of a long micropylar tube (Hughes, 1961b; Brenner, 1967; Reymanovna, 1968). It must therefore be attributed to a gymnospermous plant despite its virtually "tricolpate" appearance in end view of the grain; this statement does not prejudge the as yet unexplained detailed functions of this pollen.

Although a few difficult cases such as the seed *Carpolithus* from the Valanginian (Chandler, 1958) remain undiscussable, they are only those in which solitary specimens are involved. In all the cases of numerous specimens, investigation has more certainly confirmed their lack of angiosperm characters.

Of the fossils described for the first time since 1960, the palms cited from the Jurassic of Utah (Tidwell, Rushforth, and Simper, 1970; Tidwell et al., 1970) have been shown to come from a Tertiary geological mélange (Scott et al., 1972); palms are of course already known from the early Tertiary.

A number of strange new fruits and seeds from the late Jurassic Morrison Formation (Chandler, 1966) are not angiosperms although that seemed possible at the start of the investigation.

Among pollen grains, Schulz (1967) claimed that the Barremian to Albian *Clavatipollenites* occurred in early Jurassic strata. His article unfortunately was written before the detailed expansion and revision of the original data by Kemp (1968); Schulz's Jurassic grains can be shown to be different in several features.

Burger (1966) described *Tricolporopollenites* from the latest Jurassic strata in some Dutch boreholes. There were very few specimens, and they came from scattered samples. Such a record would not be accepted as significant in normal stratigraphic work, and the observations should therefore be repeated and amplified with many more specimens before being used as hard evidence. In the meantime much other work has been done on rocks of the same age in comparable areas without disclosing any pollen of this kind. This record of occurrence is therefore inconclusive; in the circumstances of my discussion, clear positive evidence is essential rather than a failure to disprove. It would be helpful if future cases of possible pre-Barremian angiosperm fossils could receive wide discussion while the samples are still available and the work still in progress, so that confirmatory tests could be agreed upon.

The "Upland" Theory

The "upland" theory of angiosperm origin and early development is itself secondary. It derives from a widespread attitude among biologists and even paleobiologists which may be thought of as a primary theory: that all evolution takes a long (geological) time and that therefore every major taxon must have long predated its first recorded fossil occurrence. Such a belief is still incorporated in a high proportion of the generalized range diagrams and phylogenetic dendrographs that are published; it is particularly noticeable in those relating to the Cambrian period but is just as frequently inserted in later times.

Despite great extensions in the techniques of exploration for fossils, confidence in the completeness of the record is apparently still lacking. Most paleontological specialists do not subscribe to these doubts (see Harland, 1967), which derive mainly from generalizers and biological theorists who are perhaps impelled by a desire to classify all present and past life in simple diagrams. Additionally there is sometimes distrust of evolutionary lineages on the ground that the changes observed from one fossil to a subsequent one are so complex that the necessary natural selection could not have been accomplished in the time; in fact, rapid radiation of form of organisms is a common phenomenon in the geological column.

This strong lack of confidence in paleontology has led to the upland theory, in which it is accepted that most land-plant fossils represent lowland vegetation and therefore any required but unobserved evolution can have taken place *undetected* in the uplands. The listing of geological and other points indicating the superfluous nature of this hypothesis (Hughes, 1974) does not alone convince, but there also appears to be a major, undiscussed theoretical flaw in the argument concerning the method of subsequent return of upland plants to lowland areas.

In the vegetation of the aggradational lowland in tropical areas today, all the plants can be seen to be under intense biological selection, and thus in these areas there arise the increasingly elaborate specializations such as are seen in the Orchidaceae. Although the Mesozoic vegetation was, to human appreciation, much simpler, the same kind of biologic selection pressure was likely to have existed.

In present-day higher altitudes (and latitudes) the pressure of selection subtly changes over progressively to physical environmental control by frost and other factors, with the extreme contrast in high mountains and in polar regions.

If any new kind of plant structure became selected in an upland area with a mainly physical control on selection, it is very difficult to see how its progeny would be particularly equipped to recolonize in subsequent periods the aggradational lowland near sea level, with its essentially different biological basis of selection. There appears to be no evidence even in more recent times for an upland plant in any way invading lowland habitats; such evidence as there is supports "migration" in the opposite direction (Hughes, 1973).

Although this is once again only a theoretical point, it appears to demand an answer from any keen adherent of upland theory.

Marine "Transgression"

Geological tradition and the very terms in use have for several decades encouraged the idea that marine transgressions (or upward changes of relative sea level) were spectacular and were destructive of adjacent land life, on the model of tsunamis and other great waves of the present day.

Relative rises of sea level may be of eustatic origin under control of melting land ice, but whereas this kind of process was to be expected from Gondwana glaciation in Carboniferous time, there is no such supporting evidence in Jurassic and Cretaceous time. The much more general cause of sea-level rises is a relative sinking (or epeirogenic movement) of the local or regional land surface under consideration, as a result of crustal plate movements or of general surface processes; this could be expected at any time or place. The phrase "transgression of the sea" would therefore be better changed to "local immersion of the land area."

It does not seem to be understood that such changes of relative sea level are and were continuous and gradual and subject to frequent minor reversals. At the present time, as is well appreciated by coastal defense engineers, such changes are an important if seldom publicized economic factor.

In any area the general effect of a relative sea-level rise is (owing to an altered base level of the rivers) a progressive inland, or upstream, shift of the various sectors of sedimentation and the ecologic niches dependent on them. Only where a major mountain range is close to a coast will the vegetation belts roughly related to the coast be unable to migrate gradually inland in response.

Neves (1958) described the striking differences in palynologic assemblages in successive beds of coal, nonmarine shale, and marine shale from the English Upper Carboniferous. Chaloner (1958) rapidly followed this up with a dramatic explanation of the high content of *Endosporites* (related to *Cordaites* trees) in the marine shale of Neves's example. It was due, he asserted, to the coal-swamp flora having been killed by inundation of the sea (marine shale), and thus the special marine-shale content represented pollen direct from the upland floras behind the destroyed coal forest and now adjacent to the sea. This picture of sudden flood has been copied and used for comparable situations in several other geological periods.

I believe it is necessary to challenge this whole reconstruction, because the marine shale could equally well receive its *Endosporites* pollen from plants which at all relevant times bordered the sea. This argument is purely concerned with palynological distribution and does not depend on acceptance of Cridland's reconstruction (1964) of *Cordaites* with stilt roots comparable to those of some living mangrove trees. Cridland's suggestion has the effect of removing *Cordaites/Endosporites* from the upland and also raises the question of the basis on which *Cordaites* was ever considered to be upland; there is no conclusive morphologic evidence of the tree-trunk base to support this idea. Furthermore, *Cordaites* has long been thought of as a pro-conifer, and Recent conifers belong to high altitudes and latitudes. Such reasoning breaks down in the intervening Mesozoic, in which conifers formed the main lowland and tropical vegetation; there was then no competition.

There is also the geological point that at least the Carboniferous coal seams form part of cyclothemic (repetitive) sequences in which above the marine shale, coal returned as the relative sea level fell again. Consideration of the whole cyclothem, and an attempt to make the corresponding diagram for regression as well, rapidly demonstrates that the only viable hypothesis for *Endosporites* distribution is derivation from sea-margin plants (figs. 1 and 2).

In the Jurassic the same argument has been used (Chaloner and Muir, 1968) for the plant *Hirmeriella* and the pollen *Classopollis*, which were taken to represent upland conifer vegetation behind the swamps. A contrary reconstruction suggesting a marine marginal habitat for comparable but later plants has been drawn for the Lower Cretaceous (Hughes and Moody-Stuart, 1967).

The matter is discussed here at some length only because this supposed evidence for upland plants has been used as support for the possibility of upland angiosperms in the Jurassic.

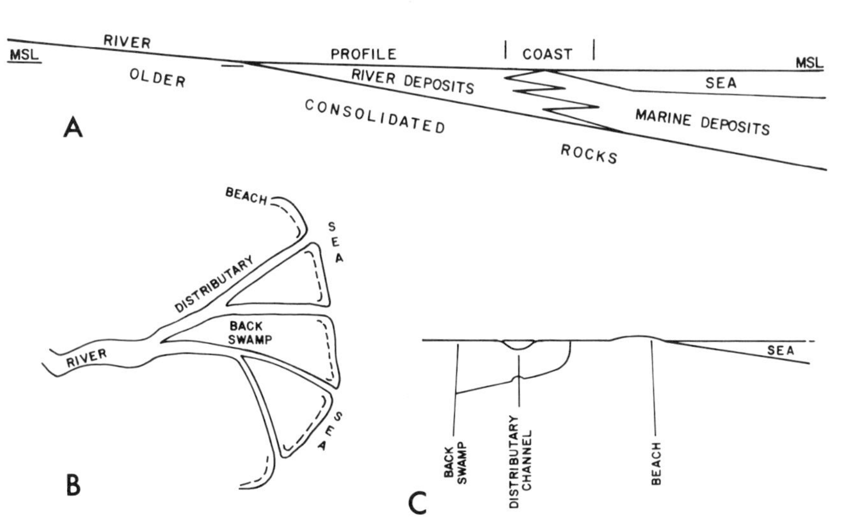

Fig. 1. Diagrams to show definition of aggradational lowland. *A,* general section of river profile and deposits; *B,* sketch maps of small area marked 'coast' in diagram *A; C,* sketch section of same. *MSL* is mean sea level.

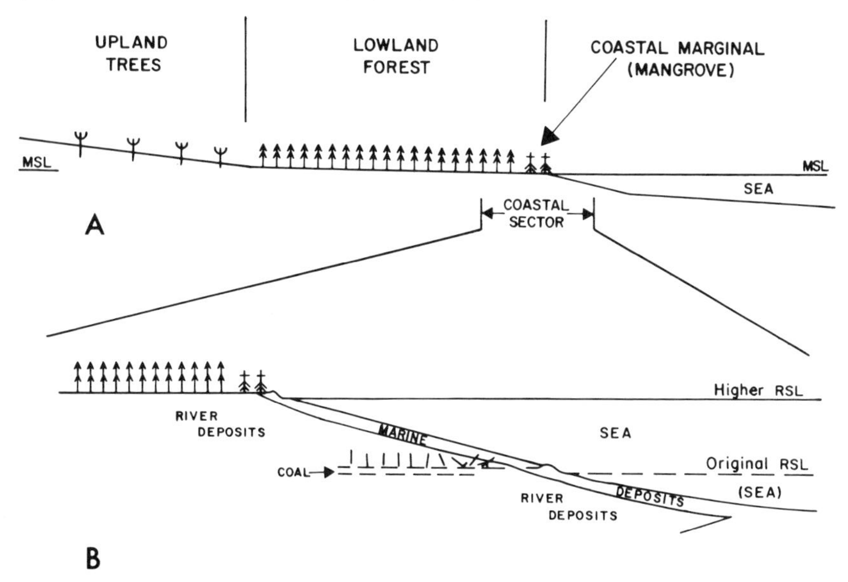

Fig. 2. Generalized diagrams of vegetation distribution. *A,* lowland, may be tens (or more) of miles wide; *B,* enlargement of coastal sector, with greatly increased vertical scale, after upward change of sea-level. RSL is relative sea level.

Paleobotanical Problems of Taxonomy

There are, in addition to the general evolutionary and geological points mentioned, some paleobotanical (or botanical) obstacles to progress.

It has been stated many times that none of the known groups of gymnosperms bears a set of characters at all suitable for consideration as a link from other gymnosperms to an angiosperm group. Working from the way in which information about gymnosperms is currently presented, the statement is difficult to refute; but because there must have been one or more progenitors among seed plants, there must also be a fallacy in the argument.

Gymnosperm groups can be considered in three categories: (a) groups with current full development, such as Pinaceae, Podocarpaceae, Cupressaceae; (b) groups erected for relatively isolated relicts, such as *Cycas, Taxodium, Ginkgo;* and (c) extinct groups, such as the Caytoniales, Benettitales, Czekanowskiales. The groups in (a) are clearly defined from many subordinate living taxa, and although their early geological history is at some points obscure they may be eliminated here because they either had not evolved or had not become sufficiently distinct from the main Jurassic conifers in early Cretaceous time to be considered. The groups in (b) are to varying degrees unsatisfactorily defined, with an extreme represented by the monotypic *Ginkgo;* in this case any group definition should be based on the maximum range of types in the widespread development of late Jurassic and early Cretaceous time, and any character based solely on *Ginkgo* should be ignored. Although it may be hard for botanists to accept, perhaps even the name 'Ginkgoales' should be abandoned as misleading. The groups in (c) are mostly based on good material from well-known mid-Jurassic or earlier floras. In most cases some early Cretaceous fossils are known, but they are usually only leaves and have contributed nothing to the taxon circumscription; for example, the state of evolution of *Weltrichia* or *Caytonia* fructifications by Valanginian time is not yet known. From this discussion it can be seen that much that has been stated about gymnosperm groups in (c), and even more about groups in (b), is irrelevant in the search for suitable *early Cretaceous* gymnosperms from which to derive earliest angiosperms.

It is also important to make clear that characterization of mid-Cretaceous angiosperms through extant families is not relevant. Probably all angiosperm fossils from the whole of the Cretaceous should be grouped in newly defined taxa based on fossil evidence *alone.* Such a suggestion usually disturbs paleoecologists, who would be deprived of their standard comparisons with living plants, but it is important for the literature to become free of extrapolations from this source even in some of the better cases such as the Platanaceae.

If neo-botanists are reluctant or unwilling to accept the disturbance implied in these paragraphs, paleobotanists can probably, in the case of the gymnosperms, make the necessary changes privately. With the angiosperms, however, the suggestions are likely to prove more fundamental in their effect on existing classifications.

Interdependence of Flora and Fauna

As seen in the context of Triassic floras, the appearance of the benettite flower *Sturianthus* was an unexpected development. Advertising the fructification was apparently the reverse of the practice of protecting the ovules from predation by insects in the Carboniferous pteridosperms (Hughes and Smart, 1967). Whether the Triassic flowers were related to insect or small-reptile activities is not known; the latter should be considered because, although the herbivorous reptiles that entered the record in the early Triassic were relatively large, gross-vegetation feeders, radiation was rapid enough to have also provided the appropriate small, active, ovule and seed eaters. If insects were involved, Coleoptera seem to have been the most likely candidates. It is not clear in this case, however, why visual rather than olfactory attraction should have been provided for a crawling animal. Olfactory attraction alone could have been provided without a perianth.

The benettite flower in the Jurassic, through *Weltrichia* and *Williamsonia* to *Cycadeoidea,* remained large and robust. The few angiosperm flowers that are known from the early Tertiary (Leppik, 1971) were already small and delicate, but by that time Lepidoptera and the aculeate Hymenoptera had also appeared. Virtually no Cretaceous fossil flowers are yet known; insect fossils are also still very rare from the Cretaceous but, judging from late Jurassic assemblages, only Coleoptera and a few early Diptera and Hymenoptera could have been available in early Cretaceous time. Birds, although present in Cretaceous time, do not appear to have been involved in this level of feeding.

The transfer or extension of function of animal visits from the initial seed dispersal to cross-pollination is of unknown timing, but it may have taken place in the mid-Jurassic benettites such as *Williamsoniella,* in which the flower was relatively small. It is also possible that the latest Jurassic *Cycadeoidea* flowers were arranged for protandry, long before angiosperm flowers had arisen.

Even in the absence of suitable Cretaceous fossils, it may be possible to interpret the progressive integration of fauna and flora from Jurassic to Paleocene time (Smart and Hughes, 1972), and thus to obtain an indication of the important Cretaceous events that led to the fundamental changes in seed plants, in pollination, gametophyte development, fertilization, and seed formation, that are regarded as original characters of angiosperms.

General Conclusion

Hitherto the search for relevant fossils has been sporadic and has been spread over five geologic periods from Carboniferous onward. If it will henceforth be concentrated on early Cretaceous rocks worldwide, there is every reason to expect success from exploration of the correct rocks to the limits of current techniques. It is perhaps surprising that this was not undertaken before; such is the powerful effect of false theory.

Fortunately there are already some younger paleobotanists working on projects which are geologically oriented for the best results. Doyle (1973) has sought out some most interesting data on early "monocotyledons," and includes further data on the palynology of the Potomac Group. Hickey and Doyle (1972) reanalyze leaf venation in the earliest angiosperm fossils in a way that is likely to change the classification radically. Krassilov (1973) concentrates on *Leptostrobus* of the Czekanowskiales (of which he has good material from the far eastern U.S.S.R.) as a possible angiosperm progenitor. The more that can be produced along these lines without the intrusion of additional theory, the better.

References

Brenner, G. 1967. The gymnospermous affinity of *Eucommiidites* Erdtman 1948. *Rev. Palaeobot. Palynol.* 5: 123–27.

Burger, D. 1966. Palynology of the uppermost Jurassic and lowermost Cretaceous strata in the eastern Netherlands. *Leidse. Geol. Mededel.* 35: 211–76.

Chaloner, W. G. 1958. The Carboniferous upland flora. *Geol. Mag.* 95: 261.

Chaloner, W. G., and Muir, M. D. 1968. Spores and floras. In *Coal and Coal-bearing Strata*, ed. D. Murchison and T. S. Westoll, pp. 127–46. Oliver, Edinburgh.

Chandler, M. E. J. 1958. Angiospermous fruits from the Lower Cretaceous of France and Lower Eocene (London Clay) of Germany. *Ann. Mag. Nat. Hist., ser. 13,* 1: 354–58.

Chandler, M. E. J. 1966. Fruiting organs from the Morrison Formation of Utah, U.S.A. *Bull. Brit. Mus. Nat. Hist.* 12: 139–71.

Cridland, A. A. 1964. *Amyelon* in American coal balls. *Palaeontology* 7: 186–209.

Davis, P. H., and Heywood, V. H. 1966. *Principles of angiosperm taxonomy.* Oliver, Edinburgh.

Doyle, J. A. 1973. Fossil evidence on early evolution of the monocotyledons. *Quart. Rev. Biol.* 48: 399–413.

Harland, W. B., ed. 1967. *The Fossil Record.* Geological Society, London.

Hickey, L. J., and Doyle, J. A. 1972. Fossil evidence on evolution of angiosperm leaf venation. *Amer. J. Bot.* 59: 661. (Abstr.)

Hughes, N. F. 1961a. Fossil evidence and angiosperm ancestry. *Sci. Progr. (London)* 49: 84–102.

Hughes, N. F. 1961b. Further interpretation of *Eucommiidites* Erdtman 1948. *Palaeontology* 4: 292–99.

Hughes, N. F. 1971. Remedy for the general data-handling failure of palaeontology. In *Data Processing in Biology and Geology* ed. J. L. Cutbill, vol. 3, pp. 321–30. Systematics Association, Academic Press, New York.

Hughes, N. F. 1973. Environment of angiosperm origins. In *Palynology of Mesophyte,* Proceedings of the 3rd International Palynological Conference (Novosibirsk), pp. 135–37. Nauka, Moscow.

Hughes, N. F. 1974. Angiosperm evolution and the superfluous upland origin hypothesis. Birbal Sahni Institute, *Paleobot. Spec. Publ.* 1: 25–29.

Hughes, N. F., and Moody-Stuart, J. C. 1967. Palynological facies and correlation in the English Wealden. *Rev. Palaeobot. Palynol.* 1: 259–68.

Hughes, N. F., and Smart, J. 1967. Plant-insect relationships in Palaeozoic and later time. In *The Fossil Record,* ed. W. B. Harland, pp. 107–17. Geological Society, London.

Kemp, E. M. 1968. Probable angiosperm pollen from British Barremian to Albian strata. *Palaeontology* 11: 421–34.

Krassilov, V. 1973. Mesozoic plants and the problem of angiosperm ancestry. *Lethaia* 6: 163–78.

Leppik, E. E. 1971. Palaeontological evidence on the morphogenic development of flower types. *Phytomorphology* 21: 164–74.

Neves, R. 1958. Upper Carboniferous plant-spore assemblages from the *Gastrioceras subcrenatum* horizon, North Staffordshire. *Geol. Mag.* 95: 1–19.

Reymanovna, M. 1968. On seeds containing *Eucommiidites troedssonii* pollen from the Jurassic of Grojec, Poland. *J. Linn. Soc., Bot.,* 61: 147–52.

Schulz, E. 1967. Sporenpaläontologische Untersuchungen rhätoliassischen Schichten im Zentralteil des Germanischen Beckens. *Paläontol. Abhandl., Abt. B,* 2: 542–633.

Scott, R. A., Barghoorn, E. S., and Leopold, E. B. 1960. How old are the angiosperms? *Amer. J. Sci.* 258A: 284–99.

Scott, R. A., Williams, P. L., Craig, L. C., Barghoorn, E. S., Hickey, L. J., and

MacGinitie, H. D. "Pre-Cretaceous" angiosperms from Utah: Evidence of Tertiary age of the palm woods and roots. *Amer. J. Bot.* 59: 886–96.

Smart, J., and Hughes, N. F. 1972. The insect and the plant: Progressive palaeoecological integration. In *Insect/Plant Relationships,* ed. H. F. Emden. *Symp. Roy. Entomol. Soc.* 6: 143–56.

Tidwell, W. D., Rushforth, S. R., and Simper, H. D. 1970. Pre-Cretaceous flowering plants: Further evidence from Utah. *Science* 170: 547–48.

Tidwell, W. D., Rushforth, S. R., Reveal, J. L., and Behunin, H. 1970. *Palmoxylon simperi* and *Palmoxylon pristina:* Two pre-Cretaceous angiosperms from Utah. *Science* 168: 835–40.

Middle Cretaceous Floral Provinces and Early Migrations of Angiosperms

GILBERT J. BRENNER, ***Department of Geological Sciences State University of New York, New Paltz***

THE TIME, LOCUS, and biological origin of flowering plants are among the major evolutionary questions remaining today. Once evolved—in which directions did angiosperms spread? Seward (1931) proposed an Arctic center of origin and dispersion. Axelrod (1952, 1959) suggested a tropical origin for the angiosperms, with subsequent poleward migrations.

Major obstacles to answering some of these questions are the scarcity and seemingly advanced nature of the first recognizable vegetative remains of angiosperms, and the problem of accurate and reliable stratigraphic dating and sampling at the intra-stage level.

During the past several years, I have examined the palynological literature and numerous palynological assemblages from Neocomian to Cenomanian beds around the world. My reasoning was that fossilized pollen, being more abundant in the stratigraphic record than are plant megafossils, could help clarify the questions of the geographic locus of the oldest angiosperm and the subsequent dispersion paths.

Monosulcate pollen grains with possible angiosperm affinities first appear in the geologic record during the Barremian to Aptian stages during the Lower Cretaceous. These grains have been called *Clavatipollenites* and are found in beds of these ages in England, Israel, Maryland (U.S.A.), and Argentina. *Clavatipollenites* has not been found in beds of the same age in Canada or the Arctic areas.

For years many palynological workers have considered the oldest tricolpate pollen to be the first definite evidence of angiosperms. The ancient angiosperms that produced this pollen were thought to have evolved rapidly during the Albian stage and spread rapidly during that age to many continents. In fact, the first appearance of tricolpate pollen is still assumed by many palynologists to mark the uppermost part (Albian) of the Lower Cretaceous. Coupled with this assumption, however, was the fact that very little was known of Lower Cretaceous palynofloras of present tropical or Arctic latitudes.

The purpose of this paper is to describe the time-latitude relationships of the earliest tricolpate pollen and the significance that floral provincialism played during the time of early migration.

During the course of this research it became apparent that Middle Cretaceous palynofloras from Barremian to Cenomanian time could be grouped into four distinct provinces with strikingly different assemblages. Climatic interpretations of these provinces agree with continental reconstructions based on plate tectonics by Dietz and Holden (1970). The provincial patterns outlined here are based on a limited number of samples in each area, but the patterns that have become apparent seem consistent with each new locality and horizon that becomes available.

Palynological Record of Early Angiosperms

Undoubted angiosperm pollen first appears in the geological record as small, smooth to simply reticulate, tricolpate grains. In the present middle latitudes of the Northern Hemisphere, tricolpates first appear within the Albian, the uppermost stage of the Lower Cretaceous. Appearances have been reported in widely separated areas such as Maryland (Brenner, 1963, 1967), western Canada (Singh, 1964, 1971; Norris, 1967), southern England (Kemp, 1968, 1970), and Kazakhstan (Bolkhovitina, 1953).

Possible angiosperm pollen that predates the first tricolpates in middle latitudes is retipilate monosulcate pollen called *Clavatipollenites* Couper, 1953. It appears for the first time in beds of Barremian age in England (Couper, 1958) and in supposed contemporaneous beds in Maryland (Brenner, 1963), and Argentina (Archangelsky and Gamerro, 1967). *Clavatipollenites* has an exine structure ranging from retipilate, with pilae free, to tectate, with the heads of the pilae fused to form a distal reticulum similar to the sexine structure found in the pollen of the Liliaceae or the more primitive genus *Ascarina* in the dicot family Chloranthaceae. Another possible pre-Albian angiosperm pollen grain is a monosulcate form, *Liliacidites peroreticulatus*, first described by (Brenner, 1963) Singh (1971) from the Patuxent and Arundel formations of Maryland. I have also found it in Barremian–Lower Aptian beds of the Zeweira Formation of the northern Negev of Israel. *Liliacidites peroreticulatus* differs from *Clavatipollenites hughesii* by its longer and more widely spaced pilae and distinct "bridgelike" structures connecting the distal capita of the pilae. Both types are considered angiospermous by Doyle (1969).

Doyle (1969) hypothesized that the tricolpate condition developed from a *Clavatipollenites* type. The monosulcate condition became trichotomosulcate, and three colpi developed by the departure of the three rays from the polar region. Although the trichotomosulcate condition is developed in some aberrant grains of *Clavatipollenites* (Hedlund and Norris, 1968; Doyle, 1969), the transitional change to the true tricolpate condition has not yet been found.

The pollen-form genus *Eucommiidites,* which is found from the Upper Triassic to Cenomanian, was considered a tricolpate grain by Erdtmann (1948). Couper (1958) demonstrated that the symmetry of the three furrows of *Eucommiidites* is dissimilar from the dicot condition. Subsequently Hughes (1961) in England and Brenner (1963, 1967) in Virginia found *Eucommiidites* in the micropylar canal of tiny gymnospermous seeds. Van Konijnenburg–Van Cittert (1971) isolated *Eucommiidites* from gymnospermous cones (possibly cycadalian).

Localities of special palynological importance (fig. 1) are listed in the Index to Localities at the end of this paper.

EARLY TRICOLPATE POLLEN FROM LOW-LATITUDE AREAS

The oldest tricolpates yet discovered in the geologic record are those that I found in the Zeweira Formation of Barremian–Lower Aptian beds of the Zohar 1 well from the northern Negev of Israel (fig. 6:10). *Clavatipollenites* was also found in the sample with these tricolpates. Tricolpates found in the Zeweira Formation (Barremian–Lower Aptian) in core sample 6 are found again higher in the well in core no. 5 in the Dragot Formation, of Upper Aptian–Lower Albian age. The beds are dated by lithocorrelation with surface sections to the south that contain ammonites, foraminifera, and lamellibranchs. To date, the tricolpates in core 6 are the oldest tricolpate pollen grains that have been reported.

Lower Albian tricolpates have been reported from the Ivory Coast by Jardiné and Magloire (1966). Müller (1966) reported the first tricolpates from Aptain beds of the Alagoas–Serigipe Basins of eastern Brazil. I have examined several samples from these Aptian horizons and have found small reticulate to scabrate tricolpates in the Alagoas and underlying Maceio formations. These formations are dated as Aptian on the basis of freshwater ostracods (Wicher, 1959; Krömmelbein, 1962; Viana, 1966). The Alagoas and Maceio formations lie unconformably just below the Albian marine beds of the Riachuelo Formation, which represents the first marine invasion subsequent to continental rifting in this area of Gondwana.

West of Brazil in the Montana region of Peru (locality 4), palynomorphs from Upper Albian to Lower Cenomanian were described by

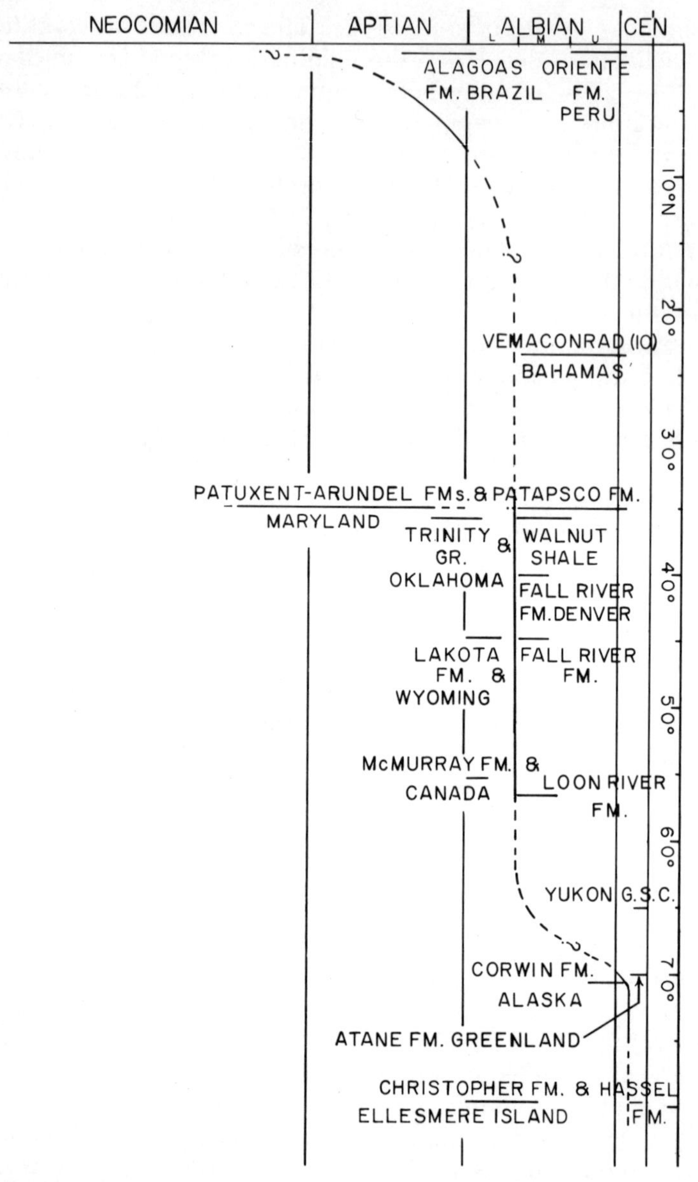

Fig. 1. Time-latitude curve for the first occurrence of tricolpate (dicot.) pollen from the equatorial region to the Arctic latitudes of the Western Hemisphere.

Brenner (1968). These sediments are presumably about the same age as the upper part of the Patapsco Formation from Maryland (Brenner, 1963). The palynomorphs are more like those described from Upper Albian assemblages from West Africa (Jardiné and Magloire, 1966) than the Patapsco material. Polyporate types similar to those in the modern genera *Thalictrum* (fig. 6:14) and *Alisma;* biporotricolpate types are found in the Peruvian material and are also described from West Africa. In North America, polyporates have not been reported earlier than the Cenomanian.

Unfortunately no Lower Cretaceous palynological studies have been made from formations south of the equator in South America other than the supposed Barremian-Aptian Baquero Formation in Patagonia (Archangelsky and Gamerro, 1967). No angiosperm pollen other than the problematical *Clavatipollenites hughesii* was found in the Baquero Formation.

Preliminary palynological investigations of Aptian-Albian horizons from the Lago San Martin plant beds did not reveal the presence of any tricolpates (Archangelsky, personal communications, 1967).

EARLY ANGIOSPERM POLLEN FROM MIDDLE LATITUDES OF THE NORTHERN HEMISPHERE

In Europe and North America there is no well-documented evidence of tricolpate pollen in beds of Lower Albian age. Kemp (1968, 1970) published a palynological study of Aptian to Upper Albian strata from southern England. The oldest tricolpate pollen was found in the Gault beds of Middle Albian age. The tricolpates were infrequent, becoming more abundant in the Upper Greensands of Upper Albian age.

At similar latitudes in the Western Hemisphere, tricolpate pollen does not appear until Middle Albian horizons. Among the best dated and palynologically best documented materials in North America are the Barremian to Cenomanian sediments in the Alberta-Saskatchewan area of western Canada, studied by Pocock (1962), Singh (1964, 1971), and Norris (1967).

The oldest tricolpates that have been found in western Canada are from Middle Albian beds of the Loon River Formation from the Peace River area (Singh, 1971). No Lower Albian beds from several sections in this area have yielded any tricolpate pollen.

In Wyoming (Davis, 1963), the Lakota Formation of Lower Albian age does not contain any tricolpates, while the Middle Albian Fall River Formation above it contains the typically small, reticulate tricolpates (locality 7). Pannella (1966) reported low frequencies of small tricolpates in the Fall River Formation of the Denver Basin (locality 6). In the overlying Upper Albian Skull Creek Formation, the diversity and frequency of tricolpates increase.

In the Oklahoma area, the Antler Sand and Walnut Shale (Hedlund and Norris, 1968) of late Middle Albian age (based on ammonites to the south) contain tricolpates very similar to those found in the middle part of the Patapsco Formation of Maryland (Brenner, 1963). Lower Albian beds of the Trinity Group below do not contain tricolpate pollen.

EARLY ANGIOSPERM POLLEN FROM HIGH-LATITUDE AREAS OF NORTH AMERICA

I have studied sediments from six Arctic areas for their possible angiosperm pollen content. Although the sample localities are rather few for such a vast area, the pattern that appears and that seems to agree with leaf studies by Smiley (1966) is that tricolpates common in Middle and Upper Albian horizons in middle latitudes to the south do not appear in the Arctic Cretaceous until Cenomanian time—a lag of several million years. The Arctic samples studied were:

1. Locality 10, upper Cenomanian. Yukon area, Geological Survey of Canada loc. 35701 (small tricolpates).
2. Locality 11, lower Upper Cretaceous. Atane Formation, Disco Island, Greenland (small tricolpates).
3. Locality 12, Upper Albian–Lower Cenomanian. Kuk River coals of the Corwin Formation (no angiosperm pollen).
4. Locality 13, Aptian–Lower Albian. Christopher Formation, Ellesmere Island (no angiosperm pollen).
5. Locality 14, Cenomanian. Hassel Formation, Ellesmere Island (small tricolpates).
6. Locality 15, upper Lower to lower Middle Albian. *Beaudanticeras* affine zone, Stang Creek, Peel Plateau, Northwest Territories. Geological Survey of Canada loc. 35257 (no angiosperm pollen).

An investigation of plant fossils in Coastal Plain deposits along the Kuk River of Alaska by Smiley (1966) indicates that the transition from gymnosperm-fern florules to angiosperm florules does not take place until Turonian time, whereas the first abundant angiosperm leaves enter the section in the late Cenomanian Chandler Formation. The sequence worked out by Smiley is supported by the palynological material from the Arctic localities listed above. The megafossils suggest conditions colder than those to the south; the palynomorphs also indicate cooler conditions than existed at the same time in western Canada and Maryland. Details of these provincial differences are discussed later in this paper.

EARLY ANGIOSPERM POLLEN FROM MIDDLE LATITUDES OF THE SOUTHERN HEMISPHERE

Unfortunately nothing has been published from Albian and Cenomanian beds from the southern latitudes of South America or Africa; therefore we must depend solely on the data published from New Zealand and Australia (Couper, 1953, 1960; Burger, 1968, 1970). Tricolpate pollen first appears in New Zealand in the Clarence Series, dated only as Albian to Cenomanian, based on marine fauna. More precise dating in Australia records the first tricolpates in the Allaru Mudstone of Middle Albian age.

SUMMARY OF MIGRATION DIRECTIONS

The time-latitude appearance of the oldest non-magnoliid angiosperm pollen, the tricolpates, can be visualized by plotting some of the data from the Western Hemisphere on a time-space latitude curve (fig. 1), modified from Axelrod (1959). The oldest tricolpates were found in Aptian beds from Brazil. These early angiosperms may not have been able to breach the barrier presented by the Caribbean Sea, a seaway that had been enlarging since rifting began in late Triassic time (Dietz and Holden, 1970). By middle Albian time, plants producing tricolpate pollen migrated into the middle latitudes of North America but did not enter the present high-latitude areas until the early part of the late Cretaceous.

Continental reconstructions by Dietz and Holden (1970) show that the areas in which the earliest tricolpates have been found in Brazil and Israel were positioned in the equatorial zone of that time and were part of the larger supercontinent of Gondwana. Sometime during the Barremian to Aptian stages the first tricolpates evolved and the plants producing this kind of pollen migrated both north and south, reaching the Australian region at about the same time as they reached the present middle latitudes of the Northern Hemisphere.

Palynofloristic Provinces During Early Angiosperm Migration

Four major floristic provinces can be discerned from the palynological assemblages mentioned in this paper; they existed with more or less modification during the extent of the Barremian-Cenomanian stages. I shall refer to these areas in such a way as to reflect the distribution of the continents prior to the more rapid drift that occurred during late Cretaceous time.

The provinces, from north to south, and their suggested climatic interpretations are:

1. *The Northern Laurasian province.* Temperate humid climate; now above 60° N latitude.
2. *The Southern Laurasian province.* Warm temperate to subtropical humid; now in the middle latitudes of the Northern Hemisphere.
3. *The Northern Gondwana province.* Tropical semiarid; now from the northern coast of South America and Africa to some undetermined latitude south of the equator.
4. *The Southern Gondwana province.* Warm temperate to subtropical humid; now in southern regions of South America and Africa, and including Australia, New Zealand, and India.

THE NORTHERN LAURASIAN PROVINCE

Aptian to Cenomanian palynofloras from several localities (localities 10–15) in the Arctic of North America are strikingly different from contemporaneous palynomorph assemblages in middle latitudes to the south.

The palynofloras of the Northern Laurasian province (figs. 2, 3) are characteristically dominated by bisaccate pollen grains of the Pinaceae and, to a lesser extent, of the Podocarpaceae.

Similar bisaccate types occur to the south in the Potomac Group of Maryland, the Fall River Formation of Wyoming and Colorado, and the Upper Blairmore of western Canada. In these areas the frequencies of the bisaccates are less than in the Arctic areas, because the gymnosperm asemblages are complemented by nonsaccate types such as *Classopollis, Eucommiidites, Araucariacites,* and *Exesipollenites.*

Classopollis, the pollen associated with the leaf genera *Brachyphyllum* and *Pagiophyllum* and the cone genus *Cheirolepis,* has not been found in the Arctic samples. This pollen-form genus is a commonly occurring subdominant in middle latitudes to the south. In South America and North Africa, *Classopollis* is the dominant gymnospermous palynomorph. In localities such as the Aptian to Albian of Brazil (fig. 4) and Peru, and the Aptian to Cenomanian of West Africa (Jardiné and Magloire, 1966), *Classopollis* makes up 60–80 percent of the total palynomorph count. Its seeming absence in the Northern Laurasian province is an identifying feature of that area.

Pollen of the Ephedraceae, which is a common constituent of

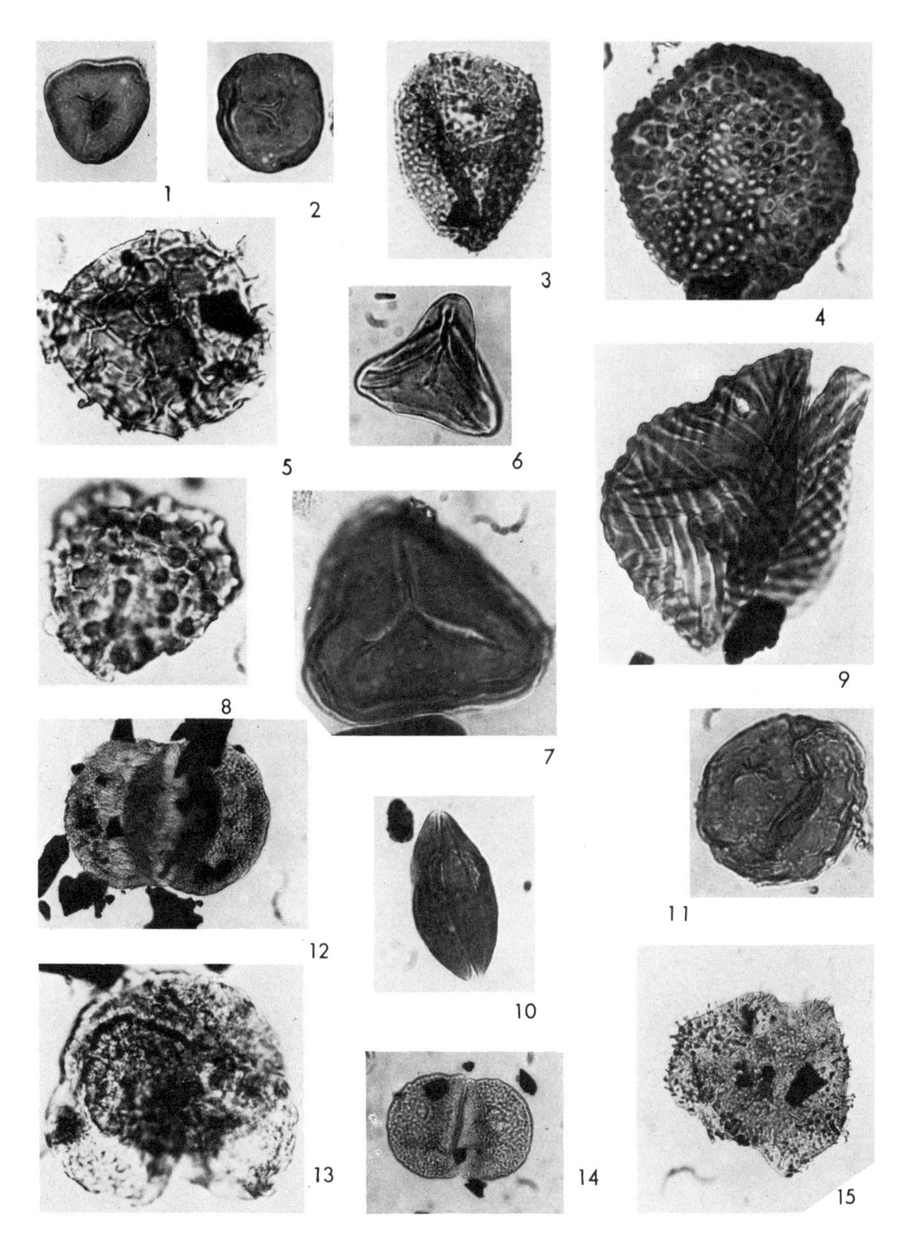

Fig. 2. Locality 13, *Christopher Formation,* Ellesmere Island, 80°40′N, 85°20′W, sec. 10–15, Collected by F. G. Stehli, 1965; Aptian to Lower Albian. All specimens × 700. (1, 2) *Stereisporites* spp. (3, 4) *Osmundacidites* spp. (5) *Lycopodiumsporites* sp. (6) *Gleicheniidites* sp. (7) *Cyathidites* sp. (8) *Converrucosisporites* sp. (9) *Cicatricosisporites* sp. (10) *Cycadopites* sp. (11) *Inaperturopollenites* sp. (12, 14) *Alisporites* sp. (13) *Abiespollenites* sp. (15) Acritarch.

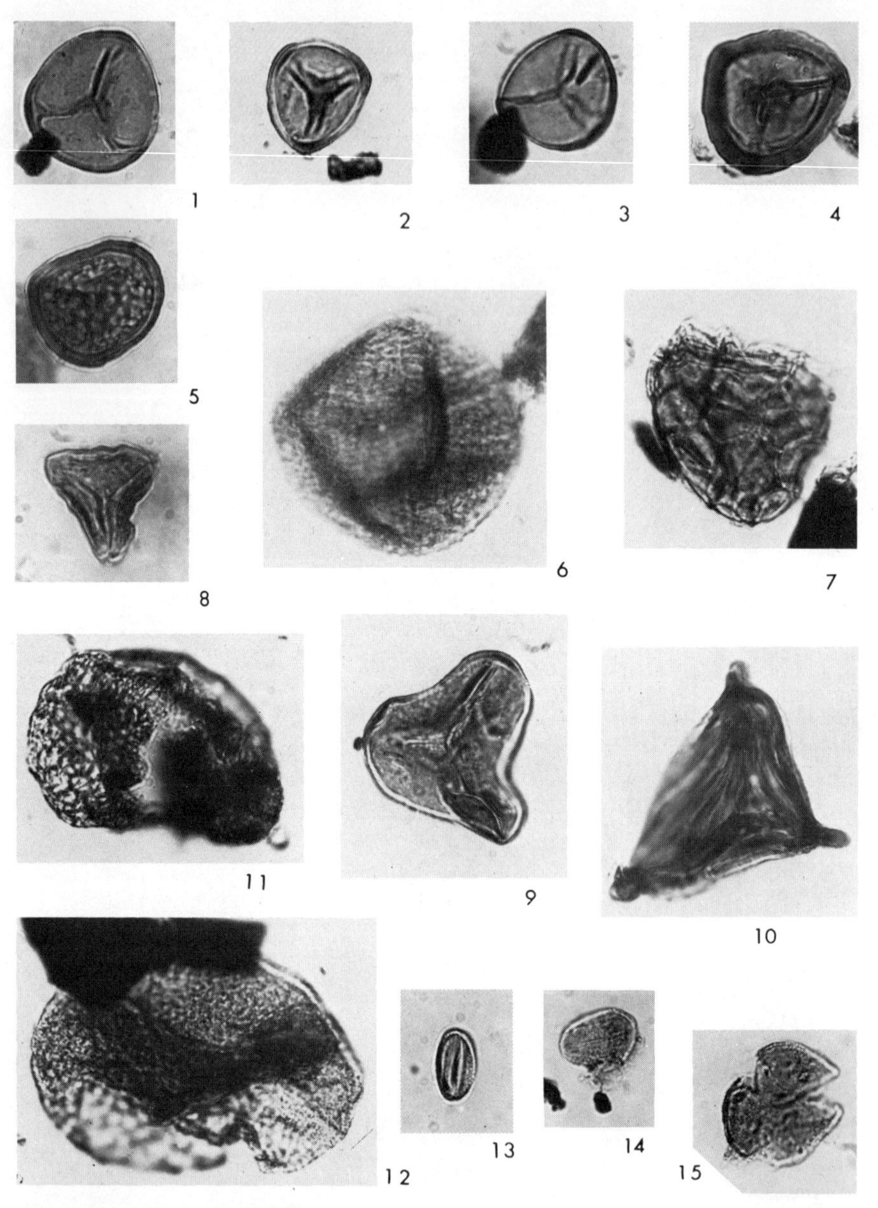

Fig. 3. Locality 10, *Geological Survey of Canada, locality 35701,* Yukon area; Upper Cenomanian. All specimens × 700. (1–5) *Stereisporites* spp. (6) *Cicatricosisporites* sp. (7) *Lycopodiumsporites* sp. (8, 9) *Gleicheniidites* spp. (10) *Appendicisporites* sp. (11, 12) *Pityosporites* spp. (13–15) *Retitricolpites* spp.

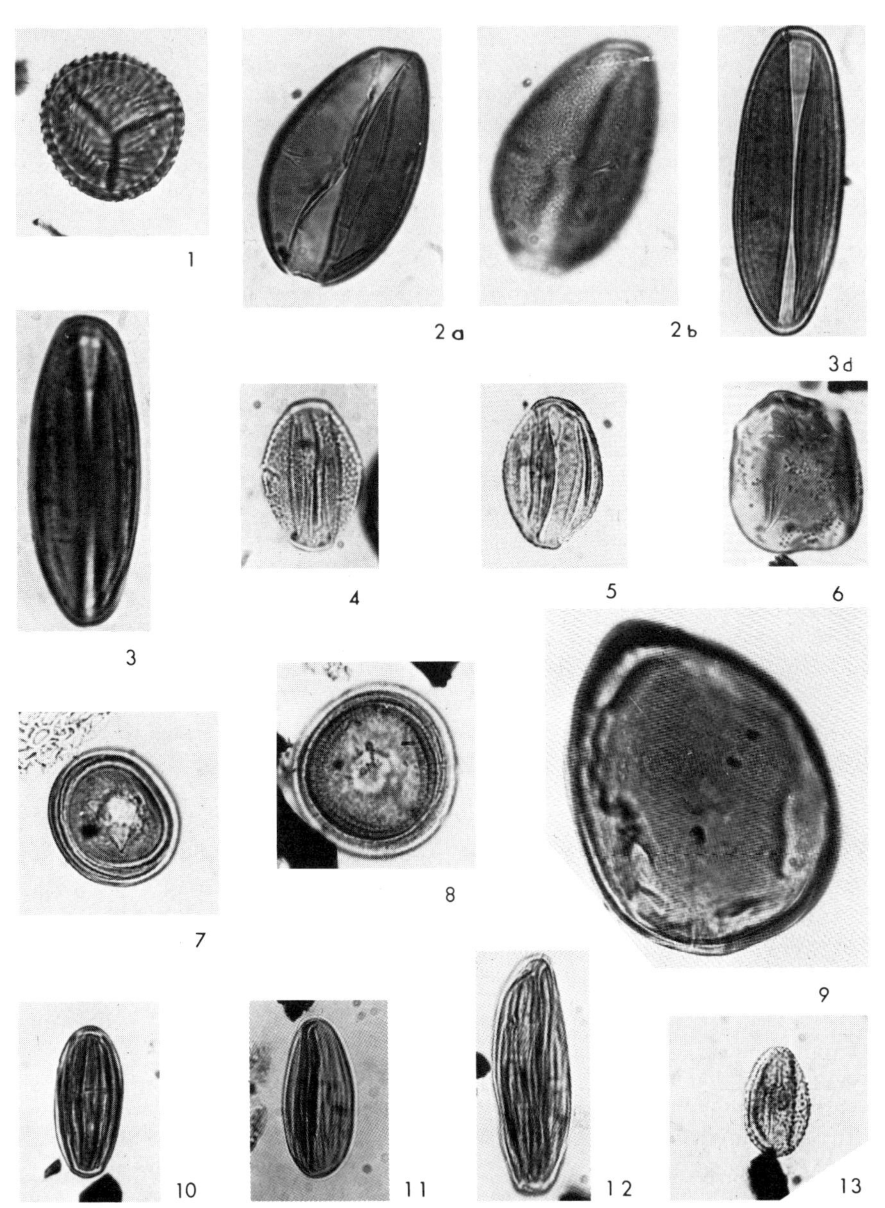

Fig. 4. Locality 1, *Alagoas Formation,* core samples from testhole Tm-1-AL core 3, area of Alagoas, Brazil; Upper Aptian. All specimens × 700. (1) *Cicatricosisporites* sp. (2, 3) *Cycadopites* spp. (4, 5) *Eucommiidites* sp. (6) *Exesipollenites* sp. (7, 8) *Classopollis* spp. (9) *Araucariacites* sp. (10–12) *Ephedripites* spp. (13) *Tricolpites* sp.

Barremian to Cenomanian horizons in the two provinces to the south, has not been found in this province. It makes up less than 1 percent of most assemblages in the middle latitudes and 2–20 percent in the tropical regions of the Northern Gondwana province. Monosulcate grains of cycadophyte affinities are also rare in the Arctic localities.

The distinctiveness of the palynomorph assemblages is also created by a very low diversity of pteridophyte spores as compared to the highly diverse spore assemblages in the middle latitudes of both hemispheres, such as those from Portugal (Groot and Groot, 1962), England (Couper, 1958; Kemp, 1970), central Russia (Bolkhovitina, 1953), and India (Venkatachala, 1968).

In the Northern Laurasian province the principal spores are the smooth trilete types found in the modern families Cyatheaceae and Gleicheniaceae and the genus *Sphagnum.* There is an evident reduction in the number and variety of the trilete spores of the Schizaeaceae, such as *Cicatricosisporites, Appendicisporites, Trilobosporites,* and *Concavissimisporites,* compared to the great number of these palynomorphs in the Southern Laurasian and Southern Gondwana provinces during Aptian and Albian times.

The high number of bisaccates, the low diversity of spores, and the paucity of spores of the Schizaeaceae and Cycadophyta seem to suggest a colder climate than existed in areas just to the south of the Northern Laurasian province. Similar characteristics of the palynospectrum of this province have been reported in high-latitude areas of Russia by Bondarenko (1958).

I would like to propose the idea that these early angiosperms did not adapt to the colder climate of this northern province until the Cenomanian stage, although to the south they already existed in early Cretaceous time. By late Albian and early Cenomanian time, in such areas as Oklahoma, Maryland, and Delaware (Brenner, 1963, 1967) the angiosperm pollen was abundant and diverse, containing tricolpates plus several types of tricolporates, triporates, and polyporates.

THE SOUTHERN LAURASIAN PROVINCE

As mentioned earlier, Middle Cretaceous palynofloras are highly diverse in the Southern Laurasian province. This diversity in the middle latitudes of both Europe and Northern America is expressed particularly in the spores of the Pteridophyta, especially the Schizaeaceae. Spores of the Gleicheniaceae are in general more abundant both in species and number in this province than in colder areas to the north and in tropical areas to the south. The pollen of

Brachyphyllum and *Pagiophyllum* is a common constituent of the present middle-latitude localities. In this province, the abundance of *Classopollis* decreases during the Aptian stage (Brenner, 1963; Kemp, 1970), whereas in the present tropical latitudes *Classopollis* continues to be a dominant element in the palynoflora until the Turonian (Müller, 1966; Jardiné and Magloire, 1966), constituting as much as 60–80 percent of the spectrum in most samples.

In addition to numerous and varied pteridophyte spores, assemblages from middle-latitude areas are rich in bisaccate pollen of the Pinaceae and Podocarpaceae.

Undoubted angiosperm pollen first appears in the Middle Albian, replacing most of the pteridophyte-gymnosperm complex by middle Upper Cretaceous time.

THE NORTHERN GONDWANA PROVINCE

Close palynological affinity between the Middle Cretaceous assemblages of eastern Peru (fig. 5, 6) and West Africa has been demonstrated (Brenner, 1968). During the late Albian and early Cenomanian, both areas contained identical species of a group of bizarre, horned palynomorphs called *Galeacornea* (Stover, 1964) and *Elaterocolpites* (Jardiné and Magloire, 1966). Numerous perinate trilete types also help to correlate the Upper Albian to Lower Cenomanian of both areas.

The palynomorphs of Northern Gondwana are similar to those of the Arctic areas in their low diversity of the pteridophyte palynomorphs, including the paucity of schizaeaceous types. As in the Arctic area, the common spore types are the cyatheaceous types such as *Cyathidites* and *Deltoidospora,* but the spores of *Sphagnum* are not as common.

As was pointed out by Couper (1964), bisaccate pollen is not found in Middle Cretaceous horizons from the tropical latitudes of Africa and South America, and I have not found any bisaccates in Israel. It appears that members of the Pinaceae and Podocarpaceae that produce bisaccate pollen did not inhabit the Northern Gondwana province. The gymnosperm pollen in Northern Gondwana is represented by araucarian and podocarpaceous, nonsaccate types, such as *Classopollis*, *Araucariacites*, and *Callialasporites* (*Applanopsis*).

The absence of bisaccate pollen gives the gymnosperm pollen complex of this province a distinctive quality. Bisaccates are present during this time in areas such as Argentina, South Africa (Scott, 1971), New Zealand, Australia, and India.

In Northern Gondwana, monosulcates belonging to a complex of

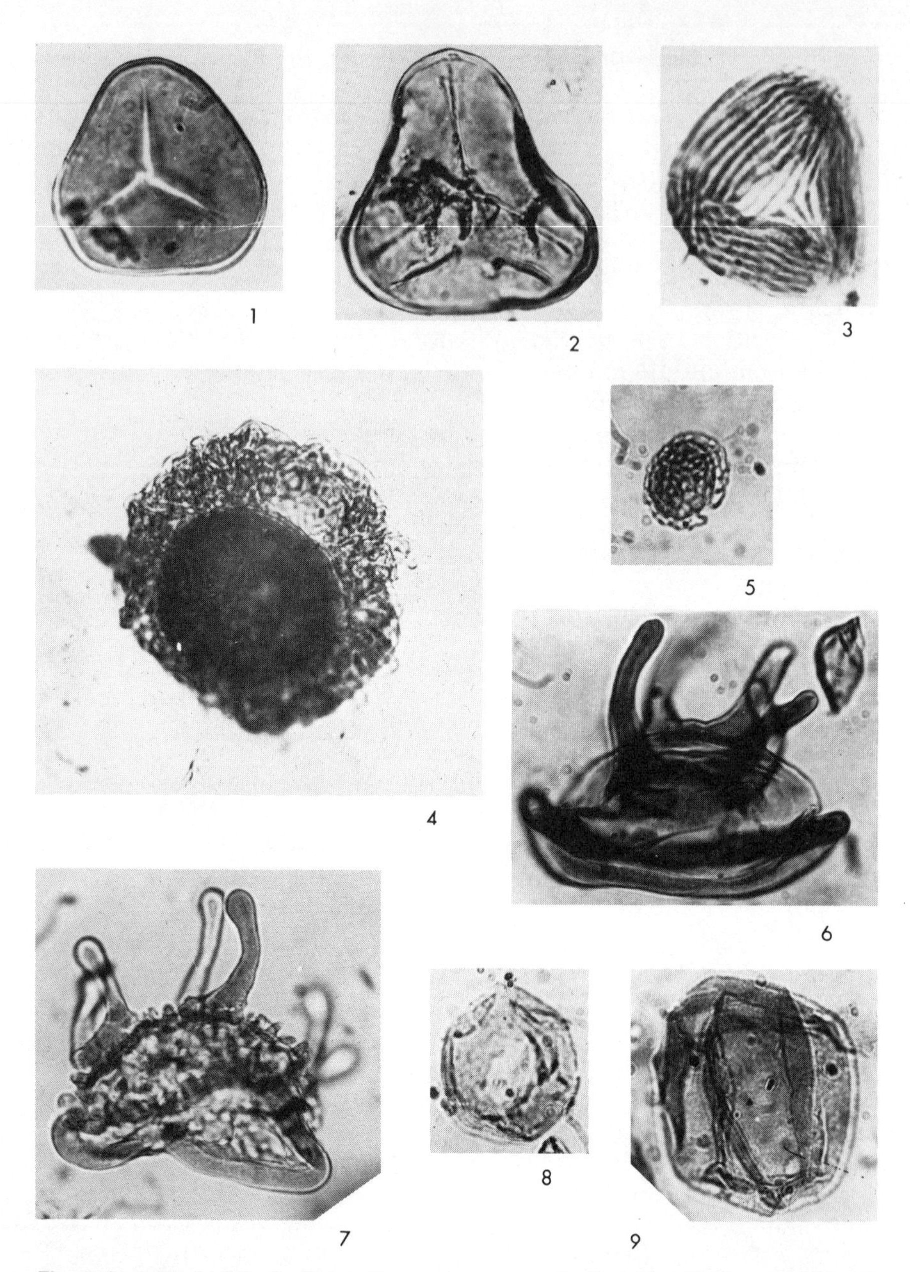

Fig. 5. Locality 2, *Oriente Formation,* outcrop sample from the Montana region of northeastern Peru, near 7°S, 75°W; Upper Albian. All specimens × 700. (1) *Cyathidites minor* Couper, 1953. (2) *Cyathidites* sp. (3) *Cicatricosisporites australiensis* (Cookson) Potonié, 1956. (4) Perotriletes pannuceus Brenner, 1963. (5) *Liliacidites peroreticulatus* (Brenner) Singh, 1971. (6) *Elaterosporites Klaszi* (Jardiné and Magloire) Jardiné, 1967. (7) *Elaterosporites protensa* (Stover) Jardiné, 1967. (8) *Inaperturopollenites dubius* (Potonié and Venitz) Pflug and Thomson, 1953. (9) *Araucariacites australis* Cookson, 1947.

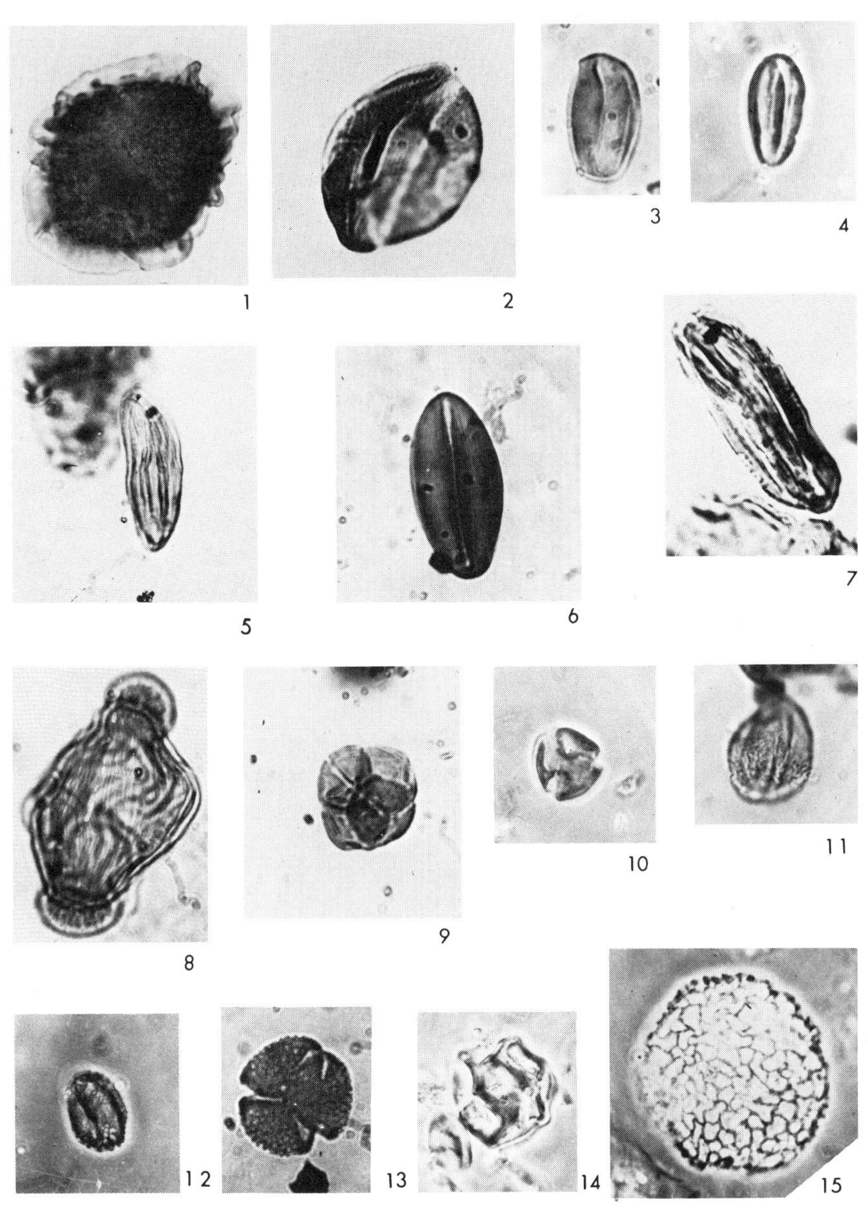

Fig. 6. Locality 2, continued. (1) *Callialasporites trilobatus* (Balme) Dev, 1961. (2) *Classopollis torosus* (Reissinger) Couper, 1953. (3) *Eucommiidites troedssonii* Erdtman, 1948. (4) *Ephedripites pentacostatus* Brenner, 1968. (5) *Ephedripites sulcatus* Brenner, 1968. (6) *Ephedripites costaliferous* Brenner, 1968. (7) *Ephedripites procerus* Brenner, 1968. (8) *Steevesipollenites dayani* Brenner, 1968. (9, 10) *Psilatricolpites tetradus* Brenner, 1968. (11) *Tricolpopollenites retiformis* Pflug and Thomson, 1953. (12) *Retitricolpites varireticulatus* Brenner, 1968. (13) *Retitricolpites* sp. B, Brenner, 1968. (14) *Multiporopollenites polygonalis* Jardiné and Magloire, 1966. (15) *Reticulatasporites jardinus* Brenner, 1968.

cycadophytes are more numerous and diverse than in any of the other provinces.

In the localities from Peru, Brazil, West Africa, and Israel the polyplicate grains of the Ephedraceae are very numerous and diverse, commonly making up 10–20 percent of the assemblages. The abundance of this family of plants in Northern Gondwana is a feature that distinguishes it from other areas. In samples from the present middle latitudes, this family is usually less than 1 percent of the assemblage and does not enter the record until late Neocomian time, and in Australia and India it does not appear until post-Aptian horizons (fig. 7).

As mentioned earlier, the oldest tricolpates were found in Barremian–Lower Aptian beds of Israel and Aptian beds from Brazil. These early tricolpates are found associated, both in Brazil and Israel with a sculptured type of *Eucommiidites* that closely approaches the tricolpate condition.

Perhaps this peculiar *Eucommiidites* belonged to a group of cycadophytes that produced the first tricolpate pollen while other members of the same group produced the monosulcate grains of *Clavatipollenites*. In this way, it would not be necessary to derive the tricolpate condition from a distal furrow; both would have evolved independently from a complex of plants approaching the angiosperm condition in other respects. The tricolpate condition could also have been derived from a monosulcate type by a sudden change in the tetrad arrangement, producing a triaperturate grain. This change may have had adaptive functions or been gene-linked to some other character.

The climate of this province is interpreted on the basis of the palynomorphs as tropical but semiarid. The low humidity is inferred from the abundance of ephedraceous pollen, the low diversity of pteridophyte spore species, and the increase in numbers of monosulcate cycadophyte types.

It is of interest to note here that Stebbins (1965) suggested that the earliest angiosperms might have evolved in semiarid conditions.

THE SOUTHERN GONDWANA PROVINCE

Very little has been published on the palynological record of strata from the middle latitudes of South America and Africa. Well-described palynomorphs from the Middle Cretaceous have been described from Australia (Dettman, 1963; Burger, 1968) and New Zealand (Couper, 1953, 1960). In southern South America the only Cretaceous material published is from the supposed Barremian-Aptian Baquero Formation of Patagonia (Archangelsky and Gamerro,

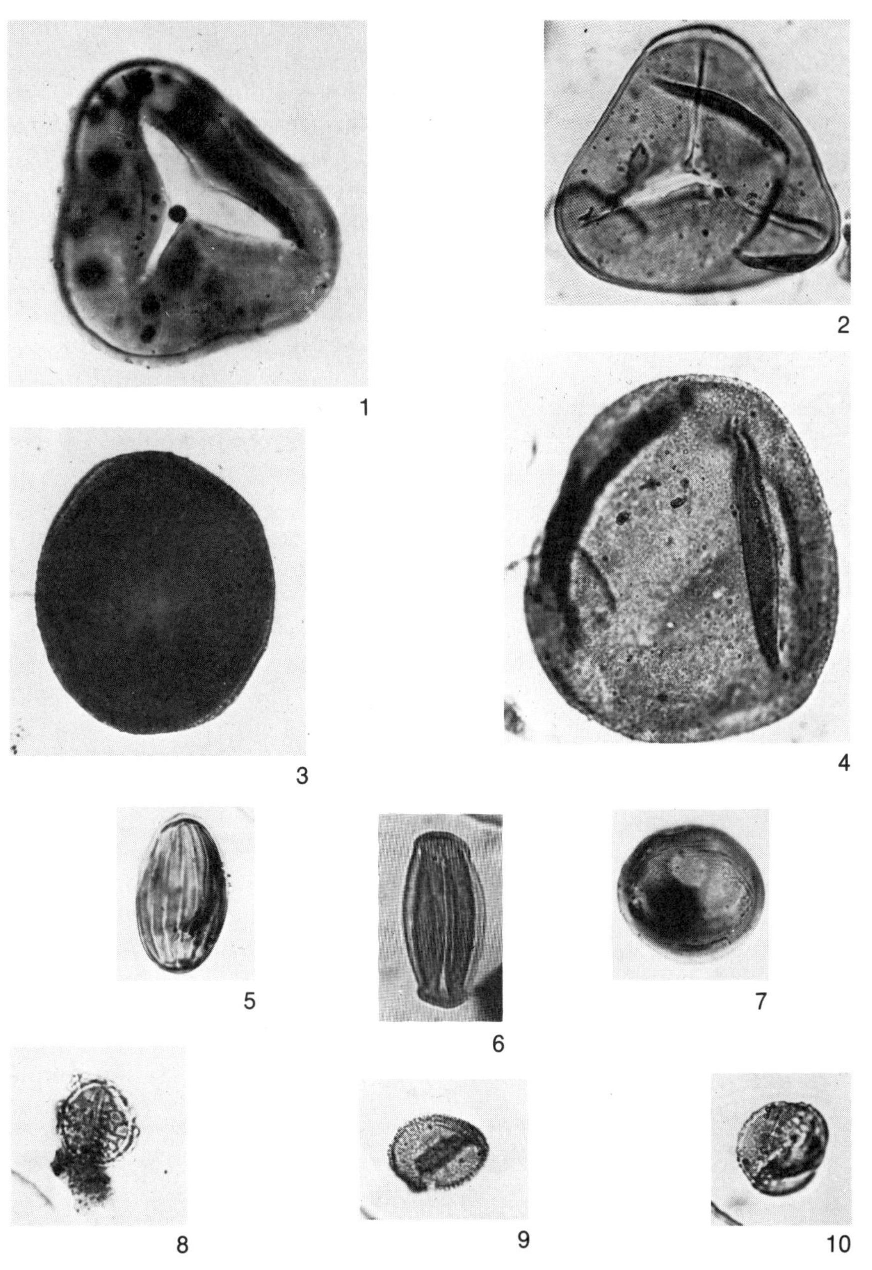

Fig. 7. *Zeweira Formation,* Zohar 1 well, core 6; northern Negev, Israel, Barremian-Aptian. All specimens are × 700. (1) *Cyathidites* sp. (2) *Deltoidospora* sp. (3, 4) *Araucariacites australis* Cookson, 1947. (5) *Ephedripites* sp. (6) *Eucommiidites troedssonii* Erdtman, 1948. (7) *Classopollis* sp. (8) *Liliacidites peroreticulatus* (Brenner) Singh, 1971. (9) *Clavatipollenities minutus* Brenner, 1963. (10) *Tricolpites* sp.

1967). In South Africa the only material published is by Scott (1971) from the Lower Neocomian Uitenhage Series.

In all the assemblages in the Southern Hemisphere, only the bisaccates of the Podocarpaceae are abundantly present. The Pinaceae appears to have been a Northern Hemisphere family during the Cretaceous. A common trisaccate form similar to those produced by the Southern Hemisphere gymnosperm *Microcachrys* is typical of Barremian to Albian assemblages of the Southern Gondwana province. The nonsaccate gymnosperms are similar to the types in Northern Gondwana, such as *Classopollis*, *Araucariacites*, and *Callialasporites* (*Applanopsis*).

Pteridophyte spores show a similar diversity to that found in the Southern Laurasian province, with numerous types belonging to the Schizaeaceae. Increasing humidity to the south of the tropical zone is indicated by these assemblage changes in the Southern regions of Gondwana.

It is interesting to note that India, although now at tropical latitudes, belongs palynologically to the Southern Gondwana province. This fact indicates the long distance northward that the Indian plate has moved from its original position in Middle Cretaceous time.

Conclusions

Palynological investigations of Barremian to Cenomanian sediments from many localities around the world suggest that the oldest tricolpate pollen is found in localities from the tropical latitudes. It has been found in Aptian beds from Brazil and Barremian–Lower Aptian beds from Israel. It now seems that tricolpates appear in younger and younger sediments and proceed into higher latitudes on both sides of the equator. A time-latitude curve from the Western Hemisphere indicates that the first tricolpates occur in the Aptian of the tropical latitudes, the Middle Albian of the present middle latitudes, and the Cenomanian of the Arctic region.

These data (although they exclude early monosulcate magnoliid types) appear to support Axelrod's contention that early angiosperm types originated in the tropics and migrated poleward; however, a pre-Cretaceous origin as proposed by Axelrod is not supported by palynological data. The time-space distribution of *Clavatipollenites* (and similar forms) as a possible early angiosperm magnoliid type should be investigated. It is possible that the tricolpate condition developed very early from magnoliid stock by a genetic change affecting the arrangement of the tetrad, without going through any gradual derivation of the tricolpate condition from a trichotomosulcate type as suggested by Doyle (1969, 1973).

Colder conditions in the Arctic regions during Albian time might have presented a barrier to newly evolved angiosperms living farther south. Adaptation to the Arctic province did not develop until Cenomanian time.

Tricolpate pollen reached the Australian area during the Middle Albian, about the same time as in the middle latitudes to the north.

Palynofloras during Middle Cretaceous time indicate four major floral provinces: the Northern Laurasian province; the Southern Laurasian province; the Northern Gondwana province; and the Southern Gondwana province.

The Northern Laurasian province includes the present Arctic region north of approximately the 60° latitude in North America. Palynological assemblages suggest temperate conditions.

The Southern Laurasian province includes the present middle-latitude areas of Europe and North America. Palynomorphs indicate warm temperate to subtropical humid conditions.

The Northern Gondwana province covers the northern parts of Africa and South America, and palynomorphs suggest semiarid conditions. The oldest tricolpates are found in this province.

The fourth major province, the Southern Gondwana province, consists of Argentina, South Africa, Australia, New Zealand, and India. Bisaccate pollen of the Podocarpaceae, absent in North Gondwana, reappears in this province. Pteridophyte spore diversity in the Southern Gondwana region suggests greater humidity than existed in the tropical latitudes.

I suggest that the tricolpate condition might have originated from a modification of sculptured *Eucommiidites* or *de novo* from monosulcate magnoliid stock by a change in tetrad arrangement. Regardless of what the precursor of the tricolpate pollen type was, these early tricolpates must reflect a major evolutionary development of dicotyledons from what was most likely a complex of archeoangiospermic stem groups developing rapidly during Middle Cretaceous time.

Finally, it should be said that the picture of angiosperm migration and floristic provinces presented here is based on material that is limited in both stratigraphic coverage and sampling localities. Many more data are needed from the southern regions of South America, from Africa, the Russian Arctic, China, and the Far East. However, I do not believe that additional data will change the basic patterns that have been recognized.

Acknowledgments

I thank the numerous scientists who cooperated in this project by supplying samples, assisting in the field collections, and offering their advice in general. Particular appreciation is due the Petroleo Brasileiro for the numerous samples they sent from the Alagoas-Serigipe basins of Brazil, the Standard Oil Company of New Jersey for the samples from Peru, and the Shell Oil Company for samples from Venezuela. The excellent material from the Negev, Israel, was generously given by M. Raab, Director of the Paleontology Division of the Geological Survey of Israel.

Special appreciation is due the National Science Foundation, which gave major financial support to the initial stages of this project (GP-5148). Additional financial support in the later stages of the work was generously given by the Research Foundation of the State University of New York, the American Philosophical Society, and the Geological Survey of Israel.

I thank Cheryl Lituchy for typing the manuscript, and Anton Friedmann for help with the drafting.

Index to Localities

* *Locality 1.* Alagoas Formation. Core samples from testhole TM-1-AL, core 3. Area of Alagoas, Brazil; Upper Aptian.

* *Locality 2.* Oriente Formation. Outcrop samples from the Montana region of northeastern Peru, near 7°S, 75°W; Upper Albian. Brenner (1968).

Locality 3. Deep-sea drilling cores from the northeast Bahamas. Vema Conrad 10: cores V22-8, RC10-281, RC10-282, RC10-283, RC10-284. Habib (1970).

* *Locality 4.* Patuxent, Arundel, and Patapsco formations. Outcrop and subsurface samples from Maryland between Baltimore and Washington, D.C. Brenner (1963).

Locality 5. Fredericksburg, Walnut Shale. Outcrop samples from Marshall County, Oklahoma. Hedlund and Norris (1968).

Locality 6. Fall River Formation. Denver basin. Pannella (1966).

Locality 7. Lakota and Fall River formations. Wyoming. Davis (1963).

Locality 8. McMurray Formation. Central Alberta plains, Lower Mannville Group. Vagvolgyi and Hills (1969).

Locality 9. Loon River Formation. Lower Peace River area, central Alberta plains. Singh (1971).

* *Locality 10.* Yukon area. Geological Survey of Canada. 35701; Upper Cenomanian.

* *Locality 11.* Atane Formation. One mile southeast of Ata delta, Nugssuak Peninsula, Greenland. Latitude 71°N, stratum 150–165 m in erosion cleft of stream. Collected by C. O. Erlanson, July 31, 1928. Cenomanian-Turonian.

* *Locality 12.* Corwin Formation. Kuk Inlet region, mine no. 1, Arctic Coastal Plain. Collected by Russel R. Dutcher and Charles L. Trotter, Pennsylvania State University, 1956. Kuk River coal section, Albian to Lower Cenomanian.

* *Locality 13.* Christopher Formation. Ellesmere Island, 80°40′N; 85°20′W, section 10-1. Collected by F. G. Stehli, Western Reserve University, 1965. Also, Geological Survey of Canada loc.

* Palynological material studied by writer.

25978, south coast of Ellef Ringis Island. Aptian to Lower Albian.

* *Locality 14.* Hassel Formation. Blacktop Ridge, near and east of Eureka, western Ellesmere Island. Collected by E. T. Tozer, 1962, Geological Survey of Canada. Cenomanian.

* *Locality 15. Beaudanticeras* affine zone. Stang Creek, Peel Plateau, Northwest Territories. Geological Survey of Canada loc. 35257. Upper Lower to lower Middle Albian.

References

Archangelsky, S., and Gamerro, J. C. 1967. Spore and pollen types of the Lower Cretaceous in Patagonia (Argentina). *Rev. Palaeobot. Palynol.* 1: 211–17.

Axelrod, D. I. 1952. A theory of angiosperm evolution. *Evolution* 6: 29–60.

Axelrod, D. I. 1959. Poleward migration of early angiosperm flora. *Science* 130: 203–7.

Bolkhovitina, N. A. 1953. Spores and pollen characteristics from the Cretaceous of the central part of the U.S.S.R. [in Russian]. *Tr. Inst. Geol. Nauk, Acad. Nauk S.S.S.R., Ed. 145, Geol. Ser.*, 61: 1–1844.

Bondarenko, N. M. 1958. Palynological characteristics of the Albian and Upper Cretaceous deposits of the Khatanga Depression. *Sb. Statei. Paleontol. Biostratigr.* 7: 43–55 (transl. from the Russian, Canadian Geological Survey).

Brenner, G. J. 1963. The spores and pollen of the Potomac Group of Maryland. *Maryland Dep. Geol. Mines Water Res.* 27: 1–215.

Brenner, G. J. 1967. Early angiosperm pollen differentiation in the Albian to Cenomanian deposits of Delaware (U.S.A.). *Rev. Palaeobot. Palynol.* 1: 219–27.

Brenner, G. J. 1968. Middle Cretaceous spores and pollen from northeastern Peru. *Pollen Spores* 10: 341–83.

Burger, D. 1968. Stratigraphy and palynology of Upper Mesozoic sections in some deep wells in the Surat Basin, Queensland. *Rec. Bur. Mineral. Geol. Geophys.* (Australia) 1968/24.

Burger, D. 1970. Early Cretaceous angiospermous pollen grains from Queensland. *Bur. Mineral Resources Geol. Geophys., Canberra, Extract no. 1,* 116: 1–10.

Couper, R. A. 1953. Upper Mesozoic and Cainozoic spores and pollen grains from New Zealand. *Bull. New Zeal. Geol. Surv.*, Paleontol. 22: 1–77.

Couper, R. A. 1958. British Mesozoic microspores and pollen grains, a systematic and stratigraphic study. *Palaeontographica* 103B: 77–179.

Couper, R. A. 1960. New Zealand Mesozoic and Cainozoic plant microfossils. *Bull. New Zeal. Geol. Surv.* 32: 5–82.

Couper, R. A. 1964. Spore-pollen correlations of the Cretaceous rocks of the Northern and Southern Hemispheres. *Soc. Econ. Paleontol. Mineral., Spec. Publ.* 11: 131–42.

Davis, P. N. 1963. Palynology and stratigraphy of the Lower Cretaceous rocks of northern Wyoming. Ph.D. thesis, Univ. of Oklahoma, Norman.

Dettman, M. E. 1963. Upper Mesozoic microfloras from southeastern Australia. *Proc. Roy. Soc. Victoria* 77: 1–148.

Dev, S. 1961. The fossil flora of the Jabalpur Series. 3. Spores and pollen grains. *Paleobotanist* 8(1,2): 43–56.

Dietz, R. S., and Holden, J. C. 1970. Reconstruction of Pangaea: Break-up and dispersion of continents, Permian to present. *J. Geophys. Res.* 75: 4939–56.

Doyle, J. A. 1969. Cretaceous angiosperm pollen of the Atlantic Coastal Plain and its evolutionary significance. *J. Arnold Arbor.* 50: 1–35.

Doyle, J. A. 1973. Fossil evidence on early evolution of the monocotyledons. *Quart. Rev. Biol.* 48: 399–413.

Erdtmann, G. 1948. Did dicotyledonous plants exist in Jurassic time? *Geol. Fören. Stockholm Förh.* 70: 265–71.

Groot, J. J., and Groot, C. R. 1962. Plant microfossils from Aptian, Albian and Cenomanian deposits of Portugal. *Communic. Serv. Geol. Portugal* 46: 133–76.

Habib, D. 1970. Middle Cretaceous palynomorph assemblages from clays near the Horizon Beta deep-sea outcrop. *Micropaleontology* 16: 345–79.

Hedlund, R. W., and Norris, G. 1968. Spores and pollen grains from Fredericksburgian (Albian) strata, Marshall County, Oklahoma. *Pollen Spores* 10: 129–59.

Hughes, N. F. 1961. Further interpretation of *Eucommiidites* Erdtmann 1948. *Palaeontology* 4: 292–99.

Jardiné, S., and Magloire, L. 1966. Palynologie et stratigraphie du Crétacé des bassins de Sénégal et de Côte d'Ivoire. *Mém. Bur. Rech. Geol. Minières.* 32: 189–245.

Kemp, E. M. 1968. Probable angiosperm pollen from British Barremian to Albian strata. *Palaeontology* 11: 421–34.

Kemp, E. M. 1970. Aptian and Albian miospores from southern England. *Palaeontographica* 131B: 1–143.

Krömmelbein, K. 1962. Zur Taxonomie und Biochronologie stratigraphisch wichtiger Ostracoden-Arten aus der oberjurassisch? unterkretazischen Bahia-Serie (Wealden-Fazies) NE-Braziliens. *Senckenbergiana Lethaea* 43: 437–528.

Müller, H. 1966. Palynological investigations on Cretaceous sediments in northeastern Brazil. In *Proceedings of the 2nd West African Micropaleontological Colloquium* (Ibadan), pp. 123–35. Brill, Leiden.

Norris, G. 1967. Spores and pollen from the lower Colorado Group (Albian-?Cenomanian) of central Alberta. *Palaeontographica* 120B: 72–115.

Pannella, G. 1966. Palynology of the Dakota Group and Graneros Shale of the Denver Basin. Ph.D. thesis, Univ. of Colorado, Boulder.

Pocock, S. A. J. 1962. Microfloral analysis and age determination of strata at the Jurassic-Cretaceous boundary in the western Canada plains. *Palaeontographica* 111B: 1–95.

Scott, L. S. 1971. Lower Cretaceous pollen and spores from the Algoa Basin (South Africa). B.Sc. thesis, University of the Orange Free State, South Africa.

Seward, A. C. 1931. *Plant Life through the Ages.* Cambridge Univ. Press, New York.

Singh, C. 1964. Microflora of the Lower Cretaceous Mannville Group, east-central Alberta. *Res. Council Alberta Can. Geol. Div. Bull.* 15: 1–239.

Singh, C. 1971. Lower Cretaceous Microfloras of the Peace River Area, Northwestern Alberta. *Res. Council Alberta Can. Geol. Div. Bull.* 28: 1–299.

Smiley, C. J. 1966. Cretaceous floras from Kuk River area, Alaska: Stratigraphic and climatic interpretations. *Bull. Geol. Soc. Amer.* 77: 1–14.

Stebbins, G. L. 1965. The probable growth habit of the earliest flowering plants. *Ann. Missouri Bot. Garden* 52: 457–68.

Stover, L. E. 1964. Some Middle Cretaceous palynomorphs from West Africa. *Micropaleontology* 9: 85–94.

Vagvolgyi, A., and Hills, L. V. 1969. Microflora of the Lower Cretaceous McMurray Formation, northeast Alberta. *Bull. Can. Petrol. Geol.* 17: 155–81.

Van Konijnenburg–Van Cittert, J. H. A. 1971. In situ gymnosperm pollen from the Middle Jurassic of Yorkshire. *Acta Bot. Neerl.* 20: 1–97.

Venkatachala, B. S. 1968. Palynology of the Mesozoic sediments of Kutch. IV. Spores and pollen from the Bhuj exposures near Bhuj, Gujarat District. *Paleobotanist* 17: 208–29.

Viana, C. F. 1966. Stratigraphic distribution of Ostracoda in the Bahia supergroup (Brazil). In *Proceedings of the 2nd West African Micropaleontological Colloquium* (Ibadan), pp. 240–57. Brill, Leiden.

Wicher, C. A. 1959. Ein Beitrag zur Altersdeutung des Recôncavo, Bahia (Brasilien). *Geol. Jahrb.* 77: 35–38.

Plate Tectonics and its Bearing on the Geographical Origin and Dispersal of Angiosperms [1]

RUDOLF M. SCHUSTER, ***Department of Botany University of Massachusetts, Amherst***

UNTIL RELATIVELY RECENTLY, dispersal and evolution of land plants were visualized as having taken place in a "steady-state" world where orientation of continents and poles was constant and where regional orogeny and climatic changes, alone, accounted for major changes in the environments of the various taxa. Such a relatively simple picture does not exist (Smith and Hallam, 1970; Hammond, 1971a, b; Schuster, 1972a). Not only have major land masses undergone extensive migration (India, in chiefly Cretaceous times; Australasia, in the Tertiary), presumably carrying with them thousands of taxa of organisms, but the splitting off of such land masses from formerly contiguous areas has caused the initiation of important disjunctions. The fact that evolution of the angiosperms occurred prior to the start of Indian plate migration, and that evolution of many modern families (and often genera) of angiosperms preceded the breakup and migration of the Australasian complex, clearly necessitates an analysis of both *origin* and *dispersal* of the angiosperms in terms of this complex geological history. In this review I attempt to analyze some presumed consequences of the dual phytogeographic results of plate movements: the massive disjunction—by breakup of supercontinents—of taxa (and floras) formerly connected, and the juxtaposition of taxa (and floras) of independent origin, coincident with collision of plates of disparate origin.

From the point of view of the student of early angiosperm history such an analysis must necessarily lean heavily on the better-documented history of the gymnosperms (Florin, 1963) and, in part,

[1] Prepared with the aid of a grant from the National Science Foundation, GB-40075. Several of the maps included here were previously published in the *Botanical Review* (Schuster, 1972a); I am indebted to Arthur Cronquist for permission to reuse them. I thank my wife, Olga M. Schuster, for her help in the preparation of this paper and James Doyle for his critical reading of the manuscript and many suggestions.

on the detailed analysis of distribution patterns in the bryophytes (Schuster, 1969, 1972a; G. L. Smith, 1972). Reasons for this lie in (1) the still very inadequate early fossil information on angiosperms, especially when based solely on study of leaf impressions; (2) the rapid rate of evolution of the angiosperms, with many instances of convergence—the difficulty of correct attribution of fossil fragments is thus often formidable; (3) the relatively modern diversification of the angiosperms, so that their history prior to the mid-Cretaceous remains highly ambiguous. Since patterns of dispersal and migration routes of gymnosperms and bryophytes are probably roughly comparable to those of the angiosperms, it seems to me valid to argue from the better-documented early history of those groups in order to "flesh out" the very lacunose early history of angiosperm migration. Furthermore, since mammals and angiosperms evolved and diversified together, and their histories are intimately linked, zoogeographic data are relevant to an understanding of early angiosperm dispersal patterns. The need is for a synthesis in which phytogeographic history is intertwined with zoogeographic history—and it is incumbent on the botanist not to neglect the latter.

Critical in any understanding of the entire problem is the time when "rafting" of continents occurred, when these migrating plates approached and collided with other plates, and when breakup of supercontinents initiated a disjunction of once-continuous floras. In my opinion the rapid diversification of angiosperms from mid-Cretaceous times onward reflects in part the opportunities and challenges presented by these geological events and related climatic changes. A. C. Smith (1972, p. 217) emphasized that "ranalean" families of angiosperms, today almost universally regarded as relicts of a complex from which all or almost all other extant angiosperms evolved, "probably date from early differentiations of ancestral angiosperms in Jurassic times, whereas comparatively highly evolved and 'modern' families may date only from late Cretaceous," although many modern and large families (e.g., Compositae) appear to have evolved as late as the Tertiary. This time sequence is reasonable on the basis of extrapolation backward from micro- and megafossils (Brenner, and Doyle and Hickey, in this volume), which almost demands a pre-Cretaceous origin of the angiosperms. Hence, not only Australasian plate migration but also Indian plate migration become important factors in comprehending modern distribution of at least some angiosperms. Even more important is recognition that until early Cretaceous times (and in several cases, until early Tertiary times) migration routes which today are closed were wide open. Indeed, massive and relatively unimpeded migrations from Laurasia to Gondwanaland and vice versa, and between regions as far apart as North America and Australasia, were feasible—and surely occurred—until at least the late Cretaceous. My analysis of the probable migrational pattern of *Nothofagus* and that of the marsupials (which is in phase with almost all zoogeographic data), strongly suggests that obstructions to relatively rapid and effective migration failed to evolve until, in many cases, late Cretaceous or early Tertiary times.

We must thus visualize a world where, until at least the Middle Cretaceous, a high level of cosmopolitanism existed in a southern supercontinent, Gondwanaland, and a similar rather extensive cosmopolitanism occurred in Laurasia.[2] At least limited biotic interchange between the two supercontinents took place: demonstrable migration from Laurasia to Gondwanaland by organisms, such as Fagaceae and Marsupialia, that were unable to engage in long-distance dispersal suggests that a relatively ineffective "barrier" was bridged between North and South America on countless occasions (as late as the end of the Cretaceous by certain placental mammals). Thus routes for effective dissemination of land organisms, and specifically the angiosperms, have changed drastically during relevant geological time. As is demonstrated later, the intrinsic ability of angiosperms to migrate has also changed dramatically, with modern taxa in most cases being much more readily dispersible over wide barriers than ancient ones. Hence, as I emphasize in the conclusion, modern, often disharmonic, distribution patterns of modern genera and families cannot be used to evaluate the former dissemination modalities and routes of early angiosperms.

It has been demonstrated (Schuster, 1972a) that primitive, mostly woody, principally ranalean angiosperm families, which in general have a relict distribution that suggests both great age and (today at least) limited dispersibility, fall into two major and numerically almost equal groups. One group of families is basically Laurasian, the other basically Gondwanalandic. Strikingly, one member of a family pair is often Laurasian (e.g., Calycanthaceae), the other Gondwanalandic (e.g., Idiospermaceae). From this situation, one is almost forced to conclude that during early stages of evolution of angiosperms, when presumably their members mostly lacked effective means of moving their disseminules (and when disseminators, such as modern mammals, were not yet on the scene in significant numbers), barriers for migration that seem significant today either did not

[2] This is clear from the history of the conifers and taxads (Florin, 1963) and hepatics (Schuster, 1972b); it becomes obvious with regard to Angiospermae by the start of the Tertiary, by which time the Arcto-Tertiary flora of Laurasia was widely dispersed.

exist or were of trivial importance.[3] From this seemingly almost equal and dual evolution of a primitive angiosperm complex in *both* Gondwanaland and Laurasia one cannot effectively deduce the origin of angiosperms in either supercontinent. As has been noted elsewhere (e.g., Schuster, 1972a), presently available data do not allow the unambiguous derivation of a "cradle" or "center of origin" for the angiosperms. Instead, we must content ourselves with drawing a limited set of conclusions dictated by recent geophysical data:

1. A basically mid- to late-Cretaceous migration of the Indian plate (McKenzie and Sclater, 1973), with contiguity of the Indian plate to Gondwanaland until at least mid-Cretaceous times.

Biogeographic conclusion: Judging from size and geological complexity alone, the Indian "raft" must have been a sort of enormous Noah's Ark on which a wide diversity of taxa could maintain themselves. Migration northward was slow enough that many if not most taxa would survive, if changed by selection pressure, as this plate migrated into and through much warmer climates. It is the phytogeographer's formidable task to recognize the descendants of these passengers and to assay the role they have played in formation of the modern Northern Hemisphere flora. In contrast to casual long-distance dispersal, "plate migration," with the potential of synchronously moving thousands of taxa entrenched in structured and cohesive communities, would tend to preserve the harmonic structure of the flora.

2. Contiguity of Australia to Antarctica until ca. 50 million years ago—thus well into the Tertiary—with survival of forests and forest communities in at least the lowlands of Antarctica until that time or later, and juxtaposition of the Australian and Asian plates quite late in the Tertiary.

Biogeographic conclusion: The Gondwanalandic elements in the flora of Australia and neighboring regions are primary and ancient; these "old" elements, cool- or cold-adapted, have been numerically depleted and submerged in a "sea" of recent immigrants from Southeast Asia that have come to occupy principally the tropical to subtropical lowlands, although some occur in the highlands (e.g., *Rhododendron, Lithocarpus, Castanopsis*). From analysis of the geological history of Australasia one arrives at the seemingly improbable

---

[3] I would not entirely negate the role of animals in dispersal of early angiosperms; the sarcotesta of *Magnolia* (and of some Mesozoic gymnosperms such as *Caytonia*), and the flesh-covered, pinkish seeds of *Zamia* argue for at least limited animal dispersal of early angiosperm and gymnosperm disseminules. However, within many groups of angiosperms there is demonstrable evolution of progressive adaptation to effective mammal and bird dispersal.

conclusion that the Fagaceae of New Guinea arrived by two very different routes, during different times: *Castanopsis* and *Lithocarpus* relatively recently, from the northwest (Asia); *Nothofagus* earlier, from South America via Antarctica and Australia.

In a few instances the biologist must continue to argue from biological necessity rather than good geological evidence. Thus we must assume the existence of "filters" and "filter bridges," to use terms coined by Simpson, during some if not most of the Cretaceous, between North and South America that apparently allowed groups as diverse as *Nothofagus*, marsupials, and early placentals to invade South America. During Cretaceous times, while the Indian raft was slowly carrying large numbers of Gondwanalandic taxa northward, it is certain that Laurasian taxa were filtering southward into Gondwanaland—or its remnants—and vice versa (Kurtén, 1969). Probably, diffuse archipelagic connections between India, Madagascar, Africa, and Australasia during early stages of Indian migration northward also constituted such filters or filter bridges. In addition, almost continuous land connections existed between Europe and Africa and, until after mid-Cretaceous times, distances between South America and Africa remained of trivial extent for many organisms.

I thus visualize a situation best described as disorderly: besides some casual, long-distance migration, opportunities for infusion of elements of one flora into another existed repeatedly during the critical times when angiosperms evolved and underwent their early dispersal. Indeed, the fact that from early to mid-Cretaceous the angiosperms were able to achieve a cosmopolitan range eloquently attests to these opportunities. A limiting factor is that throughout the first half of the Cretaceous the Angiospermae, by and large, lacked efficient, highly evolved devices that would allow rapid animal transport (either external or in the intestinal tract), and appropriate disseminators had not yet evolved.

The demise of the steady-state concept—a concept allowing that only the plant moved, not the terra firma on which it grew—has in some ways made the phytogeographer's tasks more difficult. Equally, it has forced a series of welcome constraints on phytogeographic speculation. In one swoop many "land bridges," erected mostly in the last century from "biogeographic necessity," have been swept away. Modern geophysics has also provided relatively accurate dating of geological events that must to a large degree determine the rates of migration and patterns of dispersal of organisms not fit for long-distance dispersal; the relationship between geologic events and dispersal patterns is explored in detail in a later section of this paper, on the marsupial route. The general conclusion we must arrive at, however, as Darwin (1903) stated in a celebrated letter to Hooker, is that from a biogeographic point of view the origin and early history of the flowering plants remain largely an "abominable mystery." Between wandering continents (and perhaps wandering poles) and wandering plants, we have reached the point where—in one sense—variables have multiplied. We have hardly approached a real solution to exactly where and when angiosperms evolved. Answers to these problems are likely to come only from further study of mega- and micro-fossil evidence. Rigorous adoption of constraints which the current geological revolution is imposing on phytogeographers, plus integration of fossil data, should eventually allow us to sift fact from fancy. At present, phytogeographers have to content themselves with a more restricted series of tasks, e.g., to determine specific migration routes of specific groups (such as *Nothofagus*, discussed later); the timing of such migrations; the history and interrelationships of floras. It has become clear that on the basis of living taxa one cannot accept "cradle" areas—such as Assam to Fiji or Yunnan to Queensland—as centers of origin for large and complex groups. Indeed, application of accepted geological principles automatically disproves the existence of such cradle areas, as does—by analogy—the history of the conifers (Schuster, 1972a). Concentration of generalized taxa in a specific region may represent merely a center of survival. Or, as in the case of Assam to Fiji, it may represent (Schuster, 1972a) a center of juxtaposition, where floras relatively rich in relict groups have been physically juxtaposed with subsequent but limited intermingling.

Scepticism is therefore a necessary ingredient in all phytogeographic speculation. Since, in my opinion, it is a major mistake to base phytogeographic hypotheses on study of only one group, I shall interweave data from other groups for which appropriate information is available (Bryophyta; Coniferales; land animals). As I shall develop subsequently, *early* Angiospermae, including most extant ranalean woody families, lack effective long-distance dispersal devices. Analogies to tetrapod dispersal thus are apt to prove illuminating.

Within these perimeters, I shall adress myself to certain general problems and questions:

1. From information on *extant* distribution of angiosperms can one establish a locus of origin of the Angiospermae? And did this locus lie in Laurasia or Gondwanaland?
2. Is the admittedly extraordinarily complex angiosperm flora of southeastern Asia and extending into Australasia to be ex-

plained on the basis of an origin in that region, or is this diversity due to combined persistence of old elements and commingling of disparate floras?

3. If this relict-rich flora from Assam to Fiji, the supposed cradle of the angiosperms, results in part from fusion of disparate elements of initially different floras, was this fusion due only to commingling of Australasian and east Asian elements? Or, was it due in part to an even earlier intermingling of elements present on the Indian plate with Laurasian elements?
4. Was Indian plate migration initiated sufficiently late that angiosperms could have been passively rafted to Laurasia?
5. Was stepwise migration possible, in the critical Cretaceous period, from Laurasia to Gondwanaland or its fragments?

Because accumulated geophysical evidence gives us a reasonably clear picture of how and approximately when major lithospheric plates migrated,[4] these and collateral questions can be attacked today with renewed vigor. The question is, How will we fit phytogeographical data derived from land plants and accommodate them with the geophysical evidence? No theory of origin and dispersal of the angiosperms can, today, stand close scrutiny unless it also fits the accumulated geophysical evidence.

Before I outline my interpretation of what was the likely scenario for the Angiospermae and for other land plants since the Jurassic, I shall have to deal with a series of facts, theories, hypotheses, and assumptions. A sufficent number of these are, today, incontrovertible; most of the remainder will probably be accepted without much modification, but a critical number, I think, will long remain a source of controversy. These "ground rules" fall into several categories: geological, climatic, biological. They are briefly reviewed in the next section.

---

[4] Reconstructions given by geologists need further refining. The timing assigned to events is especially subject to minor—and sometimes major—correction. A glaring example of dramatic significance: Dietz and Holden (1970) indicate that the Indian plate initiated its migration northward about 180 million years ago, at the end of the Triassic; McKenzie and Sclater (1973), on the basis of more convincing and recent paleomagnetic data, claim this occurred only about 75 million years ago, thus in the Upper Cretaceous! I accept here a conservative form of the most recent time sequence. In general the geologist has tended to speed up migration rates—a decade ago migration rates of 5 cm per year were regarded as almost maximal; today migration rates of 16 cm per year for certain plates are accepted as accurate, at least for portions of the time intervals.

Ground Rules and Approach

GEOLOGICAL GROUND RULES

I assume that concepts of plate tectonics and continental drift are now rapidly moving from the domain of theory to that of established fact. No phytogeographer remains on tenable ground by clinging to theories that are not reconcilable with recognized concepts of plate tectonics. For that reason, as detailed in Schuster (1972a), it is impossible to uphold theories (A. C. Smith, 1967, 1970; Takhtajan, 1969) that visualize a single center of origin for angiosperms extending from Southeast Asia or Assam to Australasia. Similarly, concepts of land bridges (Croizat, 1952; van Steenis, 1962, 1971) are today no longer tenable. Specific "dicta" I accept are as follows:

1. Until about the start of Jurassic times (180–195 million years ago) there is evidence for considerable cosmopolitanism as regards the flora and shallow-water marine organisms; this suggests that Wegener (1924) and Dietz and Holden (1970) are right in assuming a single Paleozoic continent, Pangaea. In the Southern Hemisphere, specifically, there had long been total contiguity of presently scattered land masses. Sahni (1939) notes that the idea of a "single more or less continuous life-province in the Southern Hemisphere during the Permo-Carboniferous period" arose from the uniformity of the "widely distributed *Glossopteris* flora," whose identity demands "free immigration."

2. At about that time, and perhaps somewhat earlier, this land mass began to break up into northern and southern continents, Laurasia and Gondwanaland, respectively (fig. 1). Gondwanalandic reconstructions differ in detail but agree in their general configurations (Dietz and Holden, 1970; Schopf, 1970a; Smith and Hallam, 1970). The degree of physical isolation of these two land masses remains conjectual.[5] It is clear from fossil evidence (Schopf, 1970a, b) that as early as the Permian little infusion of northern elements into southern Gondwanaland flora occurred (notable exceptions are treated later); distribution of fossil gymnosperms (Florin, 1963) also suggests that as early as the Carboniferous distinct northern and southern floras existed. The question is, Did these evolve in physical isolation—did they evolve on separate land masses, divided from each other by water barriers, or were they merely environmentally

[5] If, before the Upper Triassic, Africa abutted eastern North America, and the North Atlantic then started to open parallel to the appropriate sector of the Mid-Atlantic Ridge, there was long continued contact between the two continents at the shear zone between North Africa and southwestern Europe.

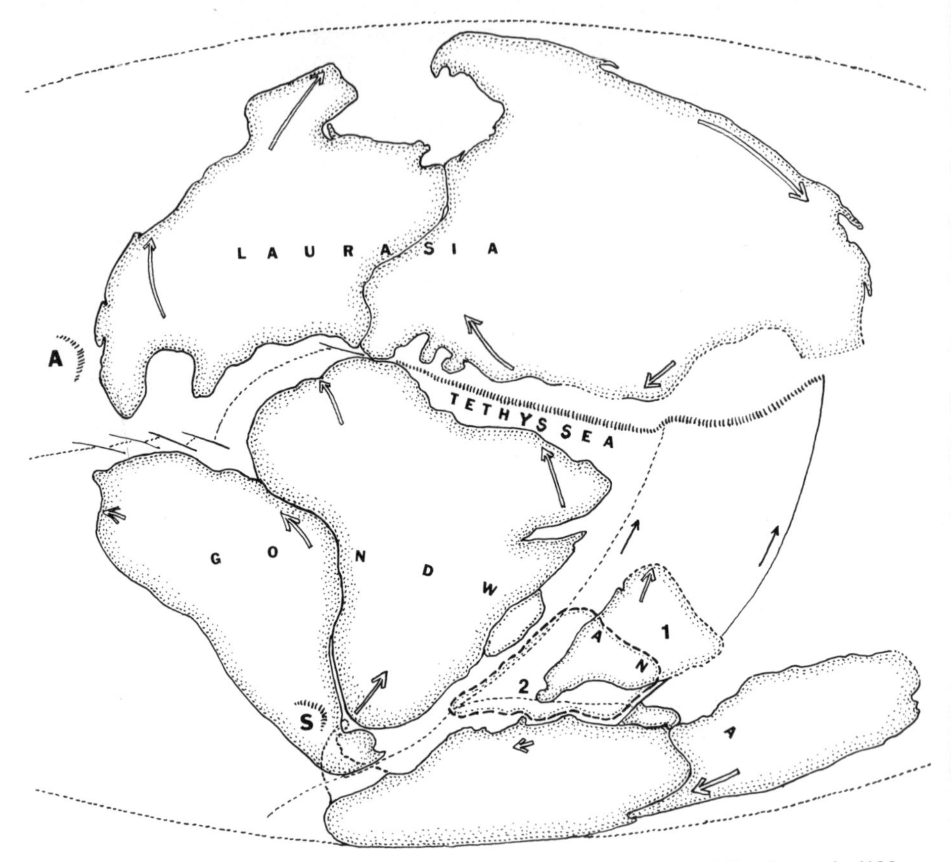

Fig. 1. Reconstruction of Laurasia and Gondwanaland at start of the Jurassic (180 million years ago). Rift lines are indicated by dotted lines; the Tethyan Trench is shown by hatching; solid arrows and solid lines denote megashears, zones of lateral slippage along plate boundaries; hollow arrows indicate rotation since the breakup of Pangaea. The Antilles and Scotia arcs (A and S) are indicated as modern reference points. Dietz and Holden (1970) indicate a position for India, 1, that is already slightly divorced from Antarctica, but according to McKenzie and Sclater (1973) a position near 2 seems likely. (Modified from Dietz and Holden, 1970.)

separated by a hot, tropical equatorial zone they could not penetrate? The latter was surely the case in the first half of the Cretaceous when angiosperms clearly were able to cross this zone repeatedly.

3. The southern land mass showed rather early rift, beginning in the Jurassic, into West Gondwanaland (South America + Africa) and East Gondwanaland (India + Antarctica + Australasia). According to Dietz and Holden (1970), a rift supposedly split the latter about the same time or shortly after Laurasia and Gondwanaland separated, with the result that India was cut free and initiated its migration

northward (fig. 1). If the Dietz and Holden reconstruction and their maps are reasonably accurate, then in a strict sense no unified continent of Gondwanaland ever existed for an extensive period. Paleobotanists such as Sahni (1939) and Schopf (1970a, b) have emphasized, however, the identity or near-identity of many elements throughout the Permo-Carboniferous *Glossopteris* flora, suggesting that a Gondwanalandic continent of considerable continuity must have existed. Indeed the continuity of a Gondwanalandic province—if not physically isolated region—is emphasized by McElhinny, Giddings, and Embleton (1974), who state that Gondwanaland existed at least 750 million years ago. The late Cretaceous distribution of other organisms lacking ability to traverse sea barriers (e.g., dinosaurs) suggests that, even that late, at least diffuse and temporary land connections must have existed between South America, Africa, and India (Hallam, 1972).[6] More weighty than arguments based on the fossil record are recent data of McKenzie and Sclater (1973), who claim that East Gondwanaland remained a single unit until India broke away "from Antarctica 75 million years ago." The Dietz and Holden reconstruction was derived prior to the existence of adequate paleomagnetic data on the Indian Ocean. Hence—and this point is critical—if McKenzie and Sclater's data are even roughly correct, then Indian plate migration was initiated only after the mid-Cretaceous; see figure 2. Extrapolating from their map, which shows the position of the magnetic anomalies in the Indian Ocean (McKenzie and Sclater, 1973, p. 66), one can only conclude that the earliest that India could have broken away from the remainder of East Gondwanaland is about 85–90 million years ago. At that time a wide diversity of angiosperms existed, including taxa belonging to many extant families. We shall return to the implications subsequently.

4. India, migrating northward some 5000 km (fig. 3), at an average rate of 7.5 cm per year, with rates varying from 16 cm per year to less than 6 cm per year, reached Laurasia some 40–45 million years ago, during the first third of the Tertiary; see also footnote 17.

5. The remainder of East Gondwanaland, Australasia and Antarctica, separated only in the Tertiary, about 45–47 million years ago, although there had been earlier, independent rifting—hence drift—along the eastern arc of the Australasian plate, so that New Caledonia, New Zealand, and peripheral islands represent the remnants of Cretaceous spreading-away along an arc paralleling eastern and

[6] Briden (1967) and McElhinny (1967) both present data, from diverging polar wandering curves, that indicate that significant dispersal—as distinct from initial rifting—of Gondwanalandic plates began only in mid-Cretaceous times, about 100 million years ago.

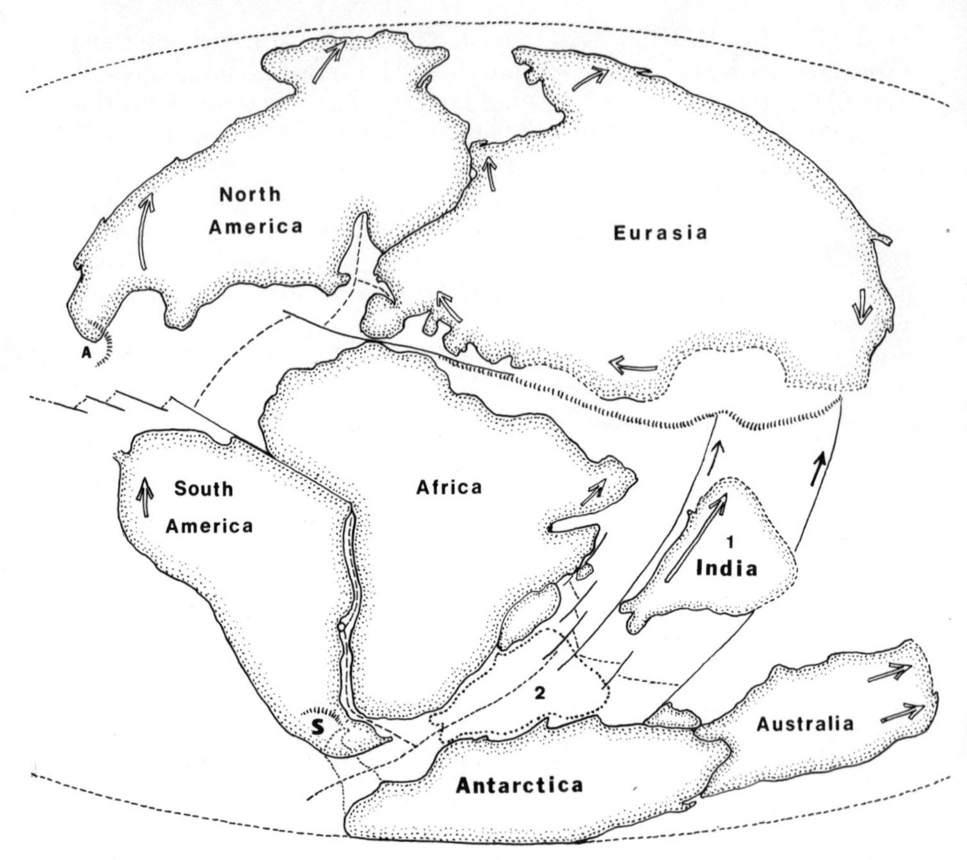

Fig. 2. The breakup of Gondwanaland after 65 million years of drift, at the start of the Cretaceous (135 million years ago). The North Atlantic and Indian oceans, according to Dietz and Holden, have begun to form; a slight rift indicates the very start of formation of the South Atlantic; Laurasia remains firmly united; according to Dietz and Holden (1970) the Indian plate has undergone about half its northward migration (1), but according to McKenzie and Sclater (1973) it still lies far to the south, perhaps as far south as 2; an intermediate position is most likely. (Modified from Dietz and Holden, 1970; symbols as in fig. 1.)

northeastern Australasia (fig. 28). The Australasian main plate—Australia with the Gondwanalandic sectors of New Guinea—probably did not approach the Eurasian plate until late in the Tertiary, perhaps later than 10–12 million years ago.

Although numerous details are uncertain, and a major item (the date of separation of the Indian plate) remains controversial, the general picture is one the biologist must live with. Other questions that carry with them major biogeographic uncertainties are whether, as Ridd (1971) and Hurley (1971) suggest, portions of Southeast Asia

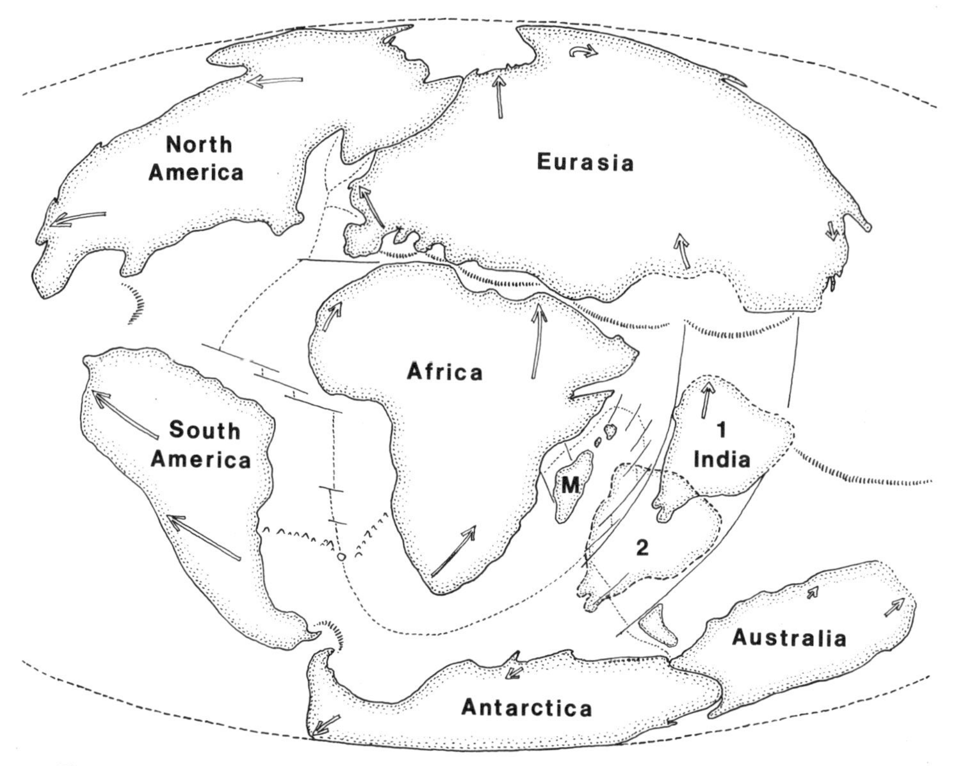

Fig. 3. Continental plates and their disposition at the start of the Tertiary (65 million years ago). The South Atlantic is fully open; Africa is isolated except from Laurasia, a rift having separated Madagascar from Africa. The position of India is questionable: at 1 according to Dietz and Holden (1970), but possibly at or near 2, extrapolating from McKenzie and Sclater (1973). By this time, former connections between South America and North America and Antarctica have been sundered. (After Dietz and Holden, 1970.)

were actually part of Gondwanaland; and whether the northern fringes of New Guinea incorporate Laurasian-derived fragments.

CLIMATIC GROUND RULES

The cliché that climates have varied in the past is, in effect, often overlooked by plant geographers. Migratory routes that now are open may have been closed in times past for purely climatic reasons; thus climatic barriers may simulate physical ones. Several general observations are pertinent to this central thesis.

1. Stepwise dispersal of organisms from one "fit" climatic zone to another is often impossible if an intervening "unfit" climatic zone exists. This principle may be critical in explaining why austral components in the present-day flora of India seem rather difficult to rec-

ognize, except perhaps in the Himalayan uplands. Migration of India through a tropical belt probably served to eliminate many cool- or cold-adapted taxa that could not evolve to adapt themselves to changing temperature stresses.

2. In this conjunction the peculiar significance of the Bryophyta in phytogeographic analysis deserves mention. These organisms occupy microhabitats, so that "their distribution is, within limits, largely a factor of the microenvironment" and, hence, we may expect them to "persist in small pockets where a suitable microenvironment persists, long after the general climate of the region has become very definitely 'inimical' " (Schuster, 1958).[7] These lines, written to explain persistence of certain elements in North America, apply equally well to the scenarios visualized for both Indian plate and Australasian plate migrations.

3. Organisms occupying microenvironments therefore are often better gauges of past phytogeographic connections than are large organisms that penetrate through several microclimatic levels. For this reason some emphasis is placed here on both the ecology and the phytogeography of such groups.

BIOLOGICAL GROUND RULES

It is unlikely that the dicta here embraced will prove acceptable to all biogeographers; I shall note the points at which there is likely to be controversy.

1. Migratory abilities of land plants are highly variable from group to group. This timeworn idea is mentioned first because unless we recognize its validity we tend to talk past each other: an example cited to make a point is countered by another example that tends to disprove it. In many cases a knowledge of the dispersibility of the organism is needed before existing patterns of dispersal become meaningful. It is useless to talk in identical terms about dispersibility of *Degeneria* and *Taraxacum*, or of *Phyllocladus* and *Hebe.* The first

---

[7] Many examples could be suggested; an obvious one is *Athalamia hyalina*, an arctic member of the Marchantiales, widespread from lat. 70–83°N., which survives as a Pleistocene relict in the "Driftless Area" of Minnesota. Here it is found not with its usual associates (*Saxifraga oppositifolia, Salix arctica, Papaver radicatum*), but on *Tilia*-forested, steep slopes and bluffs (Schuster, 1953). *Rhododendron lapponicum* in the "Driftless Area" of Wisconsin, as a Pleistocene relict, is a similar example. We must thus assume survival, in perhaps very local areas retaining appropriate microenvironments, of at least some cool-adapted taxa; these taxa, biotically depleted and stenotypic by the time of their arrival, in the case of many of the bryophytes, often survive today only as asexually propagating units. Those that did survive frequently underwent rapid speciation and diversification when they again reached appropriate climatic zones, at times (as with present-day holarctic Polytrichaceae) becoming circum-Laurasian (G. L. Smith, 1972).

member of each pair shows little or no long-distance dispersibility; the second members are relatively easily dispersed (as the Recent history of *Hebe* in subantarctic South America shows).

2. Both intrinsic and extrinsic abilities to migrate have tended to become greater with increased evolution of a group. This dictum is a most important one to keep in mind because, if we can show that it is truly valid, then we must surrender as untenable any arguments based solely on present-day migratory abilities. In brief, I contend that primitive members of any major group (order, family) often show very limited intrinsic long-distance dispersibility; they typically exist in an environment where they have not shown coadaptation with potential outside dispersers (e.g., fruit-eating birds and bats, certain insects, various mammals). Several examples follow.

With regard to bryophytes, it is clear that intrinsic ability to spread effectively is limited in "lower" groups by two important factors: unlike in *all* early tracheophytes, such as ferns, gametophytes are usually unisexual, with spores therefore carrying an X *or* a Y chromosome—hence effective dispersal must involve spores of both types landing near each other, undergoing germination, and establishing female and male populations (fig. 4). Statistical chances of this happening have been shown to be very low (see Schuster, 1969), but I also noted a significant "monoecious shift": the few bipolar Hepaticae, which obviously acquired their grossly disjunct dispersal by long-distance saltation, show high levels of monoecism. Also, lower groups rarely show asexual (gametophytic) reproduction; propagula or "gemmae" typically occur chiefly in highly evolved groups. The principle here, then, is that higher groups (such as *Bryum* in the mosses; the bisexual and gemma-producing Paradoxae in the Lejeuneaceae among Hepaticae; certain bisexual and gemmiparous Anthocerotae) have evolved devices that allow effective migration, as well as strategies (gemmae and other propagula) that facilitate dispersal asexually.[8]

With respect to seed plants, from group to group one notes similar large advances in intrinsic dispersibility. Evolution of winged seeds (as in *Pinus*, various Bignoniaceae) or fruits (*Acer, Ulmus, Ptelea*), or seeds or fruits with parachutes (*Asclepias* and *Taraxacum*, respectively), or minute and wind-borne seeds (Orchidaceae) is almost

[8] Other extraordinary devices can be cited, such as large spores (to 200 μ) which show long viability, as in *Riccia* and certain Anthocerotae; and cohesion of spores in tetrads, so that two X-carrying and two Y-carrying spores are permanently linked, as in *Riccia* spp., *Sphaerocarpus* spp., and *Cryptothallus*. Groups showing such extraordinary criteria are highly specialized in other ways as well; they often show very wide or disjunct ranges. For that reason they are not utilized here.

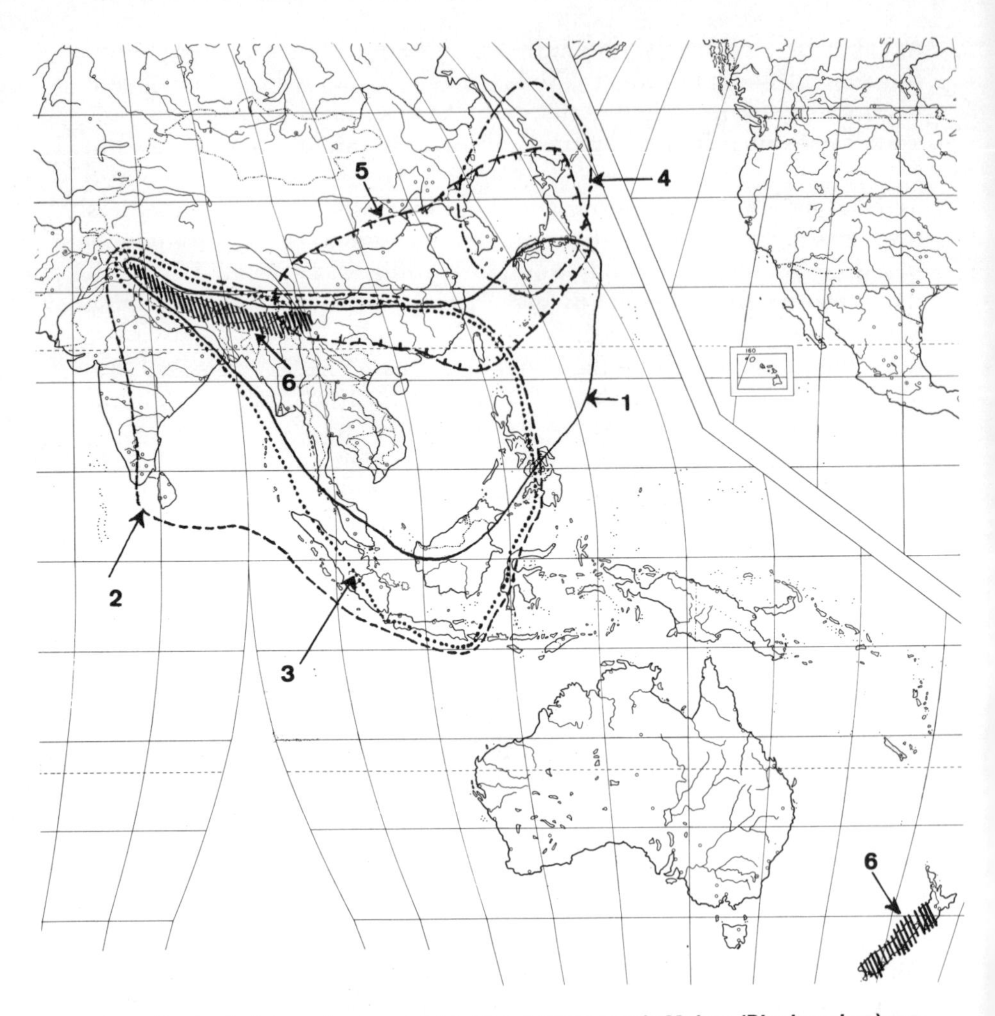

Fig. 4. Basically Laurasian ranges for Musci (Mniaceae): *Mnium (Plagiomnium) rostratum* and allies in southeastern Asia (sect. Rostratae). (1) *M. succulentum.* (2) *M. rhynchosporum.* (3) *M. integrum.* (4) *M. vesicatum.* (5) *M. maximoviczii.* (6) *M. rostratum.* Five (out of six) species are unisexual and of relatively limited range, not having extended their range beyond Wallace's Line. *Mnium rostratum,* which is synoecious, has (via casual, long-distance transport) attained New Zealand; it is also widespread in Europe and North America and is a common temperate boreal element. *Assumption:* The group is basically Laurasian, with inability to extend its range via long-distance transport, and strictly limited by unisexuality (a spore carries a Y or X chromosome but not both). *M. rostratum* illustrates the expanded possibilities available with a "bisexual" spore. *Point:* unisexual taxa generally show limited long-distance dispersibility. (After Koponen, 1972.)

always characteristic of moderately or highly advanced genera and families. Relatively bulky, nonwinged seeds (or fruits) characterize most primitive or unspecialized taxa, for example, most lower groups of angiosperms and of conifers. (There are obvious exceptions—the minute seeds of *Trochodendron* are surely an advanced character.) In general, intrinsic dispersibility of seeds and fruits shows a steady rise as we proceed through the various evolutionary stadia; this is also true of conifers (for example, *Juniperus* and *Taxus*).

Dispersibility owing to outside agents also tends to increase. Various "lower" conifers and angiosperms generally lack edible covers for seeds or fruits.[9] This is true also, in many instances, within single families or orders. In Ranunculaceae, follicles (and later achenes) are found in almost all taxa; only advanced taxa (e.g., *Actaea*) bear fleshy, highly colored, animal-dispersed fruits, or (in *Clematis*) wind-distributed fruits. In Rosaceae identical patterns recur: primitive groups (e.g., *Spiraea* and *Physocarpus*) have dry, follicular dehiscent fruits, while with evolutionary advance a fantastic range of edible, animal-dispersed fruits of all types developed (*Fragaria, Rubus, Rosa, Prunus, Malus,* to cite only a few extremes).[10] These same groups may also exhibit enhanced dispersibility owing to evolution of devices that allow dispersal on the feathers or fur of animals. The hooked achenes of taxa such as *Ranunculus* and *Acaena* are clearly evolutionary advances and are linked with wide and disjunct dispersal (e.g., on subantarctic islands).

A striking case of relative dispersibility closely linked with dia-

---

[9] Evolution of these groups occurred at a time when modern birds and mammals did not exist. James Doyle has also called my attention to the fact that in seed ferns going back to the Carboniferous, up to *Caytonia*, fleshy cupules or sarcotestas existed, as also in Pentoxylales and Cycadales. In cycads the pink, fleshy, outer seed coat of *Zamia* seems a clear—and ancient—device for animal dispersal. There is direct evidence (from coprolites) that *Caytonia* was animal dispersed—but obviously not by birds or higher mammals, presumably by reptiles ("saurochory" of van der Pijl).

I think a distinction between reptilian *seed* dispersal, evolved prior to the Cretaceous in gymnospermous groups, and "modern" bird and mammal *fruit* dispersal, evolved independently in angiosperms, must be drawn. I am sceptical of conclusions of, for instance, van der Pijl and Stebbins, that animal dispersal of seeds (as distinct from that of fruit) is primitive in angiosperms, although those taxa with a sarcotesta (e.g., *Magnolia*) may be examples of very early adaptation to animal dispersal. I think, rather, that very early diversity of primitive angiosperms occurred in this respect: certain Magnoliidae showed evolution of a sarcotesta; the presumably almost contemporary primitive Hamamelidae (e.g., Trochodendraceae, Tetracentraceae, Cercidiphyllaceae) appear never to have evolved a sarcotesta.

[10] The implied phylogenies were evolved before the phytogeographic arguments stated here.

spore type is the Empetraceae. *Corema* and *Ceratiola* are of limited range in the Northern Hemisphere; they have two and one species, respectively, and have dry fruits that release dry seeds. The third genus, *Empetrum*, has a fleshy, edible fruit; this genus has achieved (probably since late Tertiary times) a bipolar range. This and analogous cases are discussed later.

The point emphasized by such examples is that one cannot argue as to effectiveness, hence rapidity, of dispersal on the basis of modern families and genera. It is useless to try to formulate dispersal principles from the range of a modern and highly dispersible group such as the Compositae, which have effective dispersal agents like the parachute of *Taraxacum* and the barbed pappus awns of *Bidens*. Furthermore, there is a clear and obvious correlation between the time of evolution of dispersal agents (chiefly groups of birds and mammals) and evolution of modern taxa of flowering plants. Modern genera of Rosaceae, with fleshy fruits, and even genera like *Acaena*, with hooked achenes, arose as a result of coevolution with their dispersal agents [11]—and these agents are modern groups. The more primitive birds, both toothed and toothless, are almost all flesh eaters like their reptilian ancestors; seed- and fruit-eating modern taxa came later in avian evolution. Similarly, noncarnivorous mammals (which, unlike their reptilian ancestors, are often fruit eaters) evolved relatively late, and fruit-eating bats are a very late development of the mammalian line.

In general, then, with increased time and continuing evolution the modern groups, whose long-distance dispersibility becomes progressively greater, tended to arise from generalized groups with limited dispersibility. It is thus axiomatic that long-distance dispersal of, especially, primitive and generalized groups is usually unlikely, and often impossible. A corollary assumption is that early in the evolution of a specific group its long-distance dispersibility was relatively limited.

3. Migration, except under special circumstances, takes place not as individual taxa, but as communities. Except in the special case of "new" islands, or otherwise depauperate areas (e.g., the Arctic) where competition for one reason or another is reduced—and initially often lacking—extant and relatively closed communities tend

---

[11] The evolution of a wide diversity of fleshy-fruited Rosaceae in Laurasia, and their absence in Gondwanaland, appears clearly linked with evolution of modern mammals in Laurasia, and their relatively late dispersal to Gondwanalandic regions. Similarly, the wide, disjunct oceanic dispersal of *Acaena* and *Ranunculus* spp. in the Antipodes, and elsewhere, reflects achene evolution and (in part) dispersal by sea birds.

to repel invaders. Hence, casual, long-distance migration by isolated diaspores of isolated taxa tends to be far less effective than often assumed. The intruder is usually confronted by a situation in which it stands little or no chance: the ecological niche the "immigrant" demands is, if present, already occupied by one or more autochthonous taxa that effectively exploit it. Advocates of unlimited dispersibility must ponder the fact that, prior to the advent of the white man in eastern North America, none of the Eurasian weeds so rampant today were able to establish themselves in North America. Examples are *Taraxacum officinale,* various *Hieracium* species, with plumed achenes; *Solanum nigrum* and *S. dulcamara,* with fleshy fruits. *Ulex,* a pernicious weed in New Zealand, and *Opuntia,* similarly weedy in Australia, are other examples.[12] Countless others could be cited. Only after the integrity of the autochthonous communities was impaired (e.g., by overgrazing) did these invaders succeed. Yet many of these taxa possess exactly the kinds of dispersal agents that allow rapid migration. It seems inconceivable that, for example, *Taraxacum officinale* achenes were not blown across the Atlantic prior to the year 1600.[13]

A corollary of this "rule" is that casual, long-distance immigration tends to be effective only under special circumstances, such as the establishment of taxa on raw ground—as on the Hawaiian Islands. By contrast, recent long-distance dispersal, not man-induced, in other cases is often exceedingly difficult to demonstrate. Among angiosperms, a very few well-documented cases involve: a species of *Sisyrinchium* (Iridaceae; typical of eastern and middle temperate North America) in a few fjords in West Greenland; single species of *Eriocaulon* (*septangulare*) and *Spiranthes* (*romanzoffiana*), native to eastern North America, established in western Ireland and Scotland. These are surely recent immigrants; their very rarity is noteworthy.

We must recognize then that autochthonous communities tend to repel invaders—hence the effective migratory ability of an organism

[12] It surely is hardly fortuitous that the *only* genus of the almost wholly New World family Cactaceae to have spread to the Old World is *Rhipsalis,* a genus with small, fleshy fruits, easily bird-dispersed, that grows typically as an epiphyte in situations with reduced or no competition.

[13] However, if depauperization of the biota of a region occurs, a rapid flood of modern immigrants can almost be assumed. Thus it seems likely that the rather large number of Laurasian-derived taxa that have attained footholds in New Guinea and northern Australia represent principally taxa that are adapted to tropical and subtropical, high-rainfall climates. Such taxa were in short supply in Australasia, since this plate started from a far southern position, moved through the relatively dry horse latitudes, and only recently arrived (the northern section of it) in equatorial regions.

tends to be much lower than one would at first glance assume. This operating principle leads naturally to number 4.

4. Highly disjunct ranges have in most cases arisen by selective extinction and not by long-distance dispersal. The literature of systematics and biogeography is full of examples of extreme disjunctions in distribution, ranging from recent, bona-fide long-distance migrants such as those cited above, to a host of taxa that today have dissected ranges due to widespread Pleistocene extinction. Such extreme disjunction is striking in the Southern Hemisphere where, as Skottsberg (1925) and Schuster (1969) have noted, the effect of Pleistocene glaciation was catastrophic, with the biota of an entire continent destroyed. Such disjuncts are so well documented that they require no summary.

However, a basic point needs emphasis: many if not most instances of extreme disjunction are "ancient" disjunctions, achieved by extermination over much of the range of the taxon (fig. 5). The disjunct occurrences of *Shortia, Liriodendron,* and innumerable other taxa between eastern Asia and eastern North America have attracted inordinate attention from the time of Asa Gray onwards (see, especially, Li, 1971). Such disjunctions are clearly due to Tertiary and Pleistocene extinction, except for local areas with persistence of favorable climatic conditions. Such extinction, and the resultant relict distribution, is generally not so self-evident from angiosperm fossil history as from the fossil history of the gymnosperms (Florin, 1963) (figs. 7–10); but see also fig. 37.

5. Concentration of relict groups in an area does not prove that it is the center of origin of the group or groups. The high incidence of generalized taxa of Angiospermae in the region from southeastern Asia to New Caledonia, Tasmania, and New Zealand reflects the occurrence of centers of relict climatic conditions (and absence of extensive Pleistocene glaciation) rather than a center of origin (Schuster, 1972a). It has been asserted that if we plot extant ranges of primitive angiosperms they show a significant concentration in southeastern Asia, as do the gymnosperms (figs. 6, 8–10). However, if we plot the *fossil* occurrence of these gymnosperms, excluding those of Gondwanalandic origin (fig. 7), it is immediately apparent that southeastern Asia is a center of survival and not necessarily a center of origin (Florin, 1963). Most, indeed nearly all, of these gymnosperms occur today as members of angiosperm-dominated communities; we probably must assume a similar history for primitive angiosperms in the period since the middle Cretaceous.

6. Migration—and disjunction—of taxa may be achieved by mi-

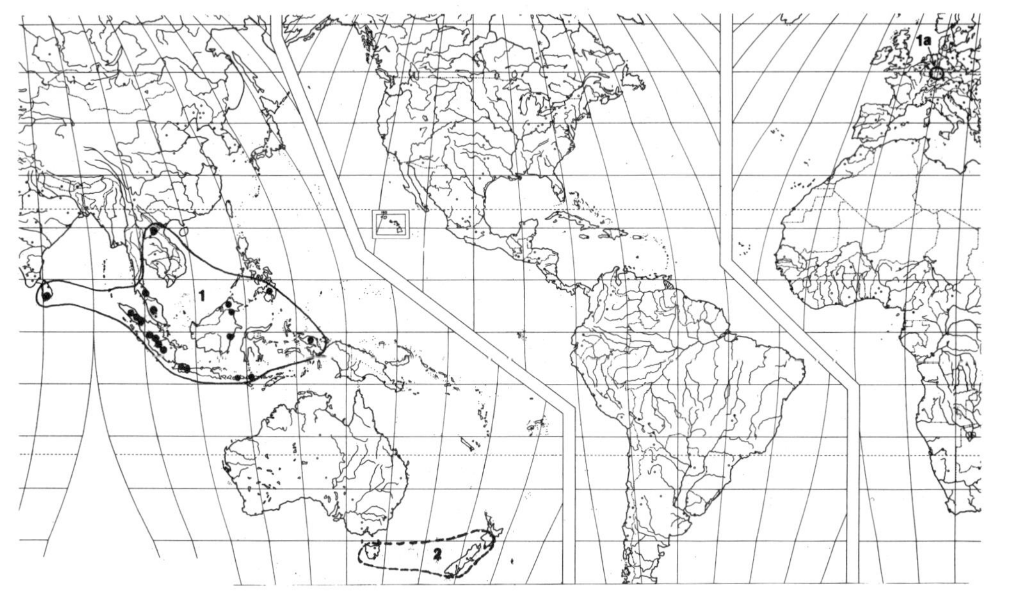

Fig. 5. Past and present total range of the specialized moss family Ephemeropsidaceae, with a single genus, *Ephemeropsis.* This is a highly derivative, "neotenic" genus like *Metzgeriopsis* and is found only on living evergreen leaves; see Meijer (*J. Hattori Bot. Lab.* 36: 552–55). Present range: (1) *E. tjibodensis* Goebl., from Ceylon and Tonkin region to New Guinea. (2) *E. trentepohlioides* (Renn.) Sainsb., on leaves in subantarctic *Nothofagus* region. *Ephemeropsis tjibodensis,* in addition to the present range (1, dots) is known from an Eocene site at Halle, Germany (1a). The combined range of (1) and (2) simulates the range of "ancient" angiosperms; it also includes the modern range of a host of conifers (*Keteleeria, Pseudolarix, Cephalotaxus, Amentotaxus*) that once were much more widespread and, in part, occurred scattered throughout much of Laurasia. *Assumption:* the range is relict; the family probably was originally Laurasian but, starting with the Tertiary, was eliminated from most of its former range. The Australasian range is secondary.

gration of land masses. An assumption central to much of what follows is that although an organism may unquestionably migrate long distances (as undoubtedly did *Empetrum* and certain species of *Carex*), such migrants are usually recognizable as foreign or disharmonic elements. Classic examples include *Calypogeia sphagnicola, Marsupella ustulata, Lophozia hatcheri,* and *Anthelia* found in antarctic or subantarctic loci as isolated disjuncts—with the bulk of their ranges (and all their proximate allies) essentially arctic (Schuster, 1969, 1972a).

Of greater significance, however, is the mass migration not of isolated taxa but of entire biotas. This has happened at least twice

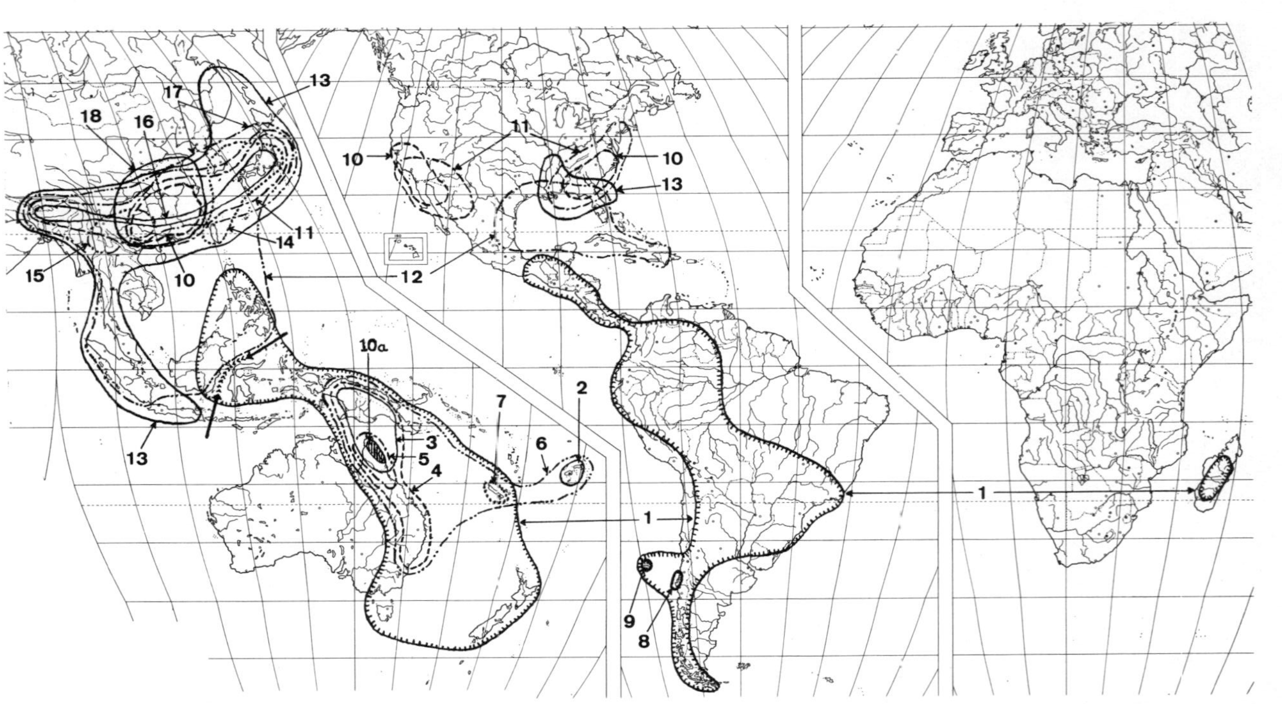

Fig. 6. Generalized distribution of 19 primitive families of woody Dicotyledonae. (1) Winteraceae. (2) Degeneriaceae. (3) Himantandraceae. (4) Eupomatiaceae. (5) Austrobaileyaceae. (6) Trimeniaceae; 1 species reaches Marquesas. (7) Amborellaceae. (8) Gomortegaceae. (9) Lactoridaceae. (10) Calycanthaceae. (10a) Idiospermataceae. (11) Saururaceae. (12) Illiciaceae. (13) Schisandraceae; *Kadsura* reaches Amboina and peninsular India. (14) Trochodendraceae. (15) Tetracentraceae. (16) Eupteleaceae. (17) Cercidiphyllaceae. (18) Eucommiaceae. See figure 36 for distribution of the Magnoliaceae. The extension of Winteraceae beyond Australasia is by means of 1 species (out of about 90 known). Arrows indicate the approximate position of Wallace's Line. Distribution is generalized and in part schematized, for maximal clarity; the families involved are often very strongly disjunct and absent from many of the intermediate areas from which they are indicated.

with regard to Asia: first the Indian plate juxtaposed a biota of unquestioned Gondwanalandic origin and a biota of Laurasian origin; later—late in the Tertiary—a second Gondwanalandic biota, in most respects very different quantitatively and qualitatively, was passively rafted on the Australasian plate northward until it came into competitive contact with a tropical Laurasian biota.[14]

In essence, long-term mass migration is visualized in which various taxa may be likened to passengers on a ship. The duration of migration is such that the passengers, in some cases, die out (as the "ship" drifts through climates ill-suited to support the taxon); in most other cases they slowly evolve. Hence, when the ship "docks," descendants of these passengers, although still showing traces of their origin, may now be very different, representing distinct species, genera, and some having even evolved into distinct families (compare the histories of the Scapaniaceae and Delavayellaceae [Schuster, 1974]; see fig. 13).

Several facts about these mass migrations are of major interest. First, both occurred well after angiosperms had evolved, and the Australasian plate migration long after many modern families were on the scene. By the mid-Cretaceous, when Indian migration started, angiosperms were already widespread. Hence we cannot assume that Indian plate migration was relevant to the initial spread of the angiosperms. Second, it seems likely that many groups of Gondwanalandic origin were introduced into the Laurasian flora, which thus acquired a high level of complexity. This is discussed later in more detail. Third, Australasian plate migration, starting well after the beginning of the Tertiary, was sufficiently late as to be irrelevant to the problem of early dispersal of the angiosperms. However, it is critical in having juxtaposed primitive relicts of the Gondwanalandic angiosperms (such as the Winteraceae, Eupomatiaceae, and Austrobaileyaceae) and Laurasian relicts (such as Magnoliaceae, Schisandraceae, Illiciaceae); see figure 6 and a following detailed discussion. The Indian plate and Australian plate migrations thus served to erase many distinctions between early Laurasian and Gondwanalandic angiosperm floras that had arisen, possibly as a result of climatic barriers, by mid-Cretaceous times. Perhaps the start of Indian plate migration was early enough (75–90 million years ago) that some old elements of Gondwanalandic origin were introduced into the Laurasian flora; for example, the Proteaceae (fig. 14) and Restionaceae (fig. 15) and possibly Lardizabalaceae, among angiosperms. It must be

---

[14] For the moment the possibility that parts of Southeast Asia formed a third such "migrant" (Ridd, 1971) is discounted as not effectively established.

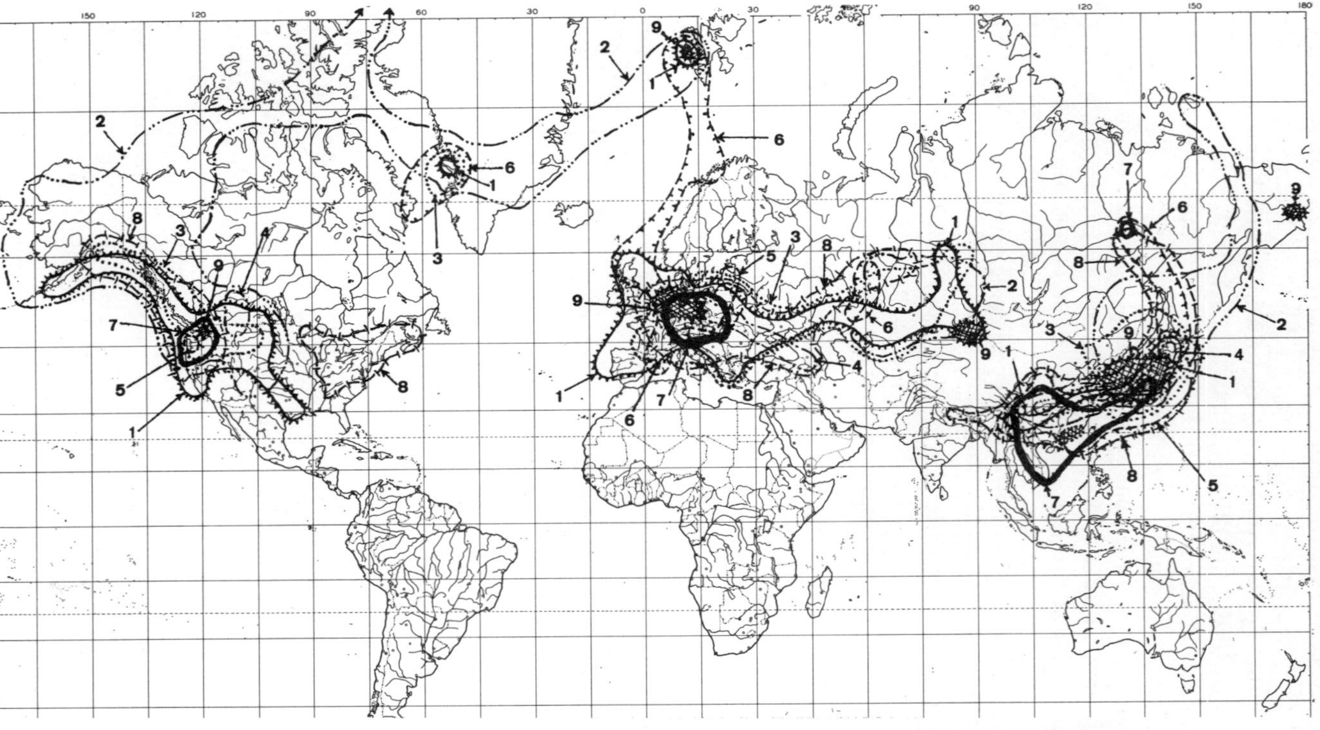

Fig. 7. Past and present range (highly generalized) of nine genera of conifers: (1) *Sequoia.* (2) *Metasequoia.* (3) *Glyptostrobus.* (4) *Cryptomeria.* (5) *Cunninghamia.* (6) *Sciadopitys.* (7) *Keteleeria.* (8) *Tsuga.* (9) *Pseudolarix.* Note that instead of one center of survival, corresponding roughly to the present range of *Keteleeria* (outlined in heavy solid line in fig. 8), there are three centers of known fossil diversity (corresponding to the heavy outline, the fossil plus present range of *Keteleeria,* in this map). Indicated absence of some (or most) of these genera from many areas probably reflects inadequate conditions for fossilization or inadequate sampling, not their actual lack of occurrence.

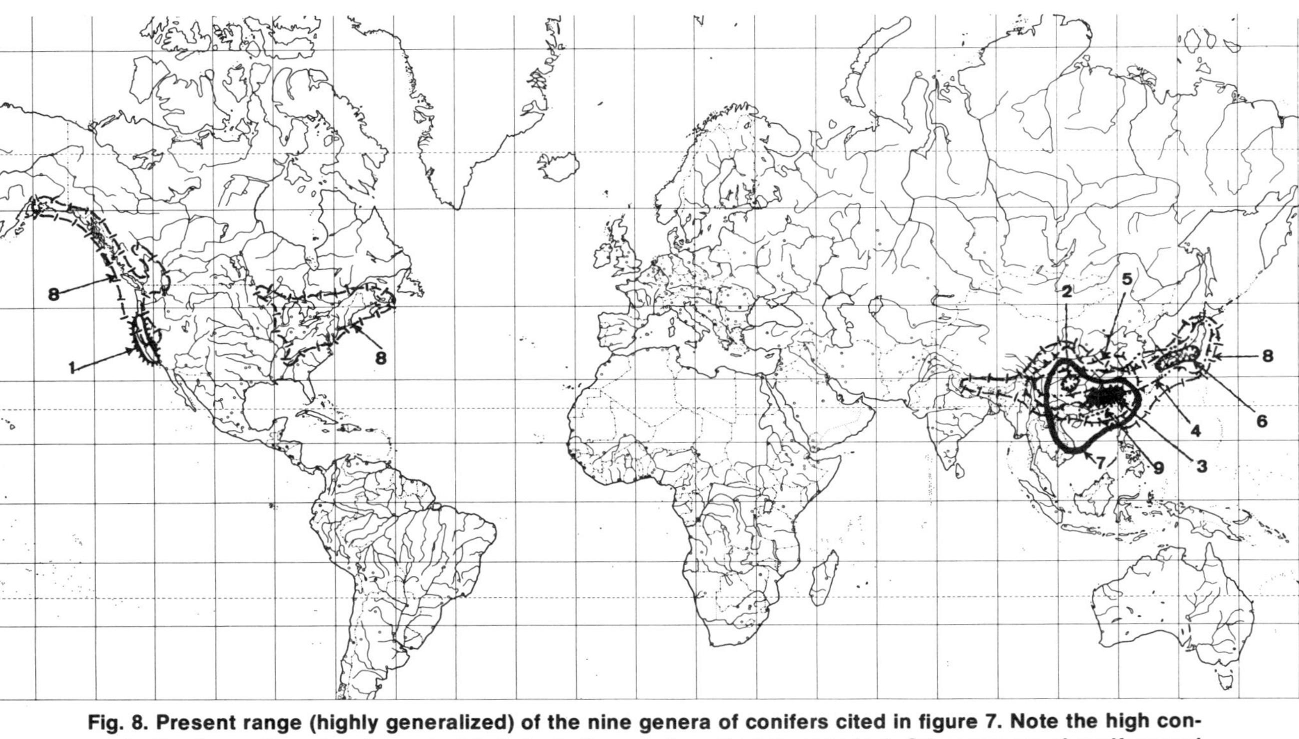

Fig. 8. Present range (highly generalized) of the nine genera of conifers cited in figure 7. Note the high concentration in southeastern to central Asia (only *Sequoia* is extinct there today). Other genera of conifers and taxads could be added to reinforce the concept of southeast Asian survival, e.g., *Cephalotaxus, Amentotaxus, Torreya, Pseudotsuga;* all these genera, at one time, had a greatly expanded range and today are clearly relict. See maps in Florin (1963). Compare distribution of Magnoliaceae and Calycanthaceae, figures 36 and 6. (Derived from data in Florin, 1963.)

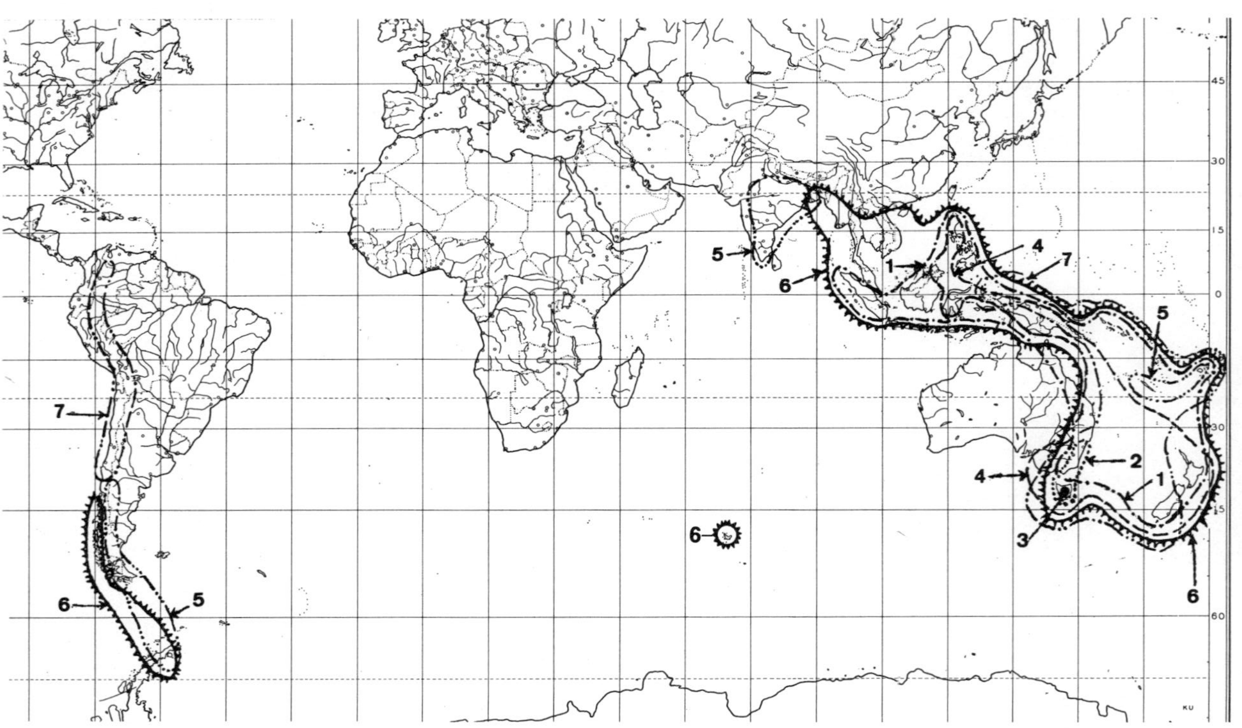

Fig. 9. Present and past range of seven Gondwanalandic taxa of conifers: (1) *Agathis.* (2) *Microstrobus.* (3) *Microcachrys.* (4) *Phyllocladus.* (5) *Acmopyle.* 6) *Dacrydium.* (7) *Podocarpus* sect. Polypodiopsis. Note known past range of two (*Acmopyle, Dacrydium*) from Antarctia and India. Note that instead of the two genera (*Agathis, Microstrobus*) present today in Australia, five of the seven were present. By contrast to the present versus past-plus-present range of Laurasian groups (compare figs. 7 and 8) these present-day ranges seem relatively slightly contracted. This is an artifact, since probably all of these genera occurred in Antarctica, whose fossil record is hardly known because of the extensive ice cover today. (Generalized from Florin, 1963.)

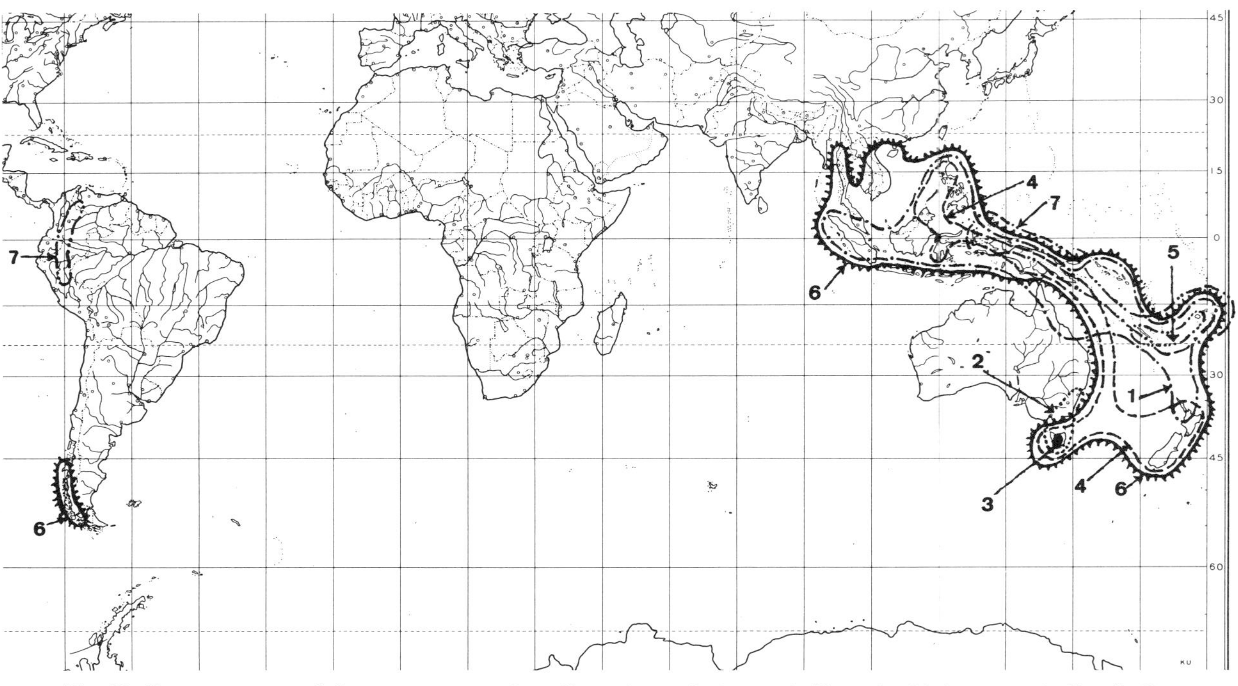

Fig. 10. Present range of the seven taxa of conifers shown in figure 9. Note the high concentration in the frontal arc of Australasia, from New Zealand and New Caledonia to Fiji, south to Tasmania and to eastern Australia (climatically very restricted occurrence of only 1 and 2). The genera that occur today in Laurasian loci (*Agathis, Phyllocladus, Dacrydium*) do not occur in non-Gondwanalandic regions in the fossil record; their current Laurasian range is believed to be recent. Compare the preceding figure. Compare also the distribution of Winteraceae (fig. 35).

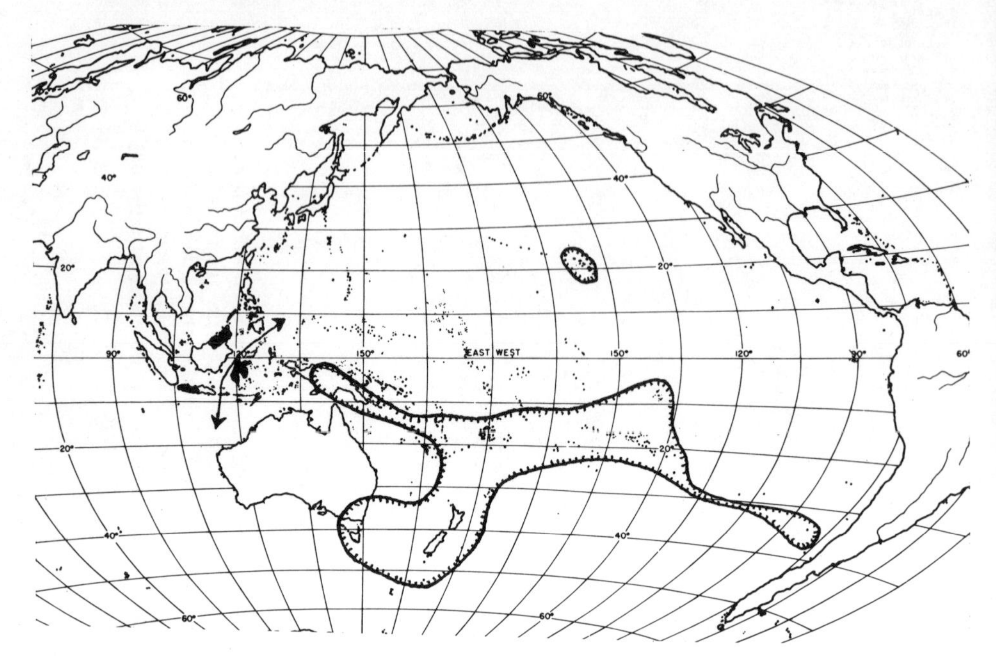

Fig. 11. Basically Gondwanalandic range of *Coprosma* (Rubiaceae). Arrows indicate Wallace's Line. Basic range enclosed in ringed areas; isolated species shown by blackened areas. *Question:* the fruits are fleshy and eaten by birds—why has there been no greater transgression of Wallace's Line? *Assumption:* invasion into the Laurasian communities (which are closed and not depauperated) is more difficult than the reverse direction, into the depauperated Australasian communities. (After Good, 1964.)

granted, however, that these plate migrations could not serve to effect the primary spread of primitive angiosperms, which was accomplished by mid-Cretaceous times. Some reasons for the rapid initial spread of the early angiosperms are touched on in the concluding section of this paper.

7. Disharmonic floras may arise by casual migration of individual taxa, by migration of land masses carrying these elements, or by a combination of the two factors. The existence of foreign elements in a flora has always caused biologists to ponder the origin of such disharmony. In the case of recent island chains, such as Hawaii (including Midway, not over 25–30 million years old; the Hawaiian Islands proper, only some 5 million years old), the entire flora is disharmonic; subantarctic elements like *Coprosma* (fig. 11) and *Gunnera* exist side-by-side with taxa derived from (probably) Australasia, such as *Metrosideros*, with others derived from southeastern Asia,

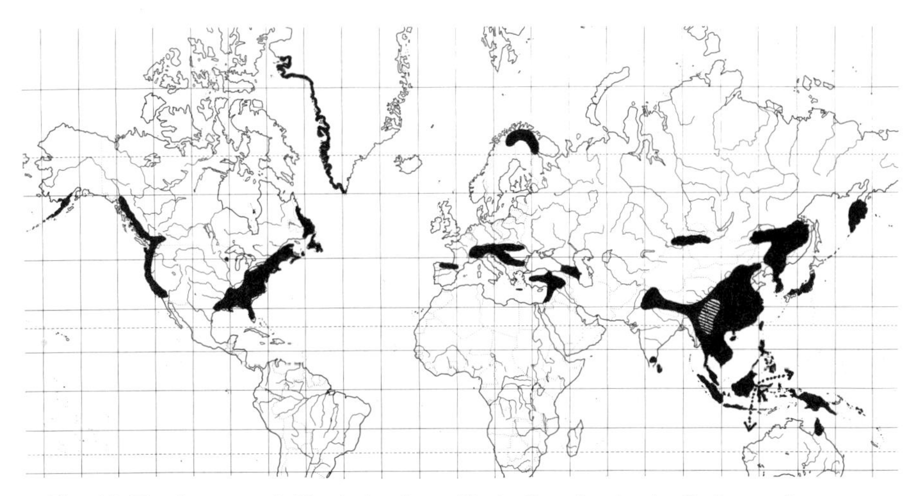

Fig. 12. Total range of *Rhododendron*, illustrating the basically Laurasian range with rapid extension (and subsequent speciation) west of Wallace's Line (dotted arrows), especially in upland New Guinea, after the uplift (in the last 1–3 million years) of the last island. Only isolated taxa reach Australia and the Solomon Islands. (Range showing late Tertiary and Pleistocene contraction: *R. ponticum*, now in southern Europe, was in Tertiary times in Britain.)

and still others from North America. In such instances long-distance immigration from a number of directions is clearly evident.

However, other disharmonic floras occur which had a much more complex history. Thus, at 8,000–10,000 ft in New Guinea one may find growing together or in neighboring sites such disparate elements as *Hebe*, *Pernettya*, *Nothofagus* (fig. 39), and *Astelia*—all subantarctic in origin—growing with or near *Castanopsis*, *Rhododendron* (fig. 12), and other genera of unquestioned Laurasian origin. The disharmonic nature of such a flora is perhaps partly the result of recent immigration (seeds of *Rhododendron* are small and winged, and perhaps easily wind-disseminated; *Pernettya* and *Astelia papuana* have pink or red, fleshy, edible fruits). Other taxa cannot be so readily explained away; for instance, *Nothofagus*, a genus which all students agree requires land for its dispersal. In New Guinea, where intermingling of highly distinctive floristic elements occurs and where the extant flora is strikingly disharmonic, it is difficult if not impossible to account for the disharmony simply on the basis of casual migration over long distances, especially since the disharmony seems clearly due to mingling from chiefly two source areas.

Similarly, with oceanic elements in the Himalayan flora, in which we find a large number of Gondwanalandic taxa, the extensive disharmony is not explicable on the basis of casual, long-distance transport.[15] In both cases the disharmony is not casual but exhibits a clear pattern. It is likely that such patterned disharmony reflects mass introduction, probably at one time. I therefore assume that such patterns have a more profound meaning than do evidently casual disjuncts.

Indian Plate Migration and its Effects

The migration northward of the initially Gondwanalandic Indian plate [16] until it became attached to Laurasia is an established fact; only its original latitude and the exact timing are debatable. This migration leads to major questions: What, if any, consequences did it have on the Tertiary history of the Laurasian flora? What plants were rafted northward on this plate? It is inconceivable that a land mass of many thousands of square kilometers underwent its migration de-

[15] With regard to the Hepaticae at least, almost all disjunct taxa present there occur as *unisexual* populations (fig. 13). The Himalayan flora does not show a trace of the monoecious shift seen when we examine the moss flora of Hawaii, or of the bipolar taxa that have relatively recently acquired scattered antarctic stations (see, e.g., Schuster, 1969, 1975). This strongly suggests that such elements, disjunct today in the Himalayas, did not arrive by long-distance transport.

[16] I refer here to the "Indian plate" without attempting to clearly establish its perimeters; Madagascar may have been attached to it. Currently accepted data indicate that this plate reached Laurasia about 45 million years ago, and its leading edge was subducted under Laurasia, resulting in the rapid and phenomenal elevation of the Himalayan system. However, inconclusive evidence has been recently offered to the effect that portions of central China (Hurley, 1971)—the critical region where many archaic families and genera of angiosperms are concentrated—and possibly also sectors of southeastern Asia (Ridd, 1971) represent Gondwanaland-derived fragments. If so—and the distribution of certain Mesozoic reptiles suggests it may have been (Colbert, 1973)—then, just possibly, a Gondwanalandic origin of various primitive families such as Magnoliaceae, Illiciaceae, Trochodendraceae, and Tetracentraceae may be postulated. The high incidence of originally Gondwanalandic taxa of Hepaticae (and migration patterns of certain Polytrichaceae) would also be more easily comprehensible if Yunnan and Szechwan and peripheral areas prove to be Gondwanalandic in origin. One reason Ridd's paper and its implications are largely discounted here is that Heirtzler, et al. (1973) show that in the area where Ridd's reconstruction positions Indochina, sea floor dated at 140 million years exists. Dietz and Holden (1971) also date the oceanic crust in the eastern Indian Ocean as "pre-Mesozoic." Burton (1970) indicates that in Paleozoic to early Mesozoic times a land-mass, presumably India, existed immediately west of the Malay Peninsula. If Indochina and Malaya were, indeed, once part of Gondwanaland, then they must have constituted a fragment that migrated north earlier than India, more than 140 million years ago. Admittedly, there is space between the eastern edge of India and western margin of Australia in various reconstructions (e.g., Smith and Hallam, 1970) into which one could tuck Indochina-Malaya. If Indochina-Malaya once occupied this position, it may have been before the day of the angiosperms, judging from sea-floor spreading.

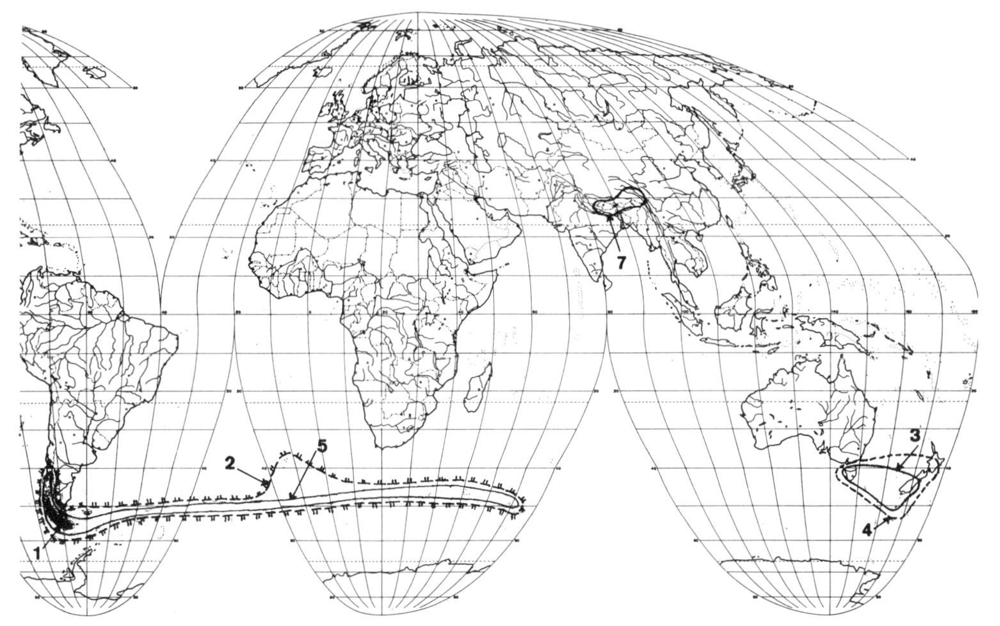

Fig. 13. Total range of Scapaniaceae subfam. Blepharophylloideae (1–6) and Delavayellaceae (7). The Blepharophylloideae, restricted to the *Nothofagus* zone except for several taxa that extend to subantarctic islands, are primitive (rhizoids in fascicles; antheridial stalk biseriate; perianth unspecialized; terminal branching retained). The group, and hence family, is presumably Gondwanalandic in range. Derived from it is the subfam. Scapanioideae (with numerous taxa, essentially all Laurasian; there is a high concentration of taxa in the Himalayas, belonging to at least two genera and six or seven subgenera; secondarily a few isolated taxa have dispersed to the Pacific and one, *Scapania gamundiae* Schust., to Tierra del Fuego; see Schuster [1969] for map) and the Delavayellaceae (sometimes considered a third subfamily of Scapaniaceae). The ancestors of both groups presumably rafted northward on the Indian plate, with dispersal outward of Scapanioideae, while the highly derivative group Delavayellaceae (monotypic) has not been able to expand its range. *Blepharidophyllum:* (1) Subg. *Isodiplophyllum* Schust., with only *B. squarrosum* (Steph.) Schust. (2, 3) Subg. *Blepharidophyllum* Ångstr. (2) *B. densifolium* (Hook.) Ångstr. (3) *B. vertebrale* (Tayl.) Ångstr. (4–6) Subg. *Clandarium* Grolle. (4) *B. xiphophyllum* Grolle. (5) *B. clandestinum* (Mont.) Grolle. (6) *B. gottscheanum* Grolle; range as in 1. (7) *Delavayella serrata* Steph.

nuded, so to speak. These questions are interrelated and, equally obviously, the answers to some extent will be based on when this migration occurred. On the basis of a supposed late Triassic start (180 million years ago) of Indian plate migration (Dietz and Holden, 1970), serious doubt is cast on the possibility that early angiosperms could have been rafted northward on the Indian plate from Gondwanalandic to Laurasian regions (Schuster, 1972a). However, recent data of McKenzie and Sclater (1973) suggest that India started its northward trek "after breaking away from Antarctica only 75 million years ago" and moved northward approximately 5000 km at an average of 7.5 cm per year.[17]

If we extrapolate from McKenzie and Sclater's map (1973, p. 66), rifting possibly initiated as much as 95 million years ago (fig. 26). If we accept a figure of 75–95 million years ago, then Indian plate migration started at a time when many modern families and even genera of Angiospermae existed. It is also evident (McKenzie and Sclater, 1973) (fig. 27) that until at least 75 million years ago there were archipelagic connections between the Indian plate, the Mascarene Plateau, Madagascar, and Africa, and perhaps diffuse ones remained to the Australasian-Antarctic plate via the Crozet Plateau and Kerguélen Plateau. (Some of these plateaus have existing islands far too young to seem relevant. One must recognize, however, that as with the proven case of Campbell Island, extant young islands often mark the neighborhood where older, eroded or drowned islands existed.)[18] Hence potential passengers could have embarked at least to this date without the need to invoke long-distance transport, and passengers already aboard could also have dropped off; others could have traversed initially limited water barriers even long after northward migration was under way. Thus certain Gondwanalandic taxa originating after Africa plus Madagascar broke off from Gondwanaland (except for South America), in Jurassic times, could have reached Africa or Madagascar via India, as late as that time—and perhaps later.

A relevant question posed by these facts is, What Gondwanalandic groups of angiosperms were likely candidates for northward migration? Another fact is also significant: in the period between 95

[17] McElhinny (1968) also states that much of the northward movement of India took place in early Tertiary, after eruption of the Deccan traps, and gives a reconstruction (1970, fig. 2b) showing that in mid-Cretaceous times India, with Madagascar attached, still lay nearly proximate to Antarctica. Le Pichon and Heirtzler (1968) state that in the lowermost Cretaceous (140 million years ago), Africa was already separating from India.

The *rate* of movement of the Indian plate can also be inferred from dating along the Ninety-East Ridge, where a stationary "plume" appears to exist. The figure on p. 234 in Sullivan (1974) is explicable only if we assume migration northward, from site 254 (rocks 75 million years old, situated ca. 30°S.) to site 217 (15 million years old, situated ca. 10°N. lat.). On the basis of these data, the Indian plate presumably moved northward at least 40° in 60 million years—a remarkable rate of migration. Both distance and rate of movement are somewhat less than McKenzie and Sclater state.

[18] Smith and Hallam (1970) state that the Upper Cretaceous marine fauna of the Coromandel Coast of India "shows strong similarities with those of Assam and South Africa," suggesting that "a continuous land barrier extended from Africa and Madagascar to central peninsular India until at least the close of the Cretaceous."

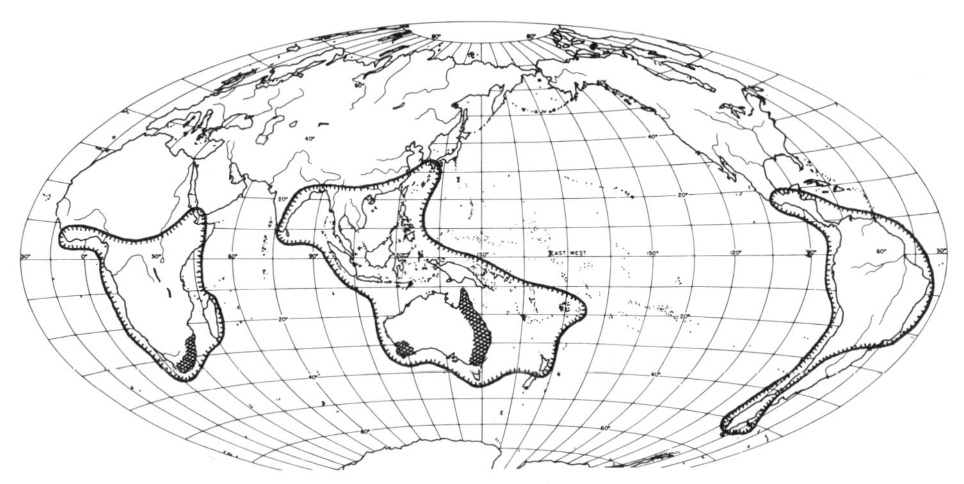

Fig. 14. Range of Proteales, family Proteaceae; the area of greatest concentration of taxa is cross-hatched. South American taxa are more allied to African than to Australian-Asian taxa. *Assumption:* range in Asia may possibly reflect two migrations from Gondwanaland, one via the Indian plate, the other, more recent, across Wallace's Line. Phylogenetic studies should be able to clarify this matter. (After Good, 1964.)

and 75 million years ago no mammals, so far as we know, existed in India, Australasia, or Antarctica, although marsupials were then diversifying in South America and the earliest placentals had reached there. Hence efficient and long-distance animal dispersal of angiosperms at that time is highly improbable; also, most modern groups of angiosperms with readily wind-carried seeds did not yet exist. Hence, relatively slow overland migration of most early angiosperms must be assumed; it is likely that late Cretaceous and early Tertiary angiosperms "walked," like the late Cretaceous dinosaurs.

Groups which are known to have existed in Gondwanaland at that time include flowering plant families like the Proteaceae (fig. 14). Convincing data from the angiosperms, to show that the Indian raft was a major vehicle for their mass transport, are largely lacking. The assumed period of migration (some 50 million years) was sufficiently long that the descendants of the initial passengers, if they survived at all, were often drastically changed.[19] Moving northward, the Indian raft encountered equatorial climates, and cool-adapted passengers

[19] An example of such change is offered by the family pair Scapaniaceae-Delavayellaceae. The primitive, purely "subantarctic" Blepharophylloideae of the first family apparently gave rise to the essentially Laurasian Scapanioideae—but also to a highly derivative, specialized monotypic family, the Delavayellaceae, now confined to northern India and adjacent China. The passenger getting on may have been like *Blepharidophyllum;* the descendant that landed was presumably well on the way to becoming *Delavayella* (fig. 13); compare also fig. 27.

probably succumbed in numbers, being outcompeted by taxa that were more malleable or had a high evolution rate. The Cretaceous Indian migration thus had very different parameters than the Tertiary Australasian migration, which showed partial canceling of the latitudinally correlated increase in temperature by a then-prevailing overall tendency toward progressive cooling of climates.

WHAT WERE THE SPECIFICS OF THE SCENARIO?

A scenario somewhat as follows is possible for the Indian plate, leaning partly on the much better known history of the Australasian plate: About or after the middle of the Cretaceous, the Indian plate broke away from East Gondwanaland. This land mass, at least 1500 km long and 900–1000 km wide, was populated at that time by a rich and diversified biota, including at least some angiosperms.[20] This land mass, furthermore, had diverse climates, ranging from probably cool to warm temperate in the southern portions to moist tropical and arid tropical; [21] on montane areas there probably survived a cold-tolerating relict biota descended from taxa that would have occurred in late Paleozoic times when the region in large part was very cold and glaciated. We must visualize a wide range of communities reflecting an already complex geography and history. It also seems highly probable that some of these elements were preadapted to the climate regime initiated with northward migration, and that selection pressures were adequate to result in the production of genotypes showing parallel, gradual adaptation to the gradual climatic changes incident to this relatively slow migration. Admittedly taxa existed then, as now, with slight ability to adapt and compete, and were soon eliminated; e.g., *Acmopyle, Dacrydium,* certain groups of *Podocarpus* which were present in India prior to mid-Cretaceous times, and

[20] Based on the fact that, by rather early in the Cretaceous, early angiosperms are represented in the microfossil flora of areas as far apart as Israel, Brazil, and the borders of the present-day North Atlantic (Virginia to New Jersey), as Brenner, and Doyle and Hickey (in this volume) show. Extrapolating from this central fact, one must almost assume that by the middle Cretaceous a diversity of recognizable angiosperms existed and had had time to become widely scattered at least in a band lying roughly between the equivalent of today's 25°N. and 25°S. lat. Since paleolatitudes (and polar position) at the middle of the Cretaceous remain uncertain (Hughes, 1973), little more than this can be currently affirmed.

[21] There are old mountains in the south of India that still attain an elevation of 8700–8841 ft, and Ceylon has elevations to 8281 ft. It seems likely that even under equatorial conditions such peaks, whose summits probably caught adequate precipitation, served as refugia for taxa not adapted to tropical conditions—in the same way that mountains of New Guinea today are refugia at elevations of 8000 ft and higher for temperate or even subantarctic taxa.

which are not known from that area later (figs. 9, 10). Thus we must assume both evolution and elimination; we must also assume that passengers got on from the Mascarenes, Madagascar, and East Africa—and, by longer-distance saltation, from elsewhere.[22]

Finally, some 40–45 million years ago, the raft arrived in a Laurasian "port." The shallows of the Tethyan Sea had already become wrinkled by the Indian raft beginning its long slide, or subduction, under Laurasia. This resulted in rapid elevation of the Himalayan chain; here we find that the northern edge of the Indian plate has slid under the Laurasian plate, with consequent major orogenic activity. The mountain building resulted in relatively rapid creation of a wide series of temperate, moist, local environments where surviving passenger taxa with such climatic preferences could find a congenial new home. The number of such survivors was probably limited, however; many had surely been eliminated by competition with more heat- and drought-tolerant taxa (e.g., Asclepiadaceae tribe Stapeliae), or had themselves given rise over thousands of generations to such tolerant genotypes.

One can find many counterarguments that would tend to invalidate portions of this scenario. Evidence in favor of its validity is available chiefly from lower plants but, inferentially, also from conifers.

EVIDENCE FROM CONIFERALES

Specific proof for Indian plate rafting northward of Coniferales is admittedly slight and based largely on sometimes ambiguous fossil evi-

---

[22] Of considerable importance is the indication by McKenzie and Sclater (1973) of a complex series of plateaus and ridges in the Indian Ocean. Volcanic islands arising from such plateaus and ridges can serve as stepping stones facilitating dispersal. The exceedingly complex topography in the Indian Ocean today, with numerous basins separating plateaus and ridges, reflects a history (figs. 26–27) rather analogous to that portrayed for the Australasian plate, where several arcs (e.g., today's New Zealand, north to New Caledonia) represent islands and plateaus formerly juxtaposed to present-day Australia. Similarly, a system of islands exists today along the Laccadive–Chagos Ridge (Chagos, Maldive Islands), the Mascarene Ridge (Seychelles, Mauritius, Réunion), the Prince Edward–Crozet Ridge (Prince Edward, Crozet Islands), the Kerguélen Plateau (Kerguélen, Heard Islands), Broken Ridge (Amsterdam, St. Paul Islands). Some of these islands are old, but many are young—20 million years old or less—and it must be assumed that other islands and plateaus, now foundered or eroded away, existed earlier. If we extrapolate backward from McKenzie and Sclater's reconstruction to early Cretaceous time, I would assume then a more southerly position for Madagascar than in the reconstructions by Dietz and Holden and by Smith and Hallam; I also assume that the Mascarene Ridge, which McKenzie and Sclater position as initially juxtaposed by its eastern flank with India, was also closely juxtaposed with Madagascar (fig. 26).

dence.[23] The fossil evidence (Florin, 1963) does show that in late Carboniferous to Permian times fossil genera like *Buriadia* occurred in India and South America and *Paranocladus* was found in South America and (questionably) India, and that *Walkomiella* occurred in the Permian in India, Australia, and Africa. This suggests, of course, geographical contiguity. None of these fossil genera is known from Laurasian localities, suggesting either a climatic or a geographical discontinuity at that time between Gondwanaland and Laurasia.

In Jurassic times, prior to the start of Indian plate migration, *Acmopyle* occurred in India (fig. 9); in Eocene-Oligocene times this genus occurred also in South America and the Antarctic Peninsula. Today the genus is known only from New Caledonia and Fiji (fig. 10), a distribution that is clearly relict and must go back at least to the Upper Cretaceous.

Dacrydium is similarly, questionably, known from Jurassic times in India, but is lacking there today; the rest of its range is basically Gondwanalandic, including Antarctica (questionably) and also Kerguélen. Associated with production of fleshy "fruits" is the apparent spread of this genus in recent times across Wallace's Line to the Philippines and extreme southeastern Asia. *Podocarpus* sectio Dacrycarpus exhibits nearly the same range, except that the group is now extinct in Tasmania and southern South America. As Florin notes, "Dacrycarpus-like fossils occurred in India throughout the Jurassic period and in early Cretaceous times." With respect to *Podocarpus* sectio Nageia the situation is more ambiguous: no fossils are known, but the section today occurs locally in southwestern India and southern Burma and extends north to southern Japan and south to the Philippines, Borneo, and Indonesia to New Guinea. Questions assert themselves: Does this almost exclusively tropical section exhibit the northward shift of other groups of Coniferales of cold-intolerant range? Is the disjunct occurrence in India (and perhaps Burma) old, reflecting a continuous presence on the Indian plate since this was attached to Gondwanaland? Is the range in Southeast Asia, north to Japan and into New Guinea a recent one, radiating out from an Indian plate center? In other words, are the suggestive simi-

[23] The most striking fact about gymnosperms and Indian plate migration—perfectly clear when the numerous distribution maps in Florin (1963) are examined—is the nearly complete lack today of any Coniferales and Taxales in the peninsula of India. Only at the northern, Laurasian edge in the Himalayas do we find intrusion of Laurasian groups such as *Abies*. The modern absence of gymnosperms stands in stark contrast to the former presence of Gondwanalandic taxa (still extant elsewhere) in Jurassic to early Cretaceous times (figs. 9, 10). This strongly suggests that as India started its long trek northward into warmer (and perhaps drier) regions, these taxa were eliminated. *Araucaria*, for example, occurred throughout India in early to late Jurassic times but is not known there after the early Cretaceous. However fragmentary, this fossil record is, I think, illuminating.

larities to the range of hepatic genera like *Gottschelia* (cf. below) nonaccidental? Similarities in range to that of the Lardizabalaceae are also suggestive.

The mystifying and apparently rather early extinction of Coniferales and Taxales on the Indian plate, apparently shortly after it began its migration northward, is surrounded by implications. It is difficult to visualize extinction of all or most of the woody flora of the region unless it was supplanted by something better adapted for survival as temperatures increased (and perhaps moisture levels decreased). Did early angiosperms, perhaps preadapted (e.g., by being able to lose their leaves during dry periods), replace the Coniferales and Taxales?

EVIDENCE FROM THE ANGIOSPERMS

Brenner and Doyle and Hickey (in this volume) show that early Cretaceous angiosperms had limited diversity, as revealed by both microfossil and megafossil evidence. Judging from leaf form, venation, and dentition, these early angiosperms did not belong to recognizable present-day assemblages or families. One must therefore conclude that if angiosperms were indeed on the Indian plate in early Cretaceous times, they could probably not be assigned to presently recognizable families. However, if migration—as distinct from initial rifting—started only in the Upper Cretaceous, then there is evidence that a wide assemblage of extant families of angiosperms existed (e.g., Proteaceae; fig. 14) and may have been rafted north on the Indian plate. If up to the late Cretaceous there were limited obstacles to migration from the Antarctic and Australasian plates onto the Indian plate, there should be corroboratory fossil evidence. Absence of adequate late Cretaceous fossil evidence from India, however, is limiting.[24] At the very least, groups like the Proteales (fig. 14), Restionaceae (fig. 15) and, less likely, the Stylidiaceae (fig. 16) and Coriariaceae (fig. 17), may have been in India by the Upper Cretaceous prior to initiation of migration. Their presence today on the Indian plate or at its periphery is admittedly equivocal evidence. If the last three families existed by Upper Cretaceous times (we know Proteaceae existed then), these groups could have dispersed to the Indian plate some time prior to 75–90 million years ago.[25] Such a

[24] Lakhanpal (1970), however, shows that the early Tertiary record in India exhibits more Gondwanalandic links than later records show.

[25] It makes little difference whether these groups "walked" into India, or, a little later, "island hopped" via archipelagic connections. The most significant consideration is that it was easier for plants—and animals—to get to India from other parts of Gondwanaland than from Laurasia.

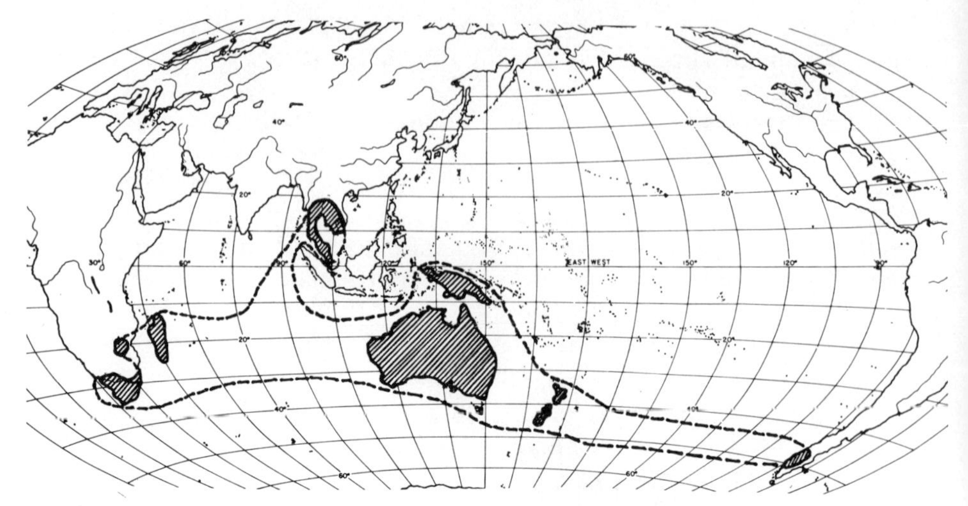

Fig. 15. Range of Restionaceae. The range possibly reflects a very ancient dispersal pattern, with presence in Africa and Madagascar not recent; possibly Cretaceous migration northward on Indian plate to Laurasia and from there spreading to southeastern Asia, or (if Ridd [1971] is right) the presence in southeastern Asia may be old. (After Good, 1964.)

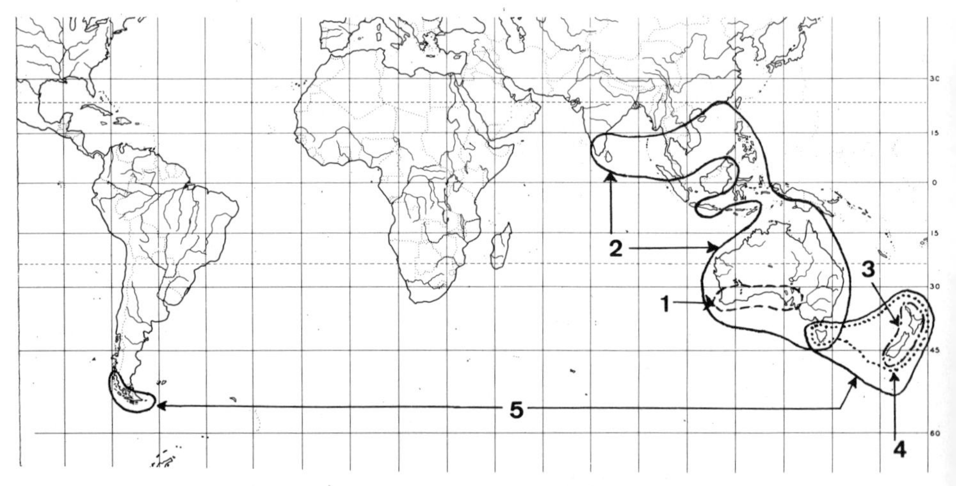

Fig. 16. Total Gondwanalandic range of the family Stylidiaceae. (1) *Levenhookia.* (2) *Stylidium.* (3) *Oreostylidium.* (4) *Forstera.* (5) *Phyllachne.* Only *Stylidium* has, in presumably relatively recent times, breached Wallace's Line. (After Good, 1964.)

scenario gives us a rather convincing explanation of why certain primitive taxa, absent in Africa, occur from New Guinea (and sometimes Australia) to the Mascarenes and/or Madagascar and/or south India. Examples are the *Bubbia*-like species from Madagascar (fig. 6) and *Gottschelia* (Schuster, 1975).

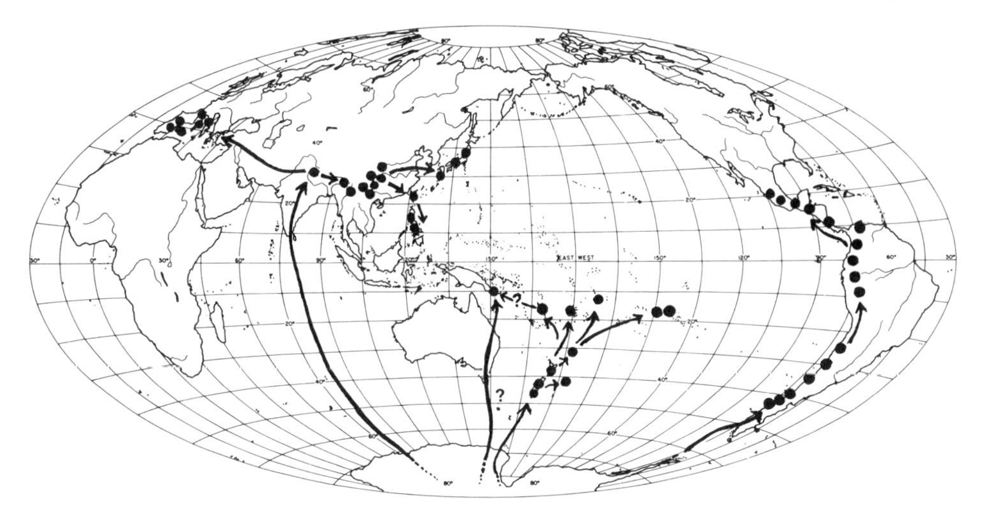

Fig. 17. Distribution of *Coriaria* (Coriariaceae; no other living genera); its affinities to Anacardiaceae are obscure, and according to Croizat (1952) it is "an offshoot from the ranunculoid" complex. Known from an Oligocene fossil from Europe that is supposedly closely similar to an extant species in the Himalayas (Croizat, 1952). Croizat assumes a dispersal outward from an origin in western Polynesia (see his map, fig. 23). However, it seems more likely that the groups are Gondwanalandic in range, with the following dispersal routes and times: migration northward from cool to temperate parts of Gondwanaland after the Andes were created to Costa Rica and Guatemala—exactly like *Desfontainia* and *Fuchsia;* migration (possibly via West Antarctica) north to New Zealand, overland in late Cretaceous times or by jumping the water gap in Tertiary times, and from there to the Chathams, New Caledonia, the Kermadecs, Samoa, the Society Islands, Fiji, New Hebrides, and New Guinea; rafting on the Indian plate, reaching the present-day Himalayan region (Sikkim) by late Eocene times and reaching Europe by Oligocene times (30–40 million years ago); fanning out from the Indian plate—and any other Gondwanalandic fragments, including possibly part of southeastern Asia and China (see Ridd, 1971; Hurley, 1971)—to reach China (Kwangsi, Kweichow, Yunnan, Hunan, Hupeh, Szechwan, Kansu, Shensi), the Philippines, Taiwan, and Japan (to Hokkaido). The possible fossil examples (Eocene) in England of Malaysian or Australasian elements like *Rhodomyrtus, Cinnamomum,* and *Tristania,* which Croizat emphasizes in this connection, may have reached Europe by rafting, initially, on the Indian plate.

THE PROBLEM OF THE LARDIZABALACEAE

The ranalean Lardizabalaceae has 8 genera (two monotypic genera, *Lardizabala* and *Boquila,* occur in central Chile; 6 genera are exclusively Asian). *Stauntonia,* with about 16 taxa occurs in the critical region from Assam and eastern Bengal to southern and western China (Yunnan, Kwangsi, Kwangtung), ranging to south China and Korea, Japan, Taiwan, and Laos. *Holboellia,* with 8 taxa, occurs in China (including Szechwan and Yunnan), India (Sikkim and Kumaun), and Tonkin. *Decaisnea* has 2 taxa, from chiefly south and west China (including Szechwan and Yunnan) and the eastern Himalayas. *Sinofranchetia* has 1 species in China (Szechwan and Hupeh). *Akebia* has 4 species, from south China to Korea and Japan. *Sargentodoxa* has 1 species, in south China to Laos. The range of the family is clearly relict and hardly explicable on the basis of chance long-distance dispersal.

Two hypotheses accounting for the anomalous present-day range suggest themselves: First, that the group was originally Laurasian—like the Fagaceae—and has died out in most of Laurasia (including North America and Europe); North American taxa migrated in Cretaceous times via the *Nothofagus*-marsupial route to South America, where they are today relict. Second, that the group was originally Gondwanalandic, and the Chilean sector of the range is relict; the group died out in most of the remainder of Gondwanaland (perhaps incident to Tertiary cooling) but its members on the Indian plate were rafted north, eventually reaching Laurasia and spreading into it to a limited extent (the hypothetical pattern of *Podocarpus* sectio Nageia). The group remained concentrated in regions clearly Gondwanalandic in origin (India) or just possibly so (Yunnan, Szechwan, Laos, Tonkin, if the hypotheses of Hurley, Ridd, and especially Colbert [1973, pp. 65, 71–72, 74, 118–19, 121, 152] are correct).[26] The range of the loosely allied hepatic genera *Blepharidopyllum* (wholly cold-antipodal in range) and *Delavayella* (Himalayas to Yunnan) might represent a similar pattern.

EVIDENCE FROM THE BRYOPHYTA

It is perhaps not coincidental that two bryologists, G. L. Smith (1972) and Schuster (1972a), the former working with Musci, the latter with Hepaticae, independently became convinced of the significance of Indian plate migration in the dissemination of Gondwanalandic taxa into Laurasia (figs. 13, 18–25).[27]

---

[26] Colbert believes that the fossil occurrence in parts of Asia outside India of otherwise exclusively Gondwanalandic Mesozoic reptiles (one related to the South American *Kannemeyeria*, the other, *Lystrosaurus*, known from the Lower Triassic of South Africa, India, Antarctica, and southeastern Asia) suggests the Gondwanalandic origin of such sectors.

[27] The peculiar relevance of Hepaticae for phytogeographic analysis is documented in detail by Schuster (1969, 1975) and briefly above. One specific point to be kept in mind, of which the student of higher plants is usually unaware: unlike gymnosperms such as *Acmopyle* (figs. 9 and 10), which tend to die out as conditions become suboptimal, many hepatics exhibit high somatic malleability and may long persist, under suboptimal conditions, as nonreproductive, purely gametophytic phases (Schuster, 1966). A well-documented, intimate link between eastern

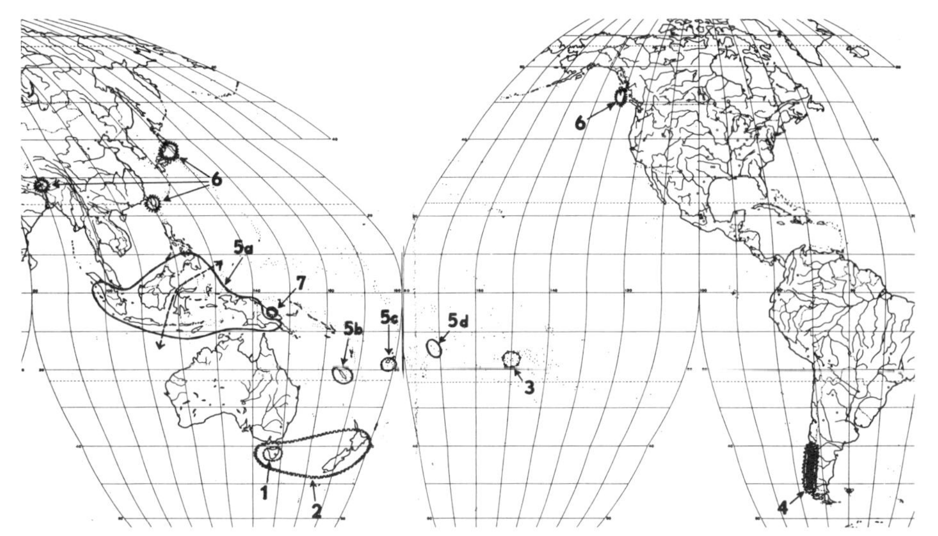

Fig. 18. Gondwanaland-derived ranges: Hepaticae, Metzgeriales, suborder Treubiinae. Total range of the family Treubiaceae, with two genera, *Treubia* Goebel (1–5) and *Apotreubia* Hatt. et al. (6–7). The group is exceedingly isolated and perhaps should form its own order. (1) *Treubia tasmanica* Schust. & Scott (the most primitive taxon). (2) *T. lacunosa* (Col.) Prosk. (3) *T. tahitensis* (Nad.) Besch. (4) *T. scapanioides* Schust. (5) *T. insignis* Goebel, including subspecies *insignis* (5a); *caledonica* (5b); *vitiensis* (5c); *bracteata* (5d). (6) *Apotreubia nana* (Hatt. & Inoue) Hatt. et al. (7) *A. pusilla* (Schust.) Schust. *Assumption: Treubia* (5a) has shown rather recent transgression of Wallace's Line (dashed line), otherwise is purely Gondwanalandic. *Apotreubia* was rafted north on the Indian plate and showed dispersal eastward and southeastward from there.

Smith (1972) postulates for Polytrichaceae that *Polytrichastrum* may have "spread into Laurasia at an early date, but since the genus was apparently not widely distributed in western Gondwanaland,

---

Asia and eastern North America, where sibling species pairs of angiosperms repeatedly recur (fig. 37), exists also with regard to the hepatics; here, however, there is apparently almost always persistence of species identity. My point is that, by and large, angiosperms and bryophytes exhibit identical or similar distributional patterns but that in the former, with a high rate of evolution and persistence of sexual reproduction, the patterns often become diffuse and not readily recognized. The slow evolution (and in the case of taxa that now reproduce purely asexually, the lack of evolution) of the hepatics often preserves in glaringly obvious fashion phytogeographic links of great antiquity. The fact that the bryophytes typically occur as members of angiosperm-dominated communities is relevant here: the two groups not only show similar phytogeographic patterns, but they often appear to have migrated as parts of integrated communities. Hence the bryophyte patterns, where better preserved, serve to document general phytogeographic patterns. This is particularly relevant in the case of Indian plate migration, which was initiated far enough in the past that, as we have seen, the angiosperm evidence is often equivocal.

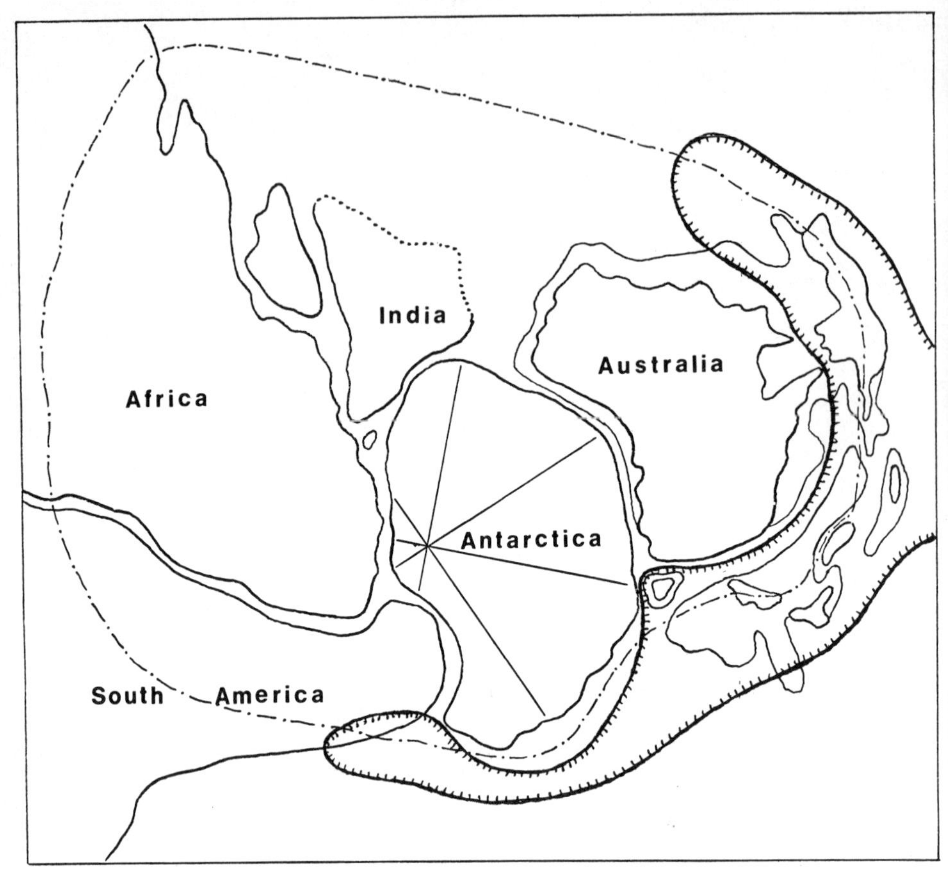

Fig. 19. Gondwanalandic range of *Treubia* Goebel (about six species), superimposed on the Gondwanaland reconstruction of Schopf (1970a); note that one species (*T. tahitensis*) has spread out beyond the range of the map; another species (*T. insignis*) has transgressed Wallace's Line to attain Sumatra and the Philippine Islands.

one may infer that it was introduced directly into Asia in the Eocene *via* the Indian subcontinent, when there would still have been some 40 million years for it to spread throughout the Northern Hemisphere." He similarly interprets the range of *Oligotrichum;* the largest concentration of taxa is in Nepal to Yunnan, where at least six species are found; in addition, the genus is found in Gondwanalandic regions (southern Chile, southeastern Brazil, southern Africa, New Zealand, and New Guinea) and also on two volcanic islands (Juan Fernandez, Tristan da Cunha). Smith believes that "*Oligotrichum* can be interpreted as a genus of Gondwanaland origin that was transported to Asia on the Indian subcontinent. Thereupon one

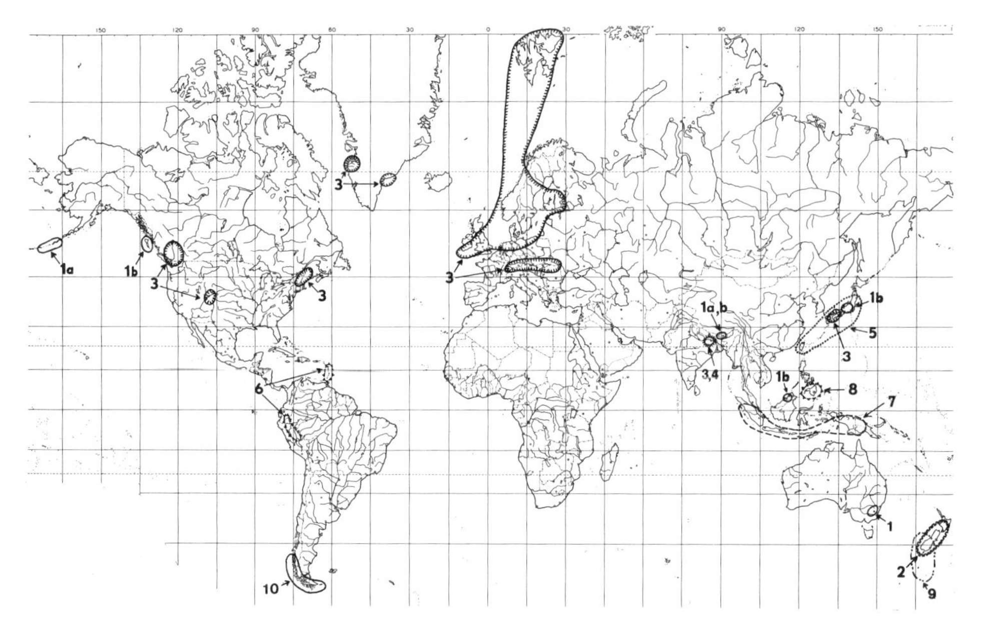

Fig. 20. Gondwanaland-derived ranges: Hepaticae, total range of Calobryales, an isolated and very primitive order (with the lowest chromosome numbers known in the Hepaticae, n = 5 and n = 4). Range of *Takakia* Hatt. & Inoue: (1a) *T. ceratophylla* (Mitt.) Grolle. (1b) *T. lepidozioides* Hatt. & Inoue. (1–10) Range of *Haplomitrium* Nees (including *Calobryum* Nees). (1) *H. intermedium* Berrie, the most primitive taxon. (2) *H. gibbsiae* (Steph.) Schust. (3) *H. hookeri* (Sm.) Nees. (4) *H. indicum* (Udar & Chand.) Schust., probably identical with *H. hookeri.* (5) *H. mnioides* (Lindb.) Schust. (6) *H. andinum* (Spr.) Schust. (7) *H. blumei* (Nees) Steph. (8) *H. giganteum* (Steph.) Grolle. (9) *H. ovalifolium* Schust. (10) *H. chilense* Schust. *Assumption:* the group is Gondwanalandic in origin—at least the Haplomitriaceae are almost certainly so. Presumably the ancestors of *Takakia* and *H. hookeri* were rafted north on the Indian plate; *Takakia* showed subsequent dispersal identical to that of *Apotreubia. Haplomitrium* showed wide Laurasian dispersal (*H. hookeri*), with Pleistocene-occasioned reduction in range. Of 12 known taxa, 10 are known today from Gondwanalandic areas, including the most primitive (*Takakia; H. hookeri = indicum, H. ovalifolium, H. intermedium*).

taxon (*O. hercynium*) became widespread in the Northern Hemisphere, and other taxa apparently spread to Malaysia, to Japan, and eventually to the Pacific coast of North America." The essential identity of Smith's interpretation of how *Oligotrichum* achieved its modern range, and the distributional mechanisms postulated by Schuster (1972a) for *Lophochaete* (fig. 21), the Calobryales (fig. 20), and the Treubiaceae (fig. 18), is striking. The role of the Indian plate in introducing Gondwanalandic elements into Laurasia, of plants at or below the level of the gymnosperms, thus seems well established.

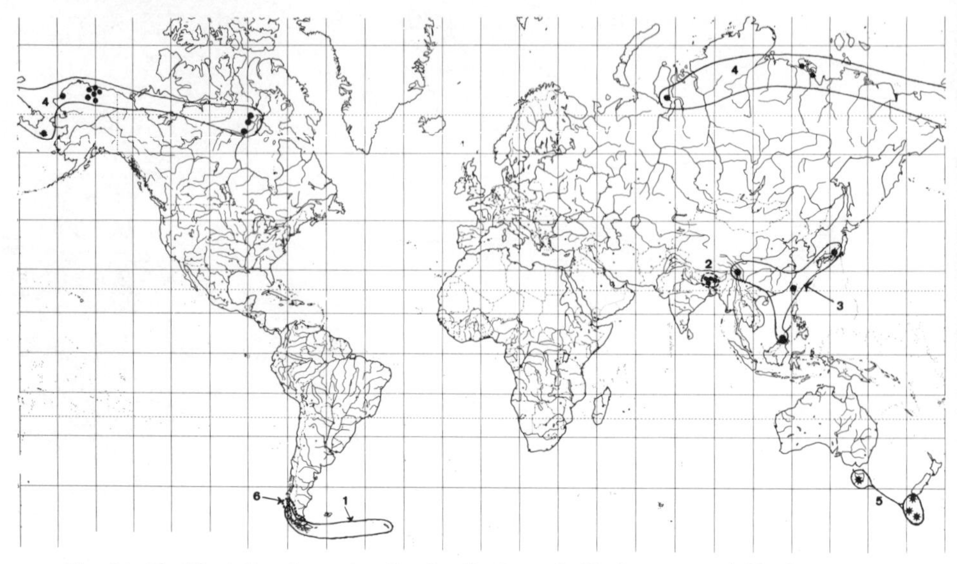

Fig. 21. Modified Gondwanalandic distribution of allied genera of Blepharostomataceae, an ancient family of Jungermanniales. (1) *Lophochaete quadrilaciniata* (Sull.) Schust., the most malleable and primitive element. (2) *L. trollii* (Herz.) Schust. (3) *L. andoi* Schust., sometimes regarded as a subspecies of (2). (4) *L. fryei* (Perss.) Schust. (5) *Isophyllaria* Hodgs., a monotypic genus of Campbell Island, Auckland Island, Stewart Island, and Tasmania. (6) *Herzogiaria* Fulf., a monotypic genus. Species (5) and (6) are phylogenetically more advanced than (1), but clearly allied. Related genera, each with three species, are *Archeophylla* Schust. (Chile, New Zealand) and *Archeochaete* Schust. (Tristan da Cunha, Tierra del Fuego, New Guinea). Of the nine species involved, only *L. fryei* (4) is monoecious; it had attained the circumpolar basin by the Tertiary, and today's range radiates out from unglaciated parts of Alaska and Siberia. *Assumption:* The five genera had a common cool-antipodal origin in oceanic loci; after segregation of *Lophochaete* from the general complex, the joint ancestor of the three allied taxa *L. trollii, L. andoi,* and *L. fryei* was rafted northward to Laurasia on the Indian plate, where *L. trollii* still occurs. Only the *L. trollii-andoi* complex is known away from ocean perimeters. *Analogy:* the range of *Anastrophyllum* (fig. 25), in which I have assumed similar rafting northward and similar evolution of one monoecious, advanced taxon peripheral to the Arctic Sea.

Smith generalized that the "predominantly Northern Hemisphere genera" of the Polytrichaceae, such as *Oligotrichum* and *Polytrichastrum*, "can be interpreted as Gondwanaland-derived members of the circumboreal Tertiary flora, or Arcto-Tertiary geoflora." The exact analogy to the new imperfectly circumboreal—and highly disjunctive—range of Hepaticae such as *Lophochaete fryei*, *Haplomitrium hookeri*, and *Acrobolbus* sectio Rhizophyllae, as well as of *Anastrophyllum* is highly suggestive. The Gondwanalandic origin of these taxa is clearly inferred (Schuster, 1969, 1972a).

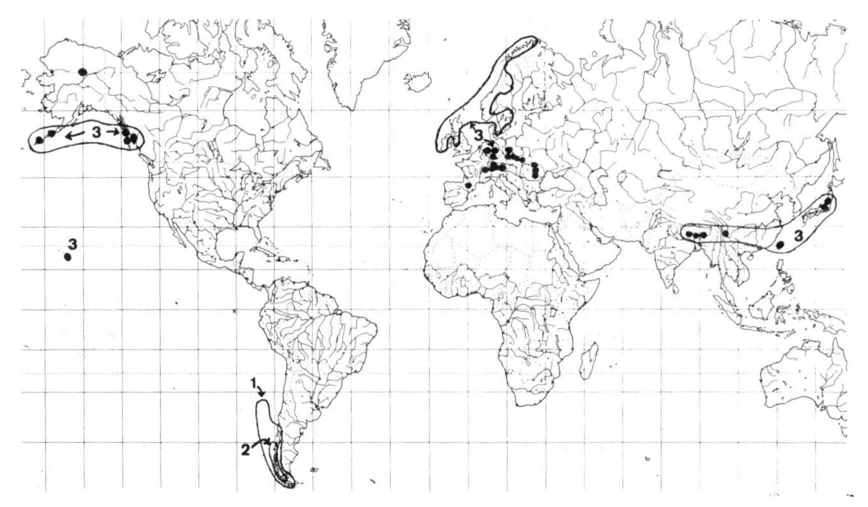

Fig. 22. Modified Gondwanalandic ranges: range of Anastrepta (Lindb.) Schiffn. (1) *A. bifida* (St.) St. (2) *A. longissima* (St.) St. (3) *A. orcadensis* (Hook.) Schiffn. Species 1 and 2 are the more primitive (they lack the ability to form gemmae) and are subantarctic in range. Species 3 occurs widely but disjunctly in the Northern Hemisphere in mostly oceanic loci, and its range impinges on Gondwanaland only in the Himalayan area; it conceivably spread widely via gemmae (sporophytes are rare). *Assumption:* the common ancestor for species 1–3 was Gondwanalandic in origin; the ancestor of 3 was rafted northward on the Indian plate and underwent dispersal from there.

A number of such Gondwanaland-derived patterns are discussed in Schuster (1972a) and are not repeated here. Others, not previously cited, reinforce the basic thesis. Among these are the following:

1. Scapaniaceae subfamily Blepharidophylloideae, the most primitive group of Scapaniaceae, is exclusively subantarctic; the derivative subfamily Scapanioideae and the highly specialized, monotypic, derivative Delavayellaceae have become Northern Hemispheric (fig. 13). Infusion into Laurasia of the blepharidophylloid ancestor of both Scapanioideae and Delavayellaceae may be assumed. Critical is the fact that today we still find in the Himalayan center not only the modern genera *Diplophyllum* and *Scapania,* but a host of isolated and primitive subgenera of *Scapania* (*Protoscapania, Ascapania, Plicatycalyx, Scapaniella,* as well as subgenus *Scapania*). All seven species of subgenus *Protoscapania* occur in the Himalayan center; only two have dispersed out of this center. It is probably not coincidental that almost all taxa are sterile and lack gemmae, except for *S. ornithopodioides,* where gemmae and rarely capsules develop—hence where long-distance dispersal remains feasible, as its presence in Hawaii indicates.

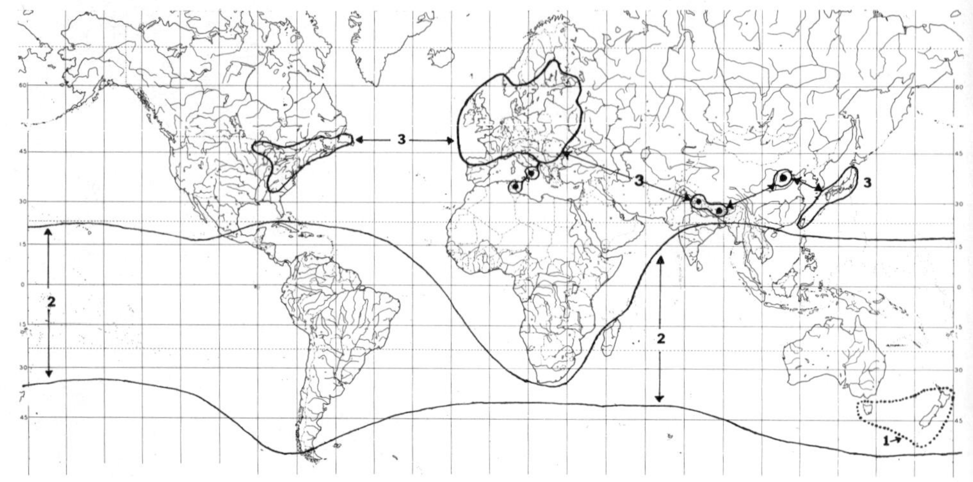

Fig. 23. Distribution of the family Trichocoleaceae Nakai (Hepaticae), a group of exclusively unisexual taxa lacking reproductive modes. There are two genera. (1) *Eotrichocolea* Schust., monotypic. (2) *Trichocolea* Dumort., with 30–40 species. (3) *Trichocolea tomentella* (Ehrh.) Dumort. occurs only in 20–70°N. This is a highly advanced taxon, lacking a perianth. All remaining taxa, including the primitive ones that retain a perianth (subg. *Leiomitra*), are basically Gondwanalandic in range; *Leiomitra* is known with certainty only from South America and Australasia and is basically nontropical. *Assumption:* the ancestors of *T. tomentella* rafted north on the Indian plate (note present occurrence there) and from there radiated out and became holarctic.

2. *Anastrepta* (fig. 22) illustrates the identical principle. The two most primitive taxa (1 and 2), which lack asexual reproduction, still occur in South America, while the most advanced (producing gemmae) is *A. orcadensis* (3); the latter occurs not only in the Himalayan center, but scattered in Laurasia, with a range suggesting widespread extinction during Tertiary and Pleistocene times.

3. The extant range of Trichocoleaceae (fig. 23), with only *Eotrichocolea* (monotypic) and *Trichocolea* (30–40 spp.) is equally instructive. The group, basically Gondwanalandic, has a single taxon, *T. tomentella,* disjunct in Laurasia and in the Himalayan center. Presumably, its ancestors were rafted northward on the Indian plate.

4. The Lepidolaeninae, with the single family Lepidolaenaceae (Schuster, 1969, map 6), has some 14 species in *Lepidolaena* and *Gackstroemia,* found under subantarctic conditions; a third genus, *Jubulopsis* Schust., a primitive monotype, occurs only in New Zealand and subantarctic islands to the south (Hamlin, 1972). Two derivative genera of a separate subfamily Trichocoleopsidoideae Schust. (Neotrichocoleaceae Inoue) occur in Japan and Korea. If the common

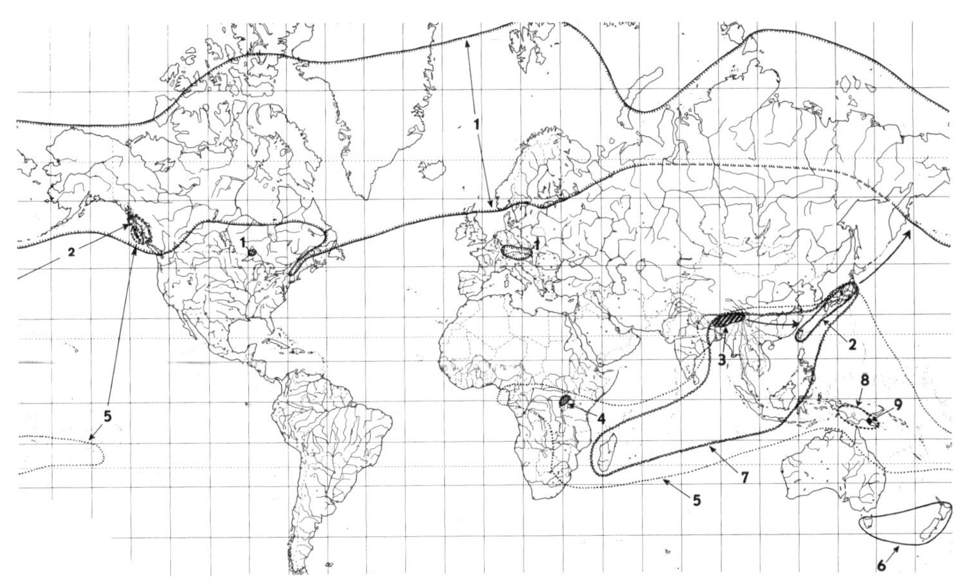

Fig. 24. Gondwanaland-derived ranges in Hepaticae (Jungermanniales): *Chandonanthus* Mitt., a genus with eight or nine species. (1–4) Range of the relatively advanced taxa of subg. *Tetralophozia;* species 2 and 3, *C. pusillus* and *C. filiformis,* are now believed to be identical. (5–9) Taxa of subg. *Chandonanthus,* the primitive, variable subgenus. Species 1–3 apparently never produce sporophytes today; species 4–9, in part, freely produce sporophytes; no asexual reproductive modes exist in any taxa. *Assumption:* the ancestors of *Tetralophozia* are to be found in subg. *Chandonanthus,* and were rafted to Laurasia via the Indian plate (where three species of both subgenera occur today). (Derived from Szweykowski, 1956; Schuster, 1960.)

ancestor of this last group arrived from Gondwanaland via the Indian plate, it migrated to eastern Asia, leaving no remnant today in the Himalayan center.

5. Identical patterns persist with *Mastigophora. Mastigophora woodsii*—known only sterile and without asexual reproductive modalities—occurs in the Himalayan center, in Scotland and Ireland, and on the Queen Charlotte Islands.

6. A less highly relict and more complex version of this pattern characterizes the ancient genus *Chandonanthus* (fig. 24; see legend)—again lacking asexual reproduction and usually sterile—in which three species occur in the Himalayan center. The common ancestor of species 1–4 (subgenus *Tetralophozia*) was presumably carried north on the Indian plate, and dispersed from there (species 1 and 2). Species 2 (presumably identical or nearly so to species 3, still in the Himalayas) dispersed to Taiwan and Japan and to the Queen Charlotte Islands, but in Tertiary times appears to have died out in

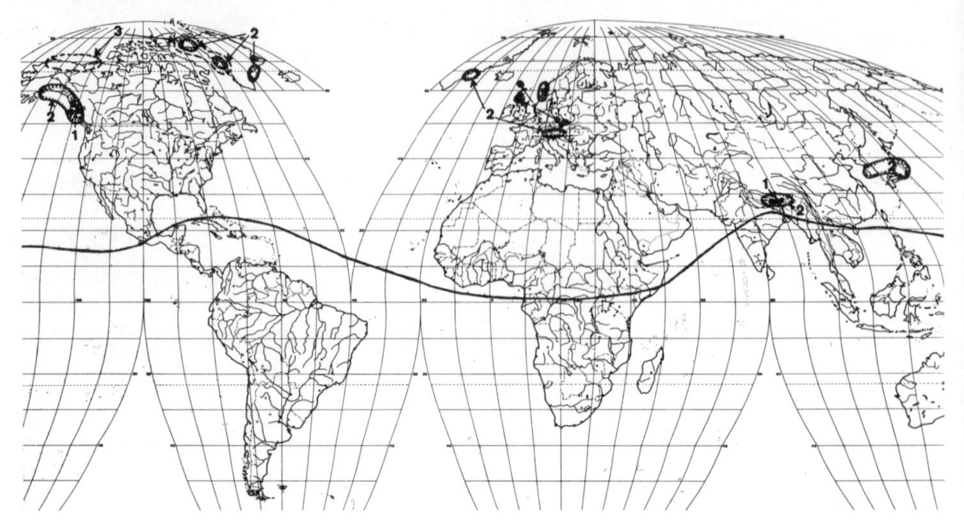

Fig. 25. Modified Gondwanalandic ranges: distribution of species of the ancient genus *Anastrophyllum*. The genus, with over 35 species, is primarily Gondwanalandic (taxa occur south to Tierra del Fuego and New Zealand), but mostly in oceanic montane districts. All taxa except *A. sphenoloboides* Schust. (3 on map) are unisexual; the latter is bisexual and produces asexual gemmae. Three other holarctic taxa (*A. minutum, A. cavifolium, A. michauxii* also produce gemmae, but none of the more ancient, tropical, and antipodal taxa do. The center of diversity is southeastern Asia to New Guinea. *Assumption:* Gondwanalandic origin, with rafting northward of ancestors of species 1–3, and infusion of the genus into Laurasia; the continued presence of species 1–2 of this "oceanic" genus in the Himalayas is suggestive. As with *Lophochaete* (fig. 21), a single, monoecious species is relict in areas now arctic; it is found in northern Alaska and the Yukon in nonglaciated areas. (1) *A. donianum* (Hook.) Steph. (2) *A. assimile* (Mitt.) Steph. (3) *A. sphenoloboides* Schust. The genus is general in range in oceanic loci south of solid line.

the Beringian region. Species 1 is widespread, and its survival seems linked to spread from arctic areas (in Alaska, Siberia, and perhaps elsewhere) in which glaciation was local or absent. Species 5, more widely distributed, may have shown the same spread from the Himalayan center to Japan and the Queen Charlotte Islands (and extinction elsewhere in Laurasia), already postulated for preceding taxa.

7. A variation of this same pattern occurs with *Anastrophyllum* (fig. 25), widespread in Gondwanalandic areas. Two ancient taxa still occur in the Himalayan center; several other species that bear gemmae (*A. minutum*, *A. hellerianum*, *A. tenue*, *A. michauxii*, *A. sphenoloboides*) have achieved ranges of varying degrees of disjunction in Laurasia—the last species is unique in the genus in being bisexual. All primitive taxa, existing in Gondwanalandic areas, are unisexual and lack gemmae.

These examples could be duplicated again and again, as with the cases of *Temnoma* (Schuster, 1969, map 14); *Metahygrobiella* ([ibid., map 13] in this genus there has been no Laurasian expansion; the single species in the Himalayan center has apparently not dispersed or speciated in the last 40 million years); *Andrewsianthus* ([ibid., map 11] again with a single species in the Himalayas, which has not speciated or dispersed; the other 17–19 species are strictly Gondwanalandic except for a few that have gotten across Wallace's Line); *Acrobolbus* ([ibid., map 8] two species, *A. wilsoni* and *A. ciliatus*, have achieved a distribution in Laurasia; the last species, found in the Himalayan center, has shown dispersal, relict today, to the Aleutians, Japan, and the Appalachians).[28]

Several significant facts are not apparent from the maps: (1) all of the taxa involved are unisexual, and almost all lack asexual bodies ("gemmae," etc.) that facilitate gametophytic dispersal; (2) the vast majority of these taxa have never been found with sporophytes; (3) hence, the disjunct ranges today appear to be relict ranges, and the former dispersal in Laurasia—in the period from 40 million years ago until the later stages of the Tertiary—was presumably in each case less or not disjunct. From these examples I would conclude that rafting on the Indian plate did occur; that, unquestionably, many other taxa crossed from Gondwanaland to Laurasia this way and then dispersed, but have died out in the Himalayan center. It is probable that such a scenario applies also to many angiosperms, with one major difference: their rate of evolution, especially under stress, is likely to be so fast that patterns which are clear when the Hepaticae

[28] In several cases the Gondwanalandic origin of a group has become obscure. Thus the Treubiaceae (fig. 18) are almost exclusively Gondwanalandic, with all taxa of *Treubia* still found in Gondwanaland. *Apotreubia* at first appears to have an anomalous, relict, and highly disjunct range: North Pacific–Himalayas–Borneo (*A. nana*) and New Guinea (*A. pusilla*). Presumably the ancestor of the specialized *Apotreubia* is *Treubia:* the latter, one can assume, evolved into *Apotreubia* during or after Indian plate migration, and then spread out, with subsequent (late Tertiary and Pleistocene) massive extinction. A similar pattern probably characterizes the small order Calobryales (fig. 20; see legend) and *Lophochaete* of the Blepharostomataceae (fig. 21; see legend). In view of the unquestionable antiquity of unisexuality in Bryophyta, with bisexual taxa derivative, it is surely significant that the only bisexual member of the Treubiaceae is *A. nana* and that in the moderately diverse Blepharostomataceae (8–9 genera in 3–4 subfamilies) we find evolution of bisexuality only in *Blepharostoma* and *Lophochaete fryei*. The sole case of bisexuality in *Anastrophyllum* is cited under example 7. (With biotype depletion under the gradually worsening climates of Tertiary-Pleistocene times, unisexual taxa, reduced to small and shrinking populations, would more readily tend to die out than would bisexual ones—the latter, reduced to very small populations, could still reproduce sexually.) Such internal evidence suggests migration patterns, and evolutionary patterns reinforce the thesis here presented. However, much more detailed analysis is needed than I have space for.

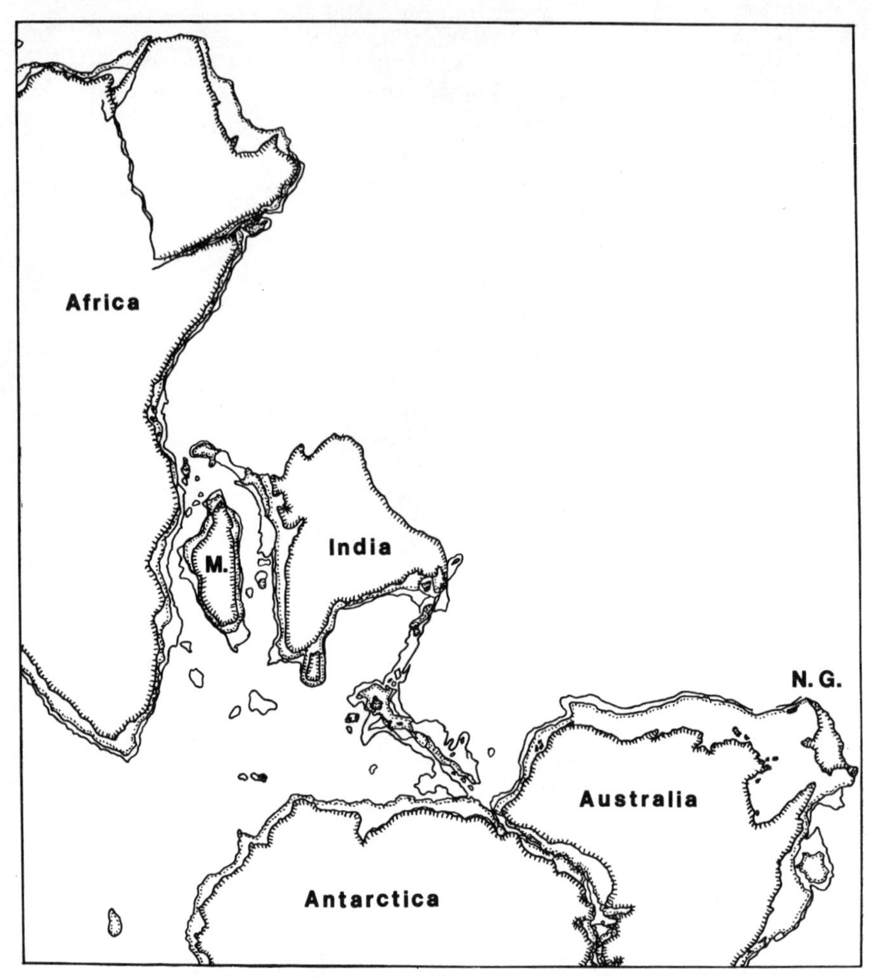

Fig. 26. Putative Middle to Upper Cretaceous configuration (90–100 million years ago). India has barely started its northward motion; putative peninsular or archipelagic connections remain between East Antarctica and Australia, India, Madagascar (*M*), and (via the Mascarene Plateau and Seychelles) Africa. Between India and Australia lies ancient sea floor (about 140 million years old) and some younger sea floor, suggesting that for at least 140 million years there has been open water between India and Australia. A reconstruction of this type would allow the stepwise migration of *Treubia, Bubbia, Gottschelia,* and other taxa with anomalous ranges today; compare text and appropriate maps. (Based on McKenzie and Sclater, 1973.)

are studied are almost obscured in the case of angiosperms. This does not imply that the migration route for the angiosperms was any less effective; it simply means that we will have to search more arduously to find unequivocal examples. Many cases may become clear only with intensive phylogenetic-morphological research.

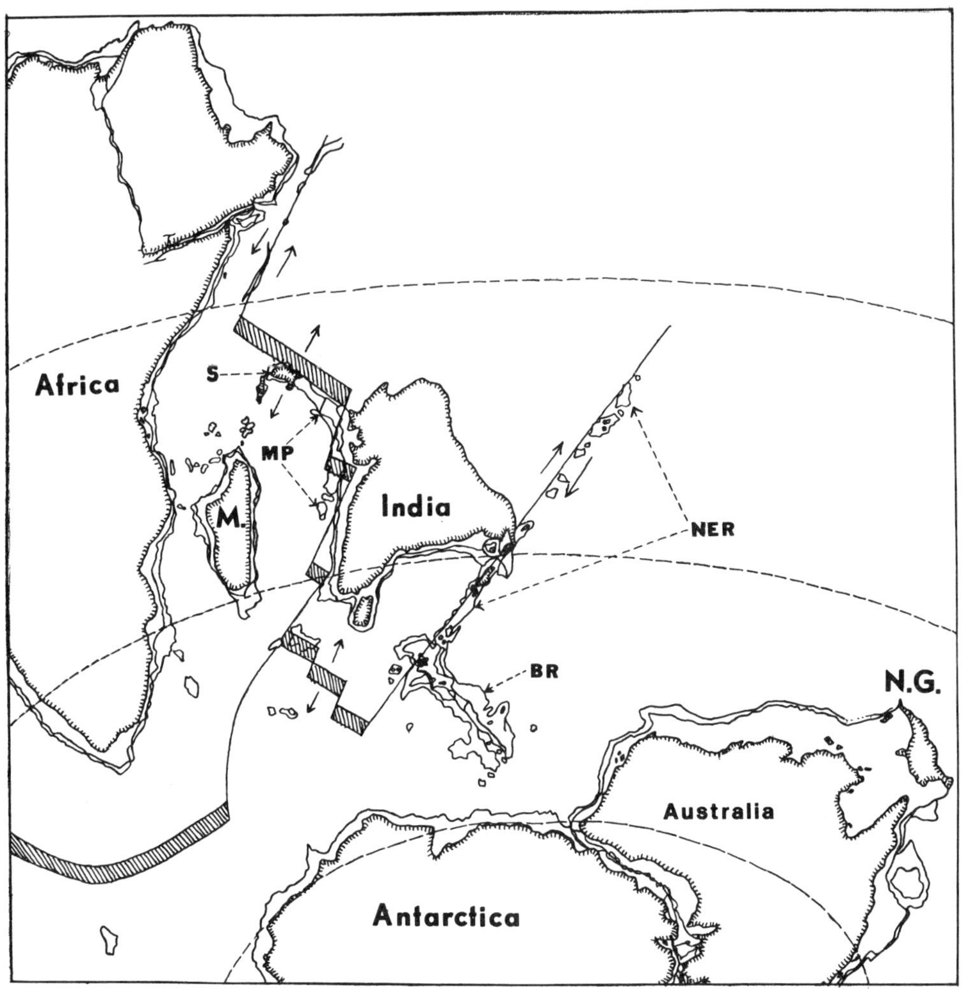

Fig. 27. Late Cretaceous configuration (75 million years ago). India has begun to move northward and northeastward. Large transform faults (at opposed arrows) enable the Indian plate to move unimpeded; new plate material is being generated in the shaded areas. *S*, Seychelles; *M*, Madagascar; *MP*, Mascarene Plateau; *NER*, Ninety-East Ridge; *BR*, Broken Ridge; *N.G.*, New Guinea. (Modified from McKenzie and Sclater, 1973.)

The pattern of gymnosperm extinction (e.g., *Dacrydium, Acmopyle*) as the Indian plate moved northward was probably matched by extinction among angiosperms. If we visualize an indented, irregular coastline extending from the land masses ancestral to New Guinea and Australia, across the then-oceanic northern edge of the Indian plate, to the Mascarenes and Madagascar, it seems likely that *Bubbia* and *Bubbia*-like Winteraceae could have occurred along this coastal region; the subsequent migration of India and the extinction of any such taxa there would explain the now-anomalous range of the *Bubbia* complex, with a large gap between New Guinea and Madagascar. A similar explanation could account for the anomalous range of the hepatic genus *Gottschelia*, found from New Guinea to Ceylon, and to Madagascar, but not in India proper. Even if genera like *Bubbia* and *Gottschelia* did not originally evolve on the Indian plate, the orientation of land masses in mid-Cretaceous times (see figs. 26 and 27) was such that migration into that area by organisms with limited dispersal powers was readily possible. Indeed, the generic identity of fossil late-Cretaceous dinosaurs such as *Laplatosaurus*, *Titanosaurus*, and *Antarctosaurus* from South America to India (Hallam, 1972) suggests that migration, even from South America via East Antarctica to Australia, India, Madagascar (and, until sometime in the Cretaceous, from there to Africa), was then readily possible, *overland.*

Australasian Plate Migration and its Effects

INTRODUCTION AND HISTORY

Long ago Joseph Hooker noted in a letter to Darwin (Nov. 1851; see Huxley, 1918, p. 445) that the floras of New Zealand (a fragment of the Australasian plate) and extratropical South America show affinities that are "very remarkable and far more than can be accounted for by any known laws of migration." He concluded that he was "becoming slowly more convinced of the Southern Flora being a fragmentary one—all that remains of a great Southern continent." Darwin wrote to Wallace in 1876 (see Hill, 1929, p. 1485) that "there must have existed a Tertiary Antarctic Continent from which various forms radiated to the southern extremities of our present continents." We now know that until about 45–50 million years ago, present-day Australia was attached to this Tertiary Antarctic continent. Until at least the Eocene, that last large southern land mass must have been wholly or very largely ice-free (McIntyre and Wilson, 1966) and, as Skottsberg (1960) eloquently phrased it, except in elevated interior portions enjoyed "a genial climate and [was] covered with vegetation: a land of lofty mountains and deep valleys, with lakes and rivers, and offering all kinds of habitats."

The history of the Australasian plate is thus a fascinating one. As the next section clearly shows, until after the end of the Cretaceous the biota of Australasia was linked with that of Antarctica and, through the agency of the juxtaposed ancestral elements of the present-day Scotia Arc, also with that of temperate South America. It is only at the start of the Tertiary that links with South America were finally severed; and apparently only in late Miocene to Pliocene times was a new link forged with the Malaysian sectors of Laurasia. These are well-established facts to which the biogeographer must fit his data.[29] The only really questionable factors of major potential significance are: first, that in mid to late Tertiary times, during its migration north, Australasia may have collided with or overridden islands whose biota was tropical-Laurasian in derivation; and second, that in its migration northward, New Guinea may have been formed by incorporation of southern areas derived from Gondwanaland, but the northern fringe may, at least in part, be of Laurasian derivation (van Bemmelen, 1949). Both of these are still highly arguable and hypothetical possibilities. Aside from these limiting factors, we must assume (Raven and Axelrod, 1972; Schuster, 1972a) that Wallace's Line, zoogeographically almost universally accepted, was formed by late-Tertiary collision of the Australasian plate with Laurasia. Until near the end of the Tertiary, New Guinea was much smaller than at present (Schuster, 1972a, p. 33) and, according to Durham (1963), the entire western third (the sector closest to the archipelagic extensions of Laurasia) was submerged. Hence we must assume the existence, until recently, of a water gap between New Guinea and the eastern part of the Malay Archipelago that was "much larger than it is today" (Schuster, 1972a).[30]

TIMING OF AUSTRALASIAN PLATE MIGRATION AND OF ACCESSORY RIFTING

The fission of the Antarctic-Australasian region of Gondwanaland (West Gondwanaland, as it is sometimes called) after the breaking off of the Indian plate is a relatively recent event.[31] Until about 47 million years ago, West Antarctica and Australia were still joined (fig. 27). The rapid northward movement of Australia and the interrelationships of Australia and New Guinea, New Caledonia, New Zealand, and the subantarctic islands region (Campbell Plateau) form a

[29] For more details on certain aspects see Schuster (1972a); space limitations preclude giving as many maps as would be desirable.

[30] The rapid and drastic reduction in number of marsupials as we go from New Guinea westward (fig. 42) clearly reflects this fact.

[31] The present outline is highly compressed because recent reviews by Schuster (1972a) and by Raven and Axelrod (1972), which complement each other, present much of the background data in more detail; they also present a more detailed bibliography. The paper by Griffiths (1971) is essential background.

complex and intricate story that is only partially clarified.[32] In essence, well before the actual initiation of the northward migration of Australia, the Australasian plate (consisting of Australia, New Zealand, New Caledonia, parts of New Guinea, etc.) began to fragment—the eastern flank fissuring to form smaller plates, on which microcontinents and island archipelagoes occur today. Until sometime rather late in the Cretaceous, Australia was still connected to a series of continents and continent fragments: East Antarctica to the south; areas underlying present-day New Zealand, New Caledonia, Norfolk Island, and Lord Howe Island, to the east; New Guinea to the north (fig. 28). In addition, rises and plateaus that existed then are now submerged or bear only small volcanic projections. The principal ones are Campbell Plateau, Lord Howe Rise, Norfolk Ridge, and Queensland Plateau (fig. 28); these are all formed of continental rocks (Cullen, 1970; Karig, 1970; Chase, 1971; Griffiths, 1971; Shor, Kirk, and Menard, 1971; van der Linden, 1971). Existing islands of these ridges and plateaus and their present position cannot be considered a valid gauge of the land areas which may have existed in the past, as Campbell Island proves. Here the oldest volcanic rocks rest in part on quartz pebbles—evidently washed in from a now-vanished land mass that apparently had eroded away. (The evolution there of the endemic composite genus *Pleurophyllum*—with one species also on Macquarie Island—suggests a considerable age. The rich and complex bryophyte flora, about 160 species of Hepaticae, on Campbell Island, which now has less than 50 sq. km, reflects its "continental" character.) As Griffiths (1971) notes, in early Eocene times (50 million years ago) the Campbell Plateau lay close to West Antarctica.

Likewise, the striking similarities between the New Zealand and New Caledonian floras (e.g., common possession of Winteraceae, *Agathis*, podocarps; Goebeliellaceae, with the sole taxon *Goebeliella cornigera; Zoopsis caledonica*, and *Geocalyx caledonica; Saccogynidium* spp., *Adelanthus piliferus* [Schuster, 1969]) suggest foundering of former connections. Griffiths (1971) notes that the Norfolk Ridge, formed of "crust of continental character," connects New Zealand and New Caledonia.

The New Guinean sector of the Australasian plate has an especially complex history; the extant biota reflects this closely. New Guinea has been only recently elevated, with perhaps much less than 50 percent of the current region above the sea until late in the Tertiary (Durham, 1963; Hermes, 1968; Davies and Smith, 1971;

[32] However, Griffiths (1971) suggests that the entire picture is explicable on the sole basis of "sea-floor spreading and the formation of new ocean basins."

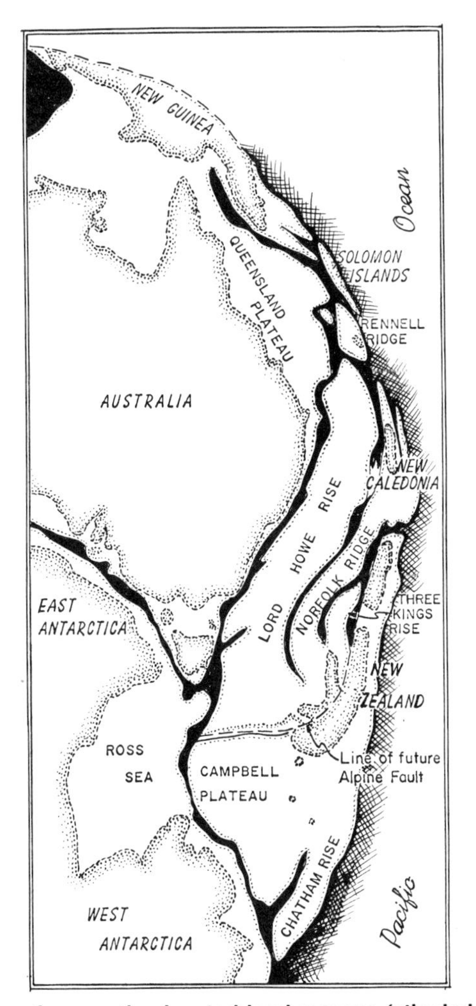

Fig. 28. Orientation of presently elevated land masses (stippled areas), rises, and plateaus that formerly were at least largely above the sea (white), and the regions where major and minor rifting began to separate these areas (black), starting about 80 million years ago. (Redrawn, after Griffiths, 1971, and Raven and Axelrod, 1972.)

Jones, 1971). Furthermore, some evidence suggests that only the southern half is surely of Gondwanalandic origin: it now appears possible that, migrating northward, the Australasian–New Guinean plate collided with (and overrode?) what may have been archipelagic extensions of Laurasia. If this is the case, it may be necessary to postulate more than 5–10 million years of intimate contact—and mutual biotic transgression—between the Gondwanaland-derived biota of Australia–New Guinea and those of Laurasia. In any event, upland

New Guinea is rich in Gondwanalandic elements (Schuster, 1972a) largely derived from Australian antecedents.

In essence, therefore, the existing distribution of land areas in Australasia reflects, first, the breakup and subsequent spreading out of a belt along the eastern, Pacific, edge of Gondwanaland, with sea-floor spreading and transcurrent faulting, and with synchronous, complex, abrasive interactions with the westward-moving Pacific plate; and second, the evolution of complex faulting, for example, that running through New Zealand. The western portions of New Zealand, "hinged" to the Australian plate, were carried northward rather rapidly, while the eastern portions, in abrasive contact with the Pacific plate, had the northward portion of their migration impeded. This must be visualized as being partly a consequence of the eastward component of the migration of New Zealand, interacting with the westward-moving Pacific plate, which was subducted along the Tonga-Kermadec Trench. The results are rises, submarine plateaus, and islands—now strikingly isolated—that seem oceanic but are really continental in origin. These rises, plateaus, and islands are oriented in arcs spreading east and northeastward from Australia, from which they are now separated by generally inactive basins (Karig, 1971). Karig notes that the Tonga-Kermadec Ridge forms a frontal arc, and the "geology of exposed islands indicates that . . . [this ridge] has been a frontal arc since the early Tertiary." The outer margins of the arc include New Zealand and, perhaps in part, Fiji, along the so-called andesite line along the Tonga-Kermadec Trench. Karig suggests that the linear ridges and troughs of the inner arc basins are "no older than latest Tertiary," thus no older than the Pliocene (Raven and Axelrod, 1972), and were evidently formed by rifting within the older frontal arc. Spreading of that frontal arc appears to have commenced about 80 million years ago, with an initial configuration as in figure 28. Hayes and Ringis (1973) state that the Central Tasman Sea between Australia and New Zealand, formed by sea-floor spreading between 60 and 80 million years ago, and the ensuing basin have been tectonically inactive for the last 60 million years.

Subsequent disruption and fission apparently also involved a movement northward of the Campbell Plateau (including the Auckland Islands and New Zealand) about 80 million years ago, in the late Cretaceous (Pittman, Herron, and Heirtzler, 1968; Heirtzler, 1971), when the Pacific-Antarctic Rise was formed.[33] The Australian–

[33] This date is suggestive: arrival of *Nothofagus*, *Leiopelma*, and the tuatara (*Sphenodon*) in New Zealand must predate this event. Presumably, ratite birds also ar-

New Guinean sector remained attached to Antarctica for a much longer period—until the Indian-Antarctic Rise was activated, about 45–50 million years ago, in the Eocene (Berggren, 1969; Dietz and Holden, 1970; Tarling, 1971). This represents the last point at which linkage existed to Antarctica, and, somewhat earlier, through it to South America. This complex history explains why land mammals are lacking from the New Zealand–New Caledonia–Fiji region; monotremes and marsupials occur in the Australia–New Guinea region, but (with a few Pliocene to Recent exceptions) no modern mammals occur there. Australia, at the time of its fission from Antarctica, was apparently positioned some 30° farther south, equal to the present distance between western Antarctica and Australia (Irving, 1964; Le Pichon and Heirtzler, 1968; Dietz and Holden, 1970; Heirtzler, 1971; Tarling, 1971; Veevers, Jones, and Talent, 1971), suggesting northward migration of the Indian-Antarctic Rise. Conversely, if the rise has been stationary, as is sometimes assumed (Wellman, McElhinny, and McDougall, 1969; Jones, 1971), there would have been about a 15° southward migration of Antarctica and equivalent migration northward of Australia. By early Oligocene, 36 million years ago (Wellman, McElhinny, and McDougal, 1969) or perhaps as late as Miocene times (McElhinny, 1970; Jones, 1971), Australia was positioned some 10° south of its current location. In contrast, perhaps impeded by its more direct, abrasive contact with the westward movement of the Pacific plate, New Zealand had moved more slowly northward, but its isolation was enhanced by sinking of the Campbell Plateau.

After late-Cretaceous to Eocene arclike spreading of the New Caledonia–New Zealand fragments, the remaining Australia–New Guinea complex started its northward movement about 45–50 million years ago. At that time northern Australia, encompassing not only Queensland but also the now-foundered Queensland Plateau, lay in relatively temperate latitudes, mostly below 40°S., at about the present latitude of Tasmania (Berggren, 1969; Dietz and Holden, 1970; Tarling, 1971). This fact has major biogeographic consequences. The New Guinean sector was largely submerged until the Miocene and later, by which time the region had rafted far enough north that the emergent lowlands lay near or perhaps within the tropics. A water barrier, persisting apparently until the Pleistocene (Jones, 1971), impeded migration from Australia to New Guinea. Ap-

---

rived before then. However, marsupials evidently crossed Antarctica after this date. Their absence from New Zealand is in accord with their arrival in or after mid-Tertiary times in Australia.

parently simultaneously with northward migration of the Australia–New Guinea complex two other events occurred: uplift, due in large part, in the case of New Guinea, to the initial approach of the Australian plate to the Asian one, but also in part to vulcanism along the New Zealand alpine fault; and rapid lowering of temperatures toward the end of the Tertiary. The roughly synchronous nature of these events tended to cancel out, in part at least, the effects of northward migration; higher peaks tended to provide refugia for cool- or cold-adapted organisms (e.g., *Nothofagus*, certain of the Winteraceae) that might otherwise have died out; the general cooling tended to diminish the effect of the northward drift.

The history of the region, from mid-Cretaceous times on, is thus complex both geologically and climatologically. The biota of the region reflects these complexities, chief of which are: initial fragmentation and isolation of subplates by sea-floor spreading at the eastern perimeter of the Australasian plate; northward drifting of the Australia–New Guinea complex; foundering of plate fragments such as the Queensland Plateau and most of the Campbell Plateau, leading to biotic isolation and creation of hyperoceanic environments; and extensive vulcanism, with elevation of high peaks, especially in New Zealand and New Guinea, and with creation of similar peaks from ocean bottom, resulting in volcanic archipelagoes (e.g., Miocene archipelagoes such as the Solomons and New Hebrides [see Karig, 1970, 1971; Hackman, 1971; Mallick, 1971; Quantin, 1971]). Collision between the Australian and Pacific plates also had other effects, such as the uplift of the Fiji Archipelago (chiefly in Eocene times), which is composed principally of ophiolites (ocean floor derived from upper mantle rocks and associated sedimentaries).

FLORISTIC CONSEQUENCES

The preceding paragraphs have suggested that certain rather definite migratory routes were open during quite specific times.

1. Until the end of the Cretaceous, migration from South America to Antarctica was possible, at least for organisms like marsupials and *Nothofagus*. By Eocene times and possibly Paleocene times the uplands of Antarctica were apparently already accumulating an ice cover, but at least until Oligocene times the more oceanic, coastal lowlands still supported forests—and could serve as migratory routes.

2. Until about 80 million years ago, step-wise migration remained feasible between South America and New Zealand, Norfolk Island, New Caledonia, etc.; after this time, rifting away of the latter regions isolated them.

3. Until about 65 million years ago, migration from South America to Australia via Antarctica was feasible, at least along the warmer, more humid coastlines; after this point, step-wise migration of organisms already in Antarctica to Australia remained feasible for another 15–20 million years, until rifting severed Australia from Antarctica. Some casual migration after these dates must be assumed since water distances initially remained small and climatic cooling had not proceeded to a critical point, at least for cold-adapted taxa.

In essence, therefore, we must assume that overland dispersal was possible for organisms without—or with strictly limited—overwater dispersibility along a route or routes between South America–Antarctica–Australasia during nearly all of the Cretaceous. This is the critical time when angiosperms first evolved and diversified, but had not yet, in most instances, evolved efficient long-distance dispersal mechanisms. It is my contention that parallels in patterns of dissemination during Cretaceous times between angiosperms and such quadrupeds as the dinosaurs (Hallam, 1967) and marsupials (Cox, 1973) are real and pertinent. It is only late in the Cretaceous, and principally in Tertiary times, that modern angiosperms with modern dispersal mechanisms evolved; it is within these same time perimeters that coevolution of efficient dispersers (e.g., fruit-eating birds and mammals) occurred. We must look at the entire picture and not its fragments. One such interrelated sector is dealt with in detail in the next section of this paper; others can only be related here briefly.

The often gross disjunctions existing between groups in the Antipodes are, in large part, demonstrably due not to long-distance dispersal but to physical fragmentation of the remnants of Gondwanaland (Schuster, 1969, 1972a). If we reassemble these fragments and elevate the foundered sectors (e.g., Queensland Plateau, Campbell Plateau), we find a mid- to late-Cretaceous geography that shows intimate connections between Australia, including New Guinea and Tasmania, the fragments to the east (chiefly New Zealand and New Caledonia), and the Antarctic regions to the south (fig. 28). Temperate sectors of this large Terra Australis—Australasia plus Antarctica of today—by late Cretaceous times were covered with forest complexes formed of various austral gymnosperms (chiefly *Araucaria, Dacrydium, Podocarpus, Agathis, Acmopyle, Phyllocladus, Arthrotaxis* [see Florin, 1963]) and principally evergreen angiosperm groups (chiefly *Nothofagus*, capsular Myrtaceae, Proteaceae, Araliaceae, Atherospermataceae, Sapindaceae, and Myrsinaceae [Schodde, 1970; Raven and Axelrod, 1972]), but presumably also included taxa of a series of primitive woody, magnolialean angiosperms (Winteraceae, Himantandraceae, Eupomatiaceae, Austrobaileyaceae, Amborellaceae, Trimeniaceae, Monimiaceae, Degeneriaceae [Schuster, 1972a]). In suitable sites a variety of other nonarborescent groups such as Gunneraceae and Epacridaceae occurred, and a wide variety of isolated groups of Hepaticae (Calobryales; Metzgeriales: Treubiinae, Phyllothalliinae; Hymenophytaceae; several suborders of Jungermanniales including: Balantiopsidinae, Perssoniellinae, Lepicoleinae, Lepidolaeninae, perhaps Ptilidiinae [Schuster, 1969, 1972a, b; also fig. 29]).

This Terra Australis retained intimate connections to South America until at least the start of the Tertiary. Forest communities persisted in Antarctica at least until Oligocene times (McIntyre and Wilson, 1966; Cranwell, 1969), but Hennig (1966) has assumed that disjunction between Australian and South American groups had become established by Oligocene-Miocene times (26 million years ago).

As a consequence, we must make certain major assumptions: (1) ancestors of those Winteraceae (*Pseudowintera, Tasmannia, Zygogynum, Belliolum, Exospermum*) present in New Zealand and New Caledonia arrived there more than 80 million years ago.[34] (2) *Nothofagus* also arrived in New Zealand and New Caledonia more than 80 million years ago. (3) Marsupials had not crossed Antarctica by that time, since they never reached New Zealand or New Caledonia; indeed, they appear to have reached South America only in the late Cretaceous, perhaps 80 million years ago (Raven and Axelrod, 1972). (4) More than one migration route existed. As is evident from the reconstruction of Schopf (1970a), after the Jurassic start of separation of the African plate, one can visualize migration of taxa from South America via East Antarctica to Tasmania-Australia, and also by a (perhaps more coastal) route from South America to New

[34] The Winteraceae (fig. 35)—phenomenally primitive except for the anomalous, coherent pollen grains—probably achieved their Gondwanalandic distribution by mid-Cretaceous times, although the pollen is not known until the Tertiary. If it is plotted on Gondwanalandic reconstructions (such as that of Smith and Hallam, 1970), the group clearly still has a range that, in toto, is nearly coherent if we reassemble the southern continents as on figures 43–44. Note the occurrence of the Winteraceae from along western South America to West Antarctica (fossil) to Tasmania, New Zealand, New Caledonia, New Guinea, and across to Madagascar (a *Bubbia*-like species!). Presumably the group occurred at one time in India and western Australia (the Cretaceous occurence of *Nothofagus*—a common associate of Winteraceae—in western Australia is relevant; figure 39). From western Australia to Madagascar is not impossibly far on the Smith-Hallam reconstruction. Such an explanation for the Madagascan occurrence of Winteraceae is much more likely than the one advanced by Carlquist (1965), who has *Bubbia* "once [extending] across mainland Africa and Asia"—for which there is neither geological nor fossil evidence.

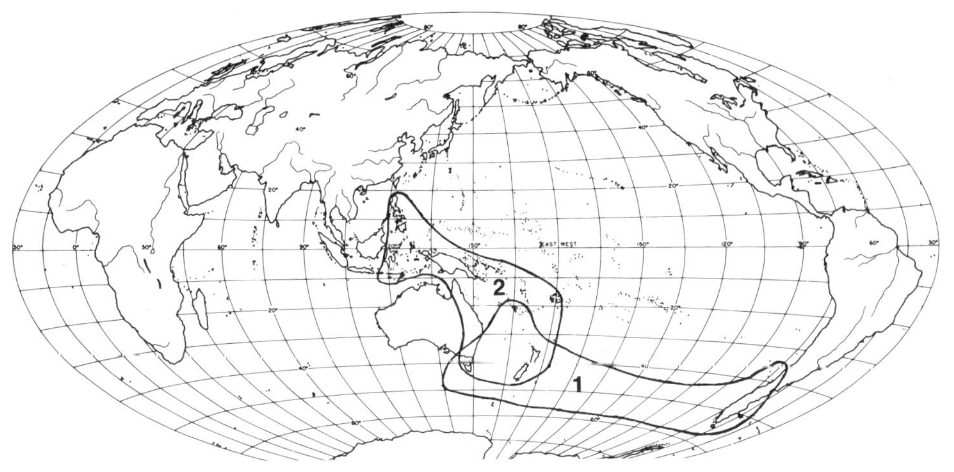

Fig. 29. Generalized range (1) of primitive (isophyllous; triradial) Jungermanniales. In this area are found the bulk of genera (and species) in the primitive orders Herbertinae (*Herberta, Triandrophyllum, Grollea, Chaetophyllopsis, Herzogianthus, Archeophylla, Archeochaete, Temnoma, Eotrichocolea, Trichocolea*), Lepidolaeninae (*Lepidolaena, Gackstroemia, Jubulopsis*), Balantiopsidinae (*Anisotachis*), Personiellinae (*Pleurocladopsis*); Ptilidiinae (*Ptilidium, Mastigophora*) and Lepicoleinae (*Vetaforma, Lepicolea*). *Assumption:* this represents the northern "leading" edge of these groups, all of which were originally cool-antipodal and Gondwanalandic. As with various gymnosperms and angiosperms (e.g., *Drimys, Nothofagus*), the remaining members were forced northward by late Tertiary times and their extant range is a relict one. The very high correspondence between Tasmania–New Zealand and Tierra del Fuego–Patagonia strongly suggests a long-continued, broad connection, which no insular or land-bridge theory could satisfy. Contrasted is the generalized range (2) of primitive, Gondwanaland-derived families of angiosperms. Note that in general these have shown more striking displacement northward; they exhibit less tolerance for the subantarctic conditions prevailing today.

Zealand via West Antarctica. Perhaps the distribution of the Eucryphiaceae (Tasmania–South America) reflects this pattern; perhaps it reflects merely the fact that *Eucryphia* underwent its overland migration after the "deadline" 80 million years ago, after which New Zealand was cut off.

LATE TERTIARY AND PLEISTOCENE HISTORY

In the next section of this paper, late-Tertiary and Pleistocene history is outlined in detail for *Nothofagus* and the marsupials; here I would like to emphasize several consequences of the Paleocene-Eocene cooling of Antarctica. Many taxa still present in Australasia and southernmost South America probably occurred across Antarctica until Oligocene times; these taxa today all possess a relict and disjunct range (figs. 30, 32–34). Also, there has been a distinct north-

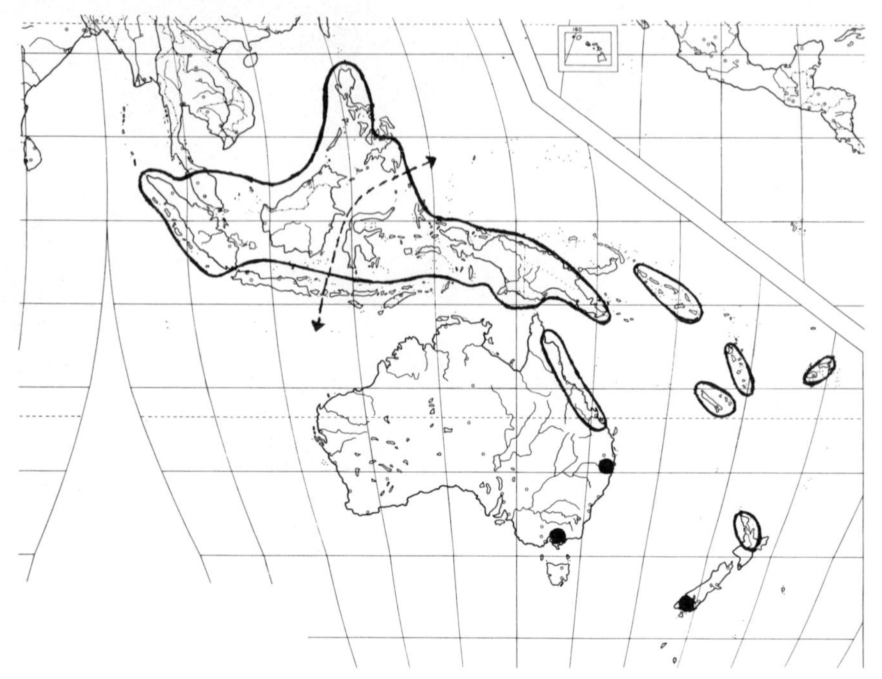

Fig. 30. Gondwanaland-derived range of *Agathis* Salisb., past (large dots; Oligocene-Miocene) and present (ringed areas). Note transgression of Wallace's Line (dashed arrow). (From Florin, 1963.)

ward shift for many taxa, and extinction southward. Thus on figure 29 we see the generalized range of primitive groups of Gondwanalandic Jungermanniales, 1, and of Gondwanalandic Angiospermae, 2; both groups were surely common on Antarctica up to early Tertiary times. The northward shift is also visible when past and present ranges of specific genera are studied, e.g., the conifers *Agathis* (fig. 30), *Araucaria,* and others (Couper, 1960). Presumably most of the primitive angiosperms now occupying region 2 on figure 29 (which corresponds rather closely to past plus present range of *Agathis*) occurred southward into Antarctica in or before the Oligocene. Finally, the northward shift was, until 47 million years ago, presumably by active migration northward; after this date the taxa, in part, were rafted northward on the Australian plate.

A consequence of this last point is that sometime relatively late in the Tertiary the Gondwanalandic elements that were shifted northward on the Australian plate came into intimate contact and competition with Laurasian-derived elements present on the archipelago now existing on the Sunda Shelf (fig. 31). Thus, relatively late in the Tertiary, a floristic "province" called Malesia (which has been a sub-

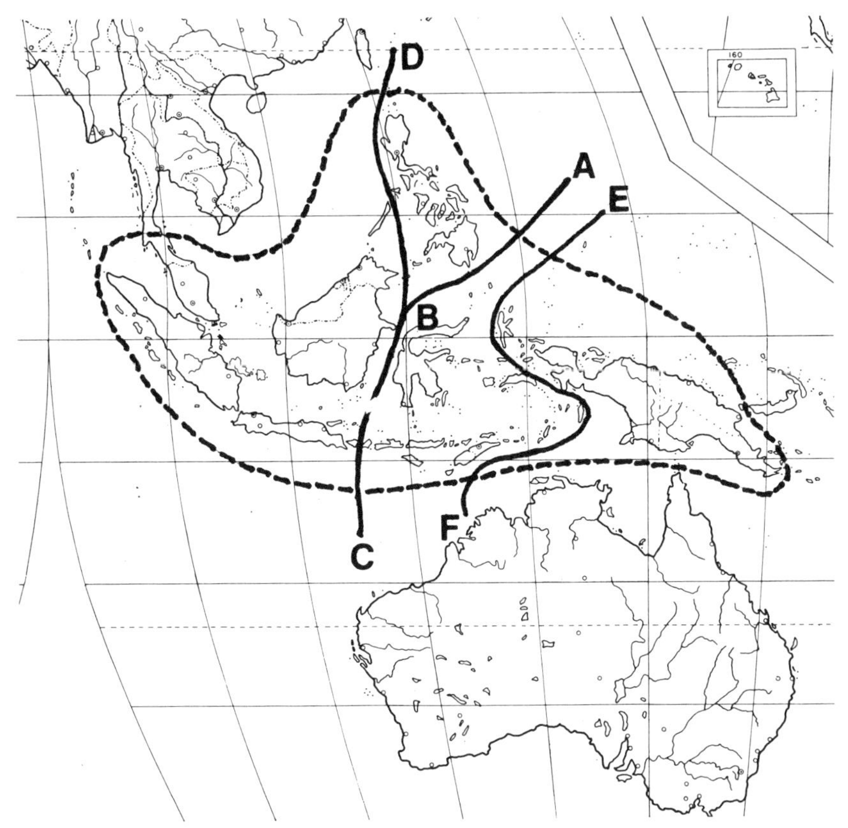

Fig. 31. Boundaries of Malesia. *A-B-C,* Wallace's Line; *D-B,* Merrill and Dickerson's modified line. Area west of D-B-C is part of the Asian (Sunda) Shelf; area east of E-F is part of the Australia–New Guinea (Sahul) Shelf. (After Keng, 1970.)

ject of contention among biogeographers for a century) came into existence. Botanists (e.g., van Steenis, 1971) have recognized such a province and have paid no attention to Wallace's Line (see Schuster, 1972a). Zoologists, by contrast, have emphasized the fact that land areas on the Sunda and Sahul shelves are fundamentally distinct and show little or no transgression by organisms not suited for long-distance dispersal (e.g., marsupials [fig. 42] or fresh-water fishes). They further emphasize that even in groups such as birds, which show theoretical wide dispersibility, we find a rapid shift, as we go eastward, from purely Laurasian faunas to mixed faunas to purely Gondwanalandic faunas. It is principally in a rather narrow band, lying between lines C–B–D and E–F on figure 31 that we find a strongly mixed fauna. Orogeny in this mixed belt is complex, and much transgression (of, for example, animals with no overwater dispersal capa-

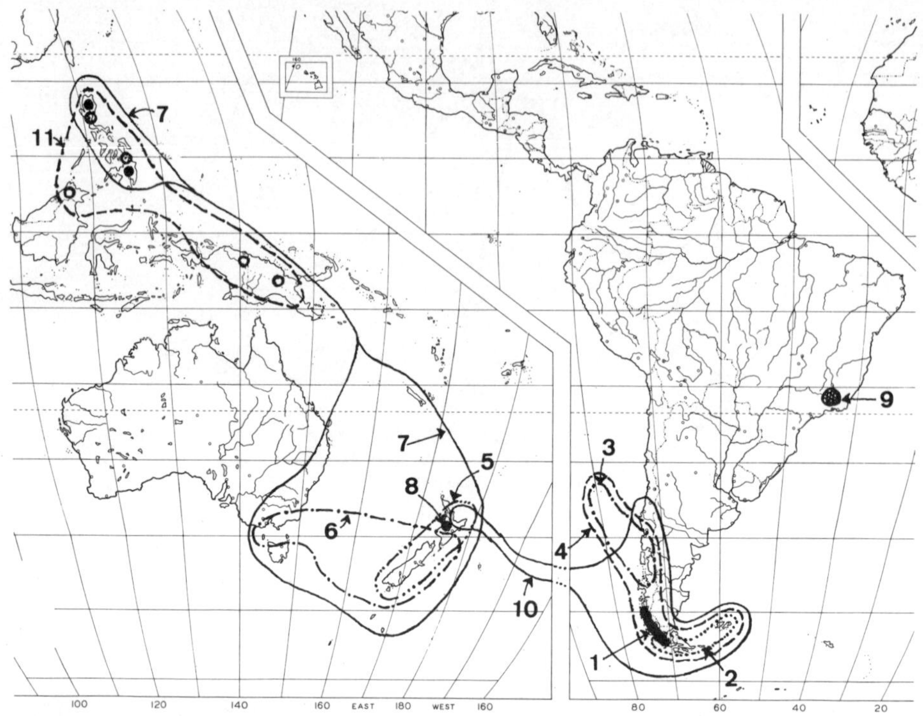

Fig. 32. Gondwanalandic ranges of unisexual Jungermanniales. Total range of the *Balantiopsis* Mitt.–*Anisotachis* Schust. complex. (1) *Anisotachis splendens* (Steph.) Schust. (2–11) *Balantiopsis* Mitt. (2) *B.* (subg. *Steereocolea*) *bisbifida.* (3) *B. asymmetrica, B. crocea,* and *B. purpurata.* (4) *B. cancellata.* (5) *B. convexiuscula.* (6) *B. tumida.* (7) *B. diplophylla s. lat.* (composite species). (8) *B. lingulata.* (9) *B. brasiliensis.* (10) *B. erinacea.* (11) *B. ciliaris.* The range recalls that of the Winteraceae (fig. 35); in both cases there is slight crossing of Wallace's Line, but no species found north and west of Wallace's Line fails to occur east of it. *Assumption:* prior to the Oligocene the group was widespread in Gondwanaland, chiefly in cooler areas; dispersal possibly occurred by mid-Cretaceous. The allied genus *Neesioscyphus* is also Gondwanalandic.

bility, such as stegodonts [see Audley-Charles and Hooijer, 1973]) is the result of early and middle Pleistocene uplift. Such transgression, often along ridge systems, followed late Pliocene–early Pleistocene plate collisions.

It is curious that plant geographers, by and large, have failed to recognize the significance of this line (or of the variants proposed [see figure 31]) while zoologists have not. Malesia is, I think, an exceptionally rich area because of the fusion and interpenetration of two floras along Wallace's Line and in part also because of the following factors: (1) As the northern end of Australia and emergent New Guinea approached the Sunda Shelf they reached tropical and

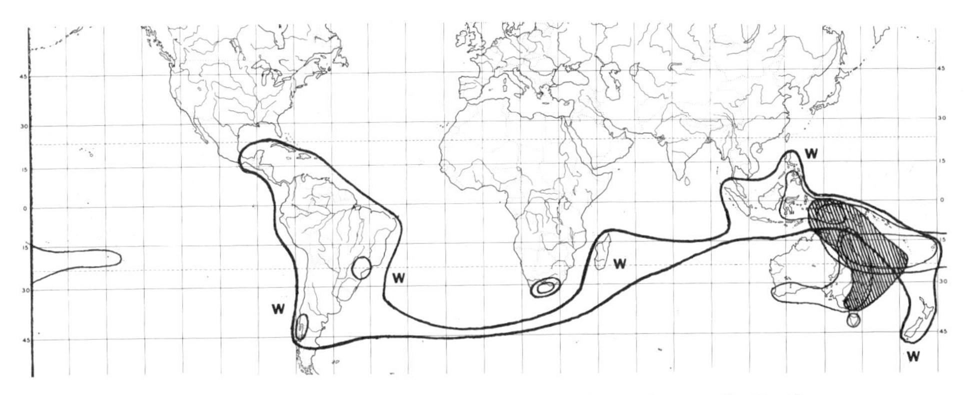

Fig. 33. Total range of the Gondwanalandic family Cunoniaceae R. Br. One genus (*Weinmannia* L., *W* on map) is widespread, and has transgressed Wallace's Line and the Isthmus of Panama. The family may have reached southern Africa in recent geological periods. The center of diversification is Australasia (shaded area), with one added genus reaching the Philippines. (After Good, 1964.)

subtropical latitudes, despite late-Tertiary and Pleistocene cooling; extinction of much of the autochthonous biota occurred, and these regions arrived biotically "depleted" in subequatorial latitudes (Schuster, 1972a). (2) Adjacent regions of southeastern Asia plus western Malesia were richly endowed with taxa adapted to such latitudes and climates. (3) The lowlands and middle elevations of New Guinea and Queensland were therefore, in Pliocene to Recent times, rapidly invaded by such Laurasian-derived elements. As a consequence, rapid and intimate admixture of floristically discrete elements occurred.

By and large, quadruped dispersal has not become more efficient with further evolution—leaving a few groups like bats out of consideration. By contrast, as already emphasized, angiosperm evolution has involved the progressive and consistent evolution, in many groups, of devices that greatly increase dispersibility of the organism. Modern groups, often (especially in tropical families) with edible, fleshy fruits or with efficient wind-dispersal mechanisms, thus can extend their ranges rapidly. This is also true of some groups such as the mangroves (see appropriate maps in van Steenis and van Balgooy, 1966).

Gondwanalandic elements also managed to infiltrate, to some extent, into western Malesia and even into Southeast Asia proper (figs. 32–34); in each case the transgression was very limited. In general, however, transgression across Wallace's Line was by *modern* genera and families, with appropriate long-distance dispersal devices. Thus the typical wide-ranging patterns found in some groups that center

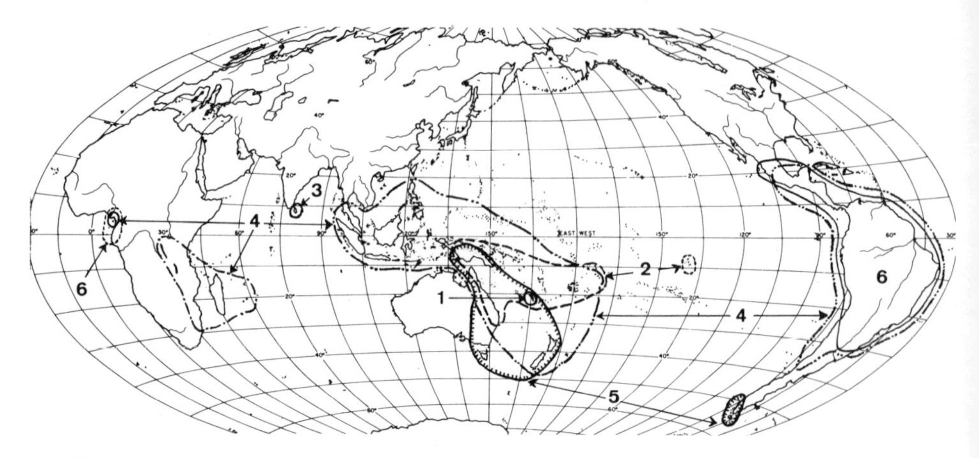

Fig. 34. Monimiaceae Juss. (*s. lat.*): the monimiaceous alliance, with about 37 genera, 540 species. Of the six families, four are present in Australasia, the center of diversity: (1) Amborellaceae Pichon. (2) Trimeniaceae (Perk. & Gilg) Gibbs. (3) Hortoniaceae. (4) Monimiaceae *s. str.* (5) Atherospermataceae R. Br. There is slight transgression of Wallace's Line by the Monimiaceae *s. str.* and of Central America by the Siparunaceae (6). Although A. C. Smith (1972) regards the group as "basically Asian-Australasian," I would regard the complex as Gondwanalandic. (After A. C. Smith, 1972.)

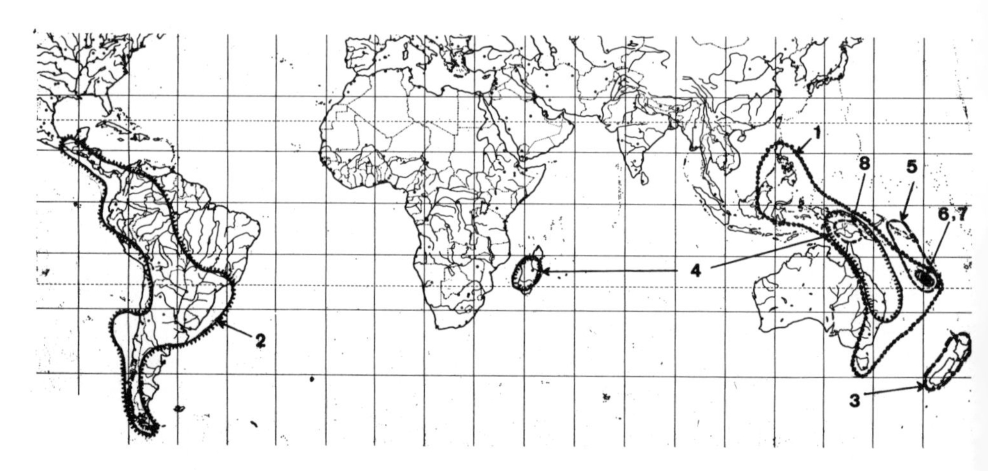

Fig. 35. Generalized extant range of the primitive, vesselless family Winteraceae. (1) *Tasmannia* R. Br., about 20 species (only 1 extends across Wallace's Line). (2) *Drimys* J.R. & G. Forst., 4 species (1 extends across the Isthmus of Panama). (3) *Pseudowintera* Dandy, 3 species (4) *Bubbia* Van Tiegh., about 30 species (5) *Belliolum* Van Tiegh., 8 species. (6) *Exospermum* Van Tiegh., 2 species (7) *Zygogynum* Baill., 6 species (8) *Tetrathalamus* Lauterb., 1 species = *Bubbia*. Fossils known from Antarctica and (doubtfully) western North America.

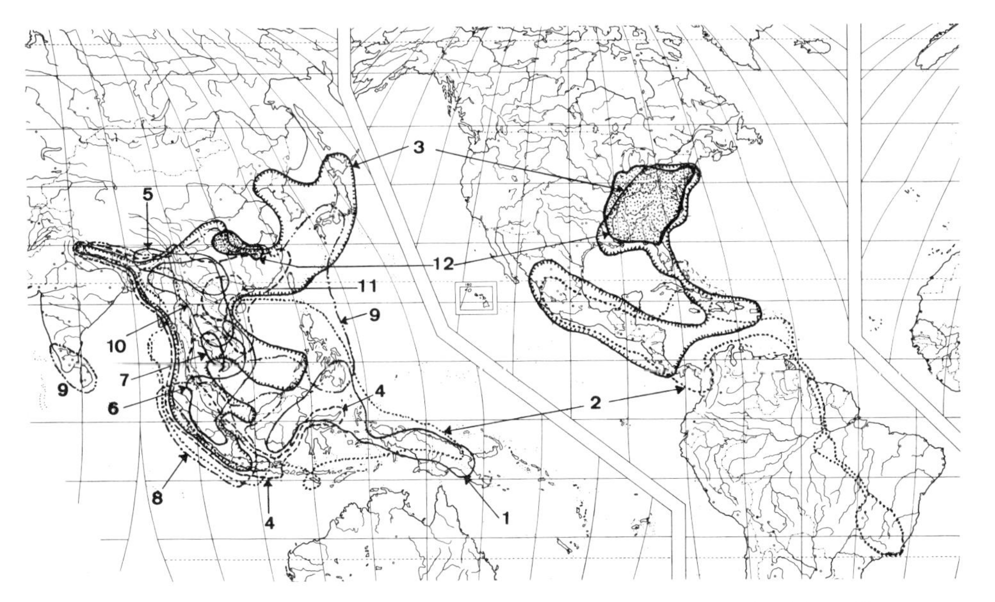

Fig. 36. Generalized extant distribution of Magnoliaceae. (1–11) Tribe Magnolieae. (1) *Elmerrillia* Dandy, 7 species. (2) *Talauma* Juss., 40 species. (3) *Magnolia* L., 80 species. (4) *Manglietia* Bl., 25 species. (5) *Alcimandra* Dandy, 1 species. (6) *Pachylarnax* Dandy, 2 species; 1 occurs as a disjunct to Assam, outside the range shown. (7) *Kmeria* (Pierre) Dandy, 2 species. (8) *Aromadendrum* Bl., 2 species. (9) *Michelia* L., 45 species. (10) *Paramichelia* Hu, 3 species. (11) *Tsoongiodendron* Chun, 1 species. (12) Tribe Liriodendreae; *Liriodendron* L., 2 species. Note the transgression of Wallace's Line by 1 species of *Manglietia* (to the Celebes) and more significantly by *Talauma* and *Elmerrillia* (to New Guinea); *Talauma* has also extended its range southward in the New World to Colombia and (two areas) Brazil. Otherwise the extant and known fossil range is purely Laurasian. (Based in large part on Dandy and Good, "Magnoliaceae," in *Die Pflanzenareale, Zweite Reihe,* fasc. 5; from Schuster, 1972a.)

on Malesia were established, e.g., *Triumfetta* (Tiliaceae), a group presumably of Laurasian origin. As van Steenis and van Balgooy note (1966), seeds in this group are water-borne and may show epizoic dispersal. By contrast, ancient groups show only slight or no transgression across Wallace's Line (figs. 6, 35, 36). Of 10 Gondwanalandic primitive angiosperm families, only one species of *Tasmannia* (Winteraceae) crosses the line (figs. 6, 35). Similarly, Laurasian families show a sharp tendency to stop short of this line or to extend to it and then stop (note, for example, the range of Illiciaceae and Schisandraceae on figure 6). Of 10 particularly primitive Laurasian angiosperm families (figs. 6, 36) only a very few taxa of the Magnoliaceae extend across Wallace's Line: *Elmerillia* and *Talauma* extend to

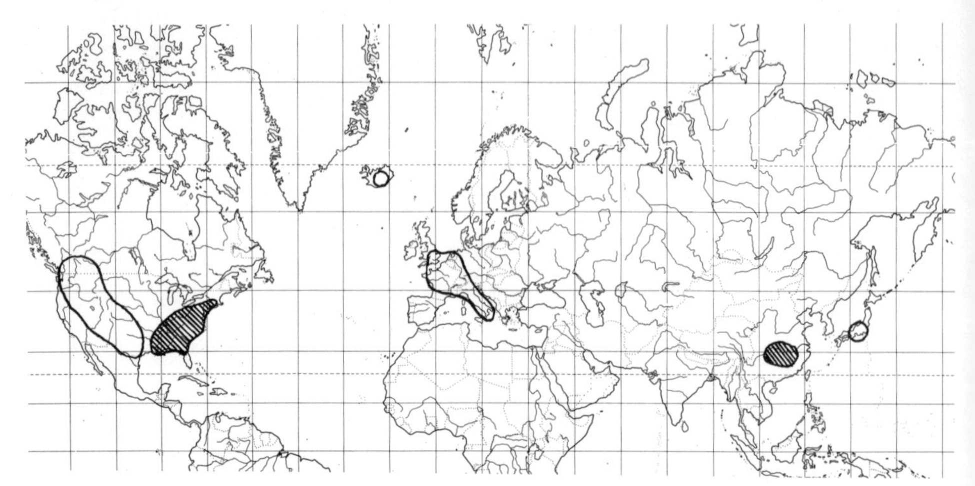

Fig. 37. Range of *Liriodendron* L. Hatched areas are present ranges; open rings, known fossil ranges. (After Good, 1964.)

New Guinea, *Manglietia* to the Celebes. In post-Miocene times none of these Laurasian families have been able to penetrate even as far as Queensland. These Laurasian families were once much more widespread. For example, the Cercidiphyllaceae, family 17 on figure 6, were common in Paleocene times in northernmost sections of Laurasia, where arctic conditions obtain today (cf. Takhtajan, 1969). Similarly, at least some Magnoliaceae, including *Magnolia* and *Liriodendron* (fig. 37), showed a much wider range in late Cretaceous to early Tertiary times, as did the Fagaceae. Distribution of the Calycanthaceae (fig. 6: 10) was also surely much more extensive in the past and is relict today; the anomalies in the distribution of that family and the allied Idiospermataceae are discussed in the section that follows.

In general one must conclude that, starting rather early in the Tertiary and extending into the Pleistocene, there was massive regional extinction of primitive and even moderately advanced angiosperm groups (see, for example, figure 37, of *Liriodendron;* compare to figure 39, showing past and present range of *Fagus*). As a consequence, centers of survival for primitive groups emerge, such as eastern North America (Li, 1971), southeastern Asia, and Australasia (including Fiji). The most important relict centers are the last two;[35] as Li has stressed with respect to the eastern Asia–eastern North America pattern, there is a rather constant tendency for eastern Asia to retain, quantitatively and qualitatively, more relict elements

[35] The reality of these centers—and the fact that they are relict centers and not centers of origin—is evident from the conifers (compare figures 7–10).

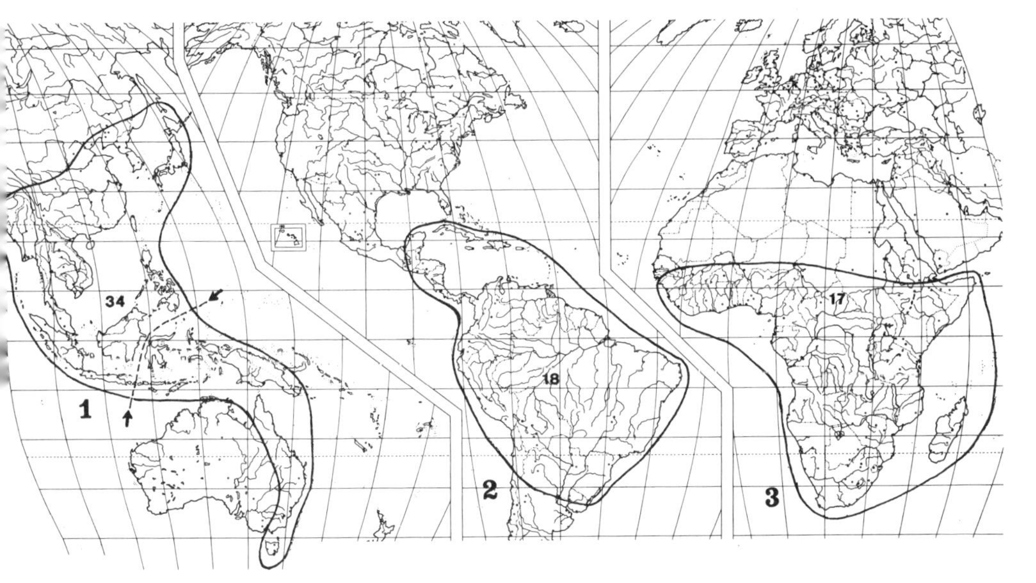

Fig. 38. Distribution of primitive families of angiosperms. The numbers indicate the number of families found in each area. Note that on each side of Wallace's Line (arrows and dashed line) some 17–18 primitive families of angiosperms occur, with only slight transgression across this line. *Assumptions:* region 1, which corresponds roughly with the area "from Assam to Fiji" that Takhtajan postulates as the "cradle of the angiosperms," represents a region of juxtaposition of a relict primitive flora west and north of the Sunda Shelf (Laurasian-derived) and another relict primitive flora east and southeast of the Sunda Shelf (Gondwanaland-derived). Regions 2 and 3 (neotropical and African realms) represent relatively homogeneous regions. (After Smith, 1967.)

than eastern North America. As has been strongly emphasized (Schuster, 1972a), in late Tertiary to Recent times the southeast Asian and Australasian centers were physically juxtaposed, with simulation of a single, large center of survival of relict groups. This center, with some 34 families (35 if we include the recently described Idiospermataceae of Blake [1972]), has sometimes been held to be the "center of origin" of the angiosperms as a whole (Smith, 1967 [see fig. 38]). As demonstrated in more detail in an earlier paper (Schuster, 1972a), this center of origin is actually not a single center at all, but has had a complex history involving recent intermingling. We clearly cannot conclude from the relict presence of numerous primitive angiosperms in the region from Fiji and Tasmania to Burma and Assam that the Angiospermae originated in such a region (Raven and Axelrod, 1972; Schuster, 1972a). Indeed, I suggest that certain primitive or relatively generalized groups may even have arrived in Australasia from Laurasia via a route involving western North America–South America–Antarctica–Australia.

The Marsupial Route, Dissemination of the Fagaceae, and Australasian Biogeography

We have seen that after Tertiary rafting northward of Australasia, there was a massive infusion of Laurasian taxa of basically east Asian derivation into Australasia in late Tertiary and Pleistocene times. Presumably at a much earlier date there was also dispersal of Laurasian taxa into Australasia via a much more complex and cumbersome route—here called the "marsupial route"—involving, basically, migration from North to South America to Antarctica and, finally, into Australasia. The introduction of Malaysian taxa and American taxa into Australasia has been confused at times (e.g., by van Steenis, 1971), and the route from North America to Australasia has not been fully documented although both *Nothofagus* and the marsupials, as well as possibly *Araucaria,* employed it.

If a genus whose ability to migrate is as limited (Cranwell, 1963, 1964; Preest, 1963) as that of *Nothofagus* could diffuse from its ancestral home in North America (presumably starting in mid-Cretaceous times) as far as New Zealand and Australia *by Upper Cretaceous times,* then floristic interrelationships of that part of the world must be assumed to be complex indeed. Correspondingly, no simple statements about "cradle areas" involving Australasia can be made. Demonstrably, both migration and survival have conspired to produce a situation so complex that no conclusions as to origin can yet be derived. A summary of the facts and purported facts is necessary for discussion.

1. The Fagaceae, with some 7–8 genera and 600 described species, are of presumed Laurasian origin.[36] Genera with a limited range today (e.g., *Fagus* [fig. 39]) were once much more widespread, including Paleocene occurrences (*Quercus*) in regions which are today Arctic (Takhtajan, 1969).

2. Van Steenis (1971) argues that the center of diversity—hence presumed center of origin—of the Fagaceae is in the area "from Yun-

[36] They belong to a group, the Hamamelididae, which includes not only the vesselless, primitive Trochodendraceae and Tetracentraceae, but also groups like Juglandaceae, Betulaceae, Hamamelidaceae, and Cercidiphyllaceae, which are all exclusively Laurasian (Cercidiphyllaceae were widespread in Laurasia at least to the Paleocene) or have a few isolated taxa (e.g., of *Quercus* and *Alnus*) that have secondarily and recently penetrated Gondwanalandic provinces.

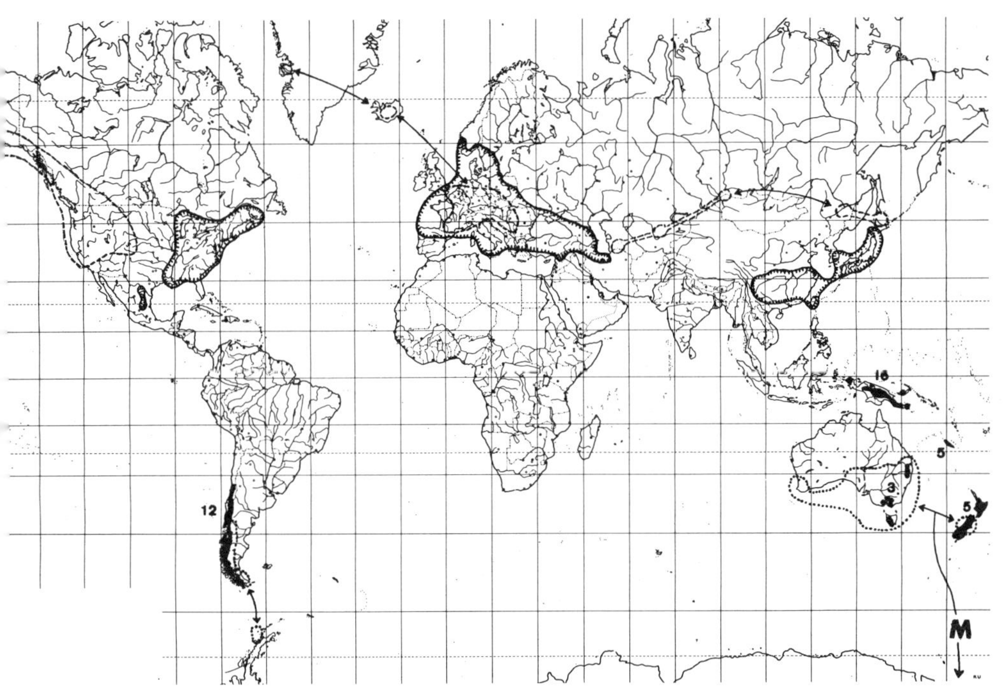

Fig. 39. Distribution of *Fagus* (Northern Hemisphere) and *Nothofagus* (Southern Hemisphere). Dashed areas, fossil range of *Fagus;* closed areas, present range. Black areas, present range of *Nothofagus* (with numbers of species indicated); dotted areas, fossil range; note fossil presence near McMurdo (*M*). It has been shown that sea transport of *Nothofagus* is virtually impossible. The common assumption is that the current range of *Nothofagus* is relict and was once basically Gondwanalandic in Cretaceous times; Australian plate migration carried it far to the northward, passively. Since the Fagaceae are Laurasian except for *Nothofagus,* I assume the the ancestor of this genus arrived in Gondwanaland by transport from North to South America; compare fig. 40. (After van Steenis, 1971.)

nan to Queensland"; this is disputable.[37] The 10 species of *Fagus* are north temperate, found today only in Eurasia and Taiwan, and eastern North America southward to Mexico (they occur as fossils in western North America). The 35–40 species of *Nothofagus* are antipodal, none crossing Wallace's Line. The nearly 300 species of

[37] Croizat (1952, fig. 27) has modern Fagaceae originating in a "Neocaledonian Center." This is even more disputable than van Steenis's ideas, although they are comparable. (Croizat's Neocaledonian center actually involves Queensland, New Guinea, and New Caledonia.) The single strongest argument in favor of such an origin is that the loosely allied, stenotypic family Balanopsidaceae (*Balanops*, with three species in New Caledonia and two in Queensland; *Trilocularia*, with two species in Fiji) is apparently exclusively restricted, today at least, to such a Neocaledonian center.

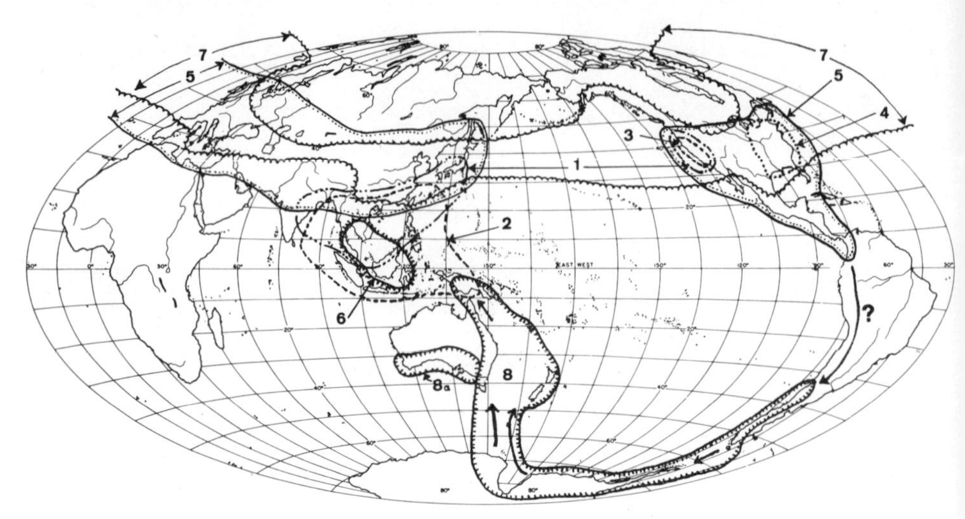

Fig. 40. Range of Fagaceae and dispersal of *Nothofagus;* see figure 39 for details on present and former range of *Fagus* and *Nothofagus.* (1) *Lithocarpus* Bl. (2) *Castanopsis* (D. Don) Spach. (3) *Chrysolepis* Helmq. (sometimes placed under *Castanopsis*). (4) *Castanea* Adans. (Eurasian sector of its range omitted for clarity). (5) *Quercus* L. (6) *Trigonobalanus* Forman. (7) *Fagus* L., past and present (highly generalized). (8) *Nothofagus* Bl.; at 8a, fossil range in western Australia. Arrows mark presumed dispersal route of *Nothofagus.* Highly generalized; several species of *Lithocarpus* extend to New Guinea.

Lithocarpus occur in eastern and southeastern Asia into Indomalaya, with 9 species (7 endemic) reaching New Guinea (Saepadmo, 1973), although Sargent (1922) refers the western North American *L. densiflora* (Hook. & Arn.) Rhed. to that genus. The 120 species of *Castanopsis* occur in tropical to subtropical Asia, with 2 species (sometimes segregated as a separate genus, *Chrysolepis* Hjelmq.) in the western United States. The 12 species of *Castanea* are Laurasian and temperate-zone, and in North America fail to occur today in the West. The 450 species of *Quercus* are widespread throughout Laurasia, barely penetrating to South America, but do not cross Wallace's Line. The 2 species of *Trigonobalanus* occur from the Malay Peninsula to Thailand and Borneo, and barely cross Wallace's Line into Celebes (fig. 40). In the general area of Malesia–Southeast Asia there are thus 6 genera; in North America, 5 genera. This corresponds very well with the situation portrayed by Li (1971), who, emphasizing the eastern Asia–North America disjunction, notes that generally there is greater survival of "remnants" of the Arcto-Tertiary flora in eastern Asia than in eastern (and also western) North America. Hence for the Fagaceae we must also accept two centers of survival, not one center of origin.

3. By late Cretaceous times, *Nothofagus,* derived either from a common ancestor with *Fagus* or, more likely, from *Fagus*-like antecedents, had spread over an area from southern South America to Antarctica and New Zealand, and by early Tertiary times, to Australia (figs. 39, 40).

4. *Nothofagus* did not reach New Guinea until Pliocene times, as far as present evidence suggests.

5. Like the marsupials, *Nothofagus* apparently never reached Africa or India.[38]

6. *Nothofagus* apparently cannot undergo sea-water dispersal (Holloway [1954] showed that the nuts sink in water and do not survive salt-water treatment) and seemingly can cover distances of only 2–3 km by wind transport (Preest, 1963). Therefore its dissemination had to be overland, or at most it could jump only narrow sea gaps.

7. The most primitive species of sectio Nothofagus, *N. alessandrii,* occurs in South America (van Steenis, 1971).

8. Biological relationships of *Nothofagus* in South America, involving symbiosis with a hemiparasitic, unigeneric family of angiosperms, the Myzodendraceae, suggest that *Nothofagus* has had a very ancient history in South America; absence of Myzodendraceae from elsewhere in the range of the genus *Nothofagus* is biogeographically highly significant. The Myzodendraceae, with only 11 species of *Myzodendron,* is probably reduced from a santalaceous ancestor that was already a hemiparasite. If *Nothofagus* had originated in Malesia or southeastern Asia in the Upper Cretaceous, it could hardly have reached South America by overland transport prior to the Tertiary—by which time the Antarctic–South American connection no longer existed. By Eocene times glaciation had begun in the

---

[38] Raven and Axelrod (1972) state that the "ancestors of *Nothofagus* . . . probably passed between Northern and Southern Hemispheres by way of Africa and Europe, since land connections were absent [in Cretaceous times] in Middle America." They cite Freeland and Dietz (1971) and Fooden (1972) in support of their claim that no land connections existed between North and South America at that time. However, Kurtén (1969) and others argue for at least tenuous connections between North and South America until the start of the Tertiary. Also, Raven and Axelrod admit that fossil evidence for the presence of *Nothofagus* in Africa is "weak" (Srivastava [1967] reports it from Nigeria). There is no confirmatory report. Supposed reports from India are stated by Darlington (1965) to be "doubtful," and he states that there is "no other evidence that *Nothofagus* ever existed in these places." The Europe–Africa–South America–Antarctica–Australasia route visualized by Raven and Axelrod seems highly unlikely on numerous grounds. Darlington (1965) also emphasizes that *Nothofagus*—like other Amentiferae—produces pollen in enormous quantities, and that "detectable quantities . . . can be carried almost unbelievable distances by prevailing winds." His review of the *Nothofagus* problem is excellent and should be studied as background for my compressed account.

Antarctic, and the filter bridge between it and South America was apparently sundered physically as well. In addition, we know that all three pollen-derived (hence phylogenetically defined?) units of *Nothofagus* were in South America by the Eocene, and the *brassii* group by Upper Cretaceous times.

From these facts, and from analogies with the distributional history of marsupials, we can draw four conclusions relevant to the basic theses of this paper: Immediate ancestors of *Nothofagus* entered Gondwanalandic areas, possibly by mid-Cretaceous times, from North America.[39] *Nothofagus* diversified in South America (pollen of all three known types, *brassii, fusca,* and *menziesii,* has been found in Eocene deposits in South America [van Steenis, 1971]) and evolved complex relationships with a fungus *Cyttaria,* and with other plants—thus coevolution led to elaboration of a unigeneric family Myzodendraceae. By Upper Cretaceous times *Nothofagus* had diffused not only to Antarctica but, more than 80 million years ago, across into Australia and New Zealand, at a time when fragmentation of the eastern sector of Australasia (New Zealand and other areas of the Campbell Plateau; New Caledonia) evidently was at most in its incipient stages; *Nothofagus*'s entry into New Caledonia and New Zealand must have occurred well before the end of the Cretaceous, because by the late Cretaceous these continental fragments had rifted free. *Nothofagus* did not reach New Guinea until well into the Tertiary; absence of its parasites, *Cyttaria* and *Myzodendron,* in New Guinea suggests that somewhere in its long migration it had left them behind; the latter possibly did not even reach Antarctica.[40]

---

[39] It might be argued that the group could not have penetrated step-wise from temperate areas of North America to temperate areas of South America. However, the genus occurs in New Caledonia at low elevations (van Steenis, 1971), and Darlington (1965) states that it may occur "and reproduce at 1000 m in tropical New Guinea," suggesting "that, in spite of its occurrence mainly in the southern cold-temperate zone, the genus is cold-tolerant rather than cold-adapted and that it might disperse through tropical or at least subtropical areas." Darlington (ibid., p. 145) notes that a common ancestor of *Fagus* and *Nothofagus* must have crossed the tropics somewhere, "and the *brassii* group of *Nothofagus* seems reasonably fitted to have done it." The fact that *N. alessandrii,* a primitive South American species, belongs to the "*brassii* group" is surely significant.

Dalziel and Elliot (1971) note that a "proto-Cordillera developed at latest in the early Mesozoic." This, if extended far enough northward, would probably have served as a satisfactory migratory route through tropical South America. From South America "the southernmost Andes once continued into West Antarctica as an essentially rectilinear Cordillera" (ibid.); this surely constituted an important migratory route. The age of the area is adequate: "southern South America, the North and South Scotia Ridges, and the Antarctic Peninsula are underlain at least in part by continental rocks certainly as old as Palaeozoic."

[40] *Cyttaria* (placed as a single genus in the family Cyttariaceae and order Cyttariales [Kobayasi, 1966; Gamundi, 1971]) includes seven South American species (Gamundi, 1971) and four species in Australia, New Zealand, and Tasmania but not

It is just possible that this scenario for *Nothofagus* also applies to the paradoxical family-pair Calycanthaceae-Idiospermataceae (fig. 6: 10, 10a). The Calycanthaceae have a relict distribution in the eastern and southwestern United States (2–3 species of *Calycanthus*) and China (3 species of *Chimonanthus;* 1 species of *Sinocalycanthus* closely allied to *Calycanthus*); the Idiospermataceae have a presumably highly relict range in northern Queensland and include only *Idiospermum* (Blake, 1972) with a single species, once placed in *Calycanthus*. The only rational alternative to the pattern outlined for *Nothofagus* is, in my opinion, that the common ancestor of these two families was Gondwanalandic; the ancestor of the Calycanthaceae was rafted north on the Indian plate and showed subsequent dispersal and speciation in the last 40 million years, with subsequent extinction in all but three surviving centers.

If the scenario of *Nothofagus* dispersal is approximately correct, certain important conclusions are unavoidable.[41] The most profound conclusion is that the mere presence of a group in a specific area is subject to gross misinterpretation. Thus the presence of a center of Fagaceae in the area from Yunnan to Queensland, involving, among others, *Fagus*, *Lithocarpus*, *Castanopsis*, *Trigonobalanus*, *Quercus*, and *Nothofagus*, led van Steenis to argue for a Malesian route by which *Nothofagus* entered New Guinea—and from there migrated across Australasia to the Antarctic to South America. In actuality, the Malesian route has been available only since, at most, Miocene or Pliocene times. The presence in New Guinea of *Castanopsis*, a Laurasian genus with a wide range (sometimes found together with *Nothofagus*), thus has a fascinating and complex explanation: the two

---

New Guinea (Kobayasi, 1966). According to Kobayasi (1966) the most primitive taxa are *C. hookeri* and *C. darwinii* of the Tierra del Fuego–Patagonia region; the genus presumably evolved there and spread, probably via West Antarctica, to Australasia. All this suggests a migration route from South America (in which *Nothofagus* has existed long enough to develop at least 18 species of parasites of the Myzodendraceae and Cyttariaceae) via West Antarctica to Australasia, with *Myzodendron* being left behind before Australasia was attained, but *Cyttaria* following the host into Australasia but left behind before the host reached New Guinea. This interpretation is clearly in phase with the late-Tertiary incidence of *Nothofagus* pollen in New Guinea. Alternatively, *Myzodendron* may have evolved after the link between South America and Antarctica was severed at the start of the Tertiary. In either case, the complex biotic relationships suggest origin of *Nothofagus* in South America.

[41] This scenario is drawn as much on the basis of what we know today to have been the migratory route of marsupials from South America to Australasia, at a somewhat later date (explaining why these animals did not get to New Zealand and New Caledonia), as it is from *Nothofagus* itself (fig. 41). Relevant is the fact that this is part of a pattern—a route open to organisms with limited powers of dispersal (e.g., ratite birds)—from probably mid-Cretaceous times until at least the start of the Tertiary.

genera arrived via drastically different routes if my interpretation is correct. The fact that *Castanopsis* has not penetrated east and south beyond New Guinea argues forcefully for its recent intrusion into the flora of that recently elevated area. The fact that, in turn, *Nothofagus* has not penetrated westward even toward Wallace's Line also argues for its recent arrival in New Guinea.[42] (The high peaks of Indonesia and Borneo appear to offer suitable habitats for *Nothofagus.*)

I have dwelt at some length on the Fagaceae, and especially on *Nothofagus,* because that group has been repeatedly cited as solid evidence for a Malesian route which, of necessity, must have been effective from Yunnan to South America by middle Cretaceous times, if not earlier. Indeed, van Steenis is forced to assume an origin of *Nothofagus* by Jurassic times in order to have it arrive in South America by the Upper Cretaceous. Since present-day geological evidence demolishes the reality of such a route, the presence of *Nothofagus* and *Castanopsis* in New Guinea—side by side—has been a major phytogeographic embarrassment *so long as we assumed they arrived via the same route.* However, if we assume, as I have here, that these genera arrived in Australasia by very different routes and at different times from Laurasia, then a scenario is at hand that meets two primary considerations: it fits the known geological evidence, and it is in phase with the known history of the marsupials.[43]

Marsupials (fig. 41) first appear in the Albian of North America (over 100 million years ago), and the first relatively complex fauna of marsupials is from the Upper Cretaceous of North America (Darlington, 1957; Cox, 1973). (The poorly known Asian genus *Deltatherium,* sometimes considered to be a Cretaceous marsupial, is probably not referable to that group [Butler and Kielan-Jaworowska, 1973].) Here were found the ancestors of the other marsupials—the Didelphidae—and one can argue that the group spread from North to South America across a filter or filter bridge formed by archipelagic connections that existed in Cretaceous times between these two land

---

[42] Speciation in *Nothofagus* (as in *Tasmannia*) appears to be recent and imperfect in New Guinea, suggesting both recent evolution (by elevation) of appropriate and isolated loci for diversification and, perhaps, relatively recent arrival.

[43] It is also in phase with known phytogeography and phylogeny of the Fagaceae: *Fagus* and *Nothofagus* are the only genera of the Fagoideae; *Fagus* today occurs in an interrupted band around the Northern Hemisphere, as far south as Mexico, and was formerly in western North America. It is just as easy to assume that the *Fagus*-like ancestor of *Nothofagus,* perhaps occurring in Mesoamerica, migrated south to South America as it is to assume that *Fagus*-like taxa that do not get anywhere near Wallace's Line migrated southward into Australasia. Crossing of Wallace's Line by *Trigonobalanus* can be matched by the crossing into northern South America of its nearest ally, and only other genus of the Quercoideae, *Quercus.*

Fig. 41. Presumed dispersal of marsupials. (1) cross-hatched; Middle and Upper Cretaceous distribution of didelphids; the earliest known fossils are from western North America, west of the Cretaceous epicontinental sea, in the Albian; about 20 million years later found in South America. (2) By the Tertiary, their range had extended to Europe and southernmost South America. (3) By Oligocene-Pliocene times their range had reached Tasmania, and beyond it, Australia; presumably the original invasion was more than 47 million years ago but less than 80 million years ago (note absence from New Zealand, New Caledonia, etc.). (4) Present range: the range in North America represents a single didelphid which appears to have arrived there after the Pleistocene (by the Pliocene no marsupials were left in North America). See figure 42 for concentration of taxa in Australasia. Arrows suggest probable migration routes; by the end of the Cretaceous no obstacles existed to migration across to Europe, or to Antarctica, or to Australia. (Based in small part on Thenius, 1972.)

masses.[44] In South America, marsupials first appear in the late Cretaceous, about 80 million years ago. The fact that fossil didelphids

[44] Both Hallam (1972) and Kurtén (1969) agree that from about the start of the Tertiary until near the end of the Pliocene, North and South America were effectively isolated and no filter existed that would allow passage of mammals. Prior to that time the marsupials and, after them, the ancestors of the strange assemblage of Tertiary South American placentals had arrived (e.g., giant sloths, notoungulates, armadillos and allies, and such ungulates as *Macrauchenia*)—clearly in the Upper Cretaceous. Entry of both marsupials and placentals into South America must have been by a Cretaceous filter bridge—even though no trace of it is shown for the period from late Jurassic to early Tertiary times in the reconstructions of Dietz and Holden. Kurtén specifically notes that as late as the "beginning of the Tertiary," South America remained "tenuously" connected to North America; he cites as evidence the reverse migration from South America of notoungulates and edentates (armadillos and allies), which apparently evolved in South America but are found in early Tertiary beds in North America.

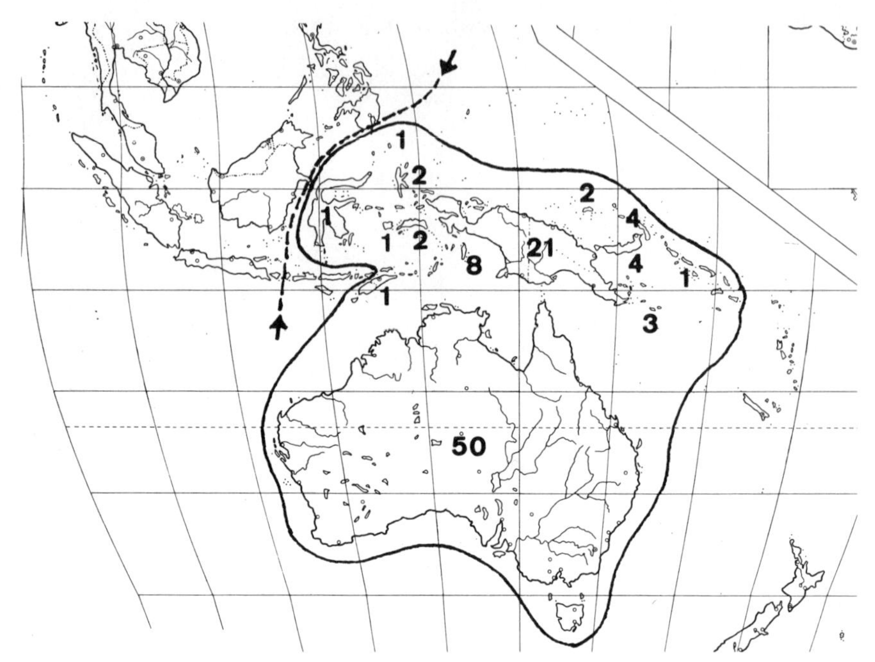

Fig. 42. Concentration of extant genera of marsupials in Australasia, with numbers of genera indicated. (After Simpson, 1961.)

and allies first appeared in some diversity 100 million years ago in North America, underwent striking diversification (clearly later) in South America, but never reached Australasia,[45] where much more specialized families of marsupials underwent the phenomenal mid- and late-Tertiary "explosion" that has been repeatedly documented (fig. 42), is consistent with the history outlined for *Nothofagus.* The much better known fossil history of the marsupials indeed lends added credence to the scheme I have outlined for the Fagaceae.

This scenario assumes that initial diversification of marsupials, present from Oligocene times onward in Australia, had already taken

[45] The classification of marsupials remains controversial. Simpson (1945) recognizes a superfamily Didelphoidea, with 12 extant genera in the most primitive family, Didelphidae, all in South America today, and only 1 species of *Didelphis* northward in North America since the start of the Pleistocene. But 11 fossil genera in two subfamilies are listed for the Upper Cretaceous for North America! In addition, only the unigeneric, Eocene, South American Caroloameghiniidae are assigned by Simpson to the Didelphoidea. Other workers (e.g., Kurtén, 1969) accept an order Marsupicarnivora, including the preceding marsupials and also the Borhyenidae (marsupial carnivores of South America, 21 genera from Paleocene to Pliocene times) and Dasyuridae (marsupial mice, cats, and wolves; some 9 genera in Australia–Tasmania–New Guinea or, according to Tate [1947], as many as 20 genera; the earliest fossils are Pleistocene). The sequence in the fossil record,

place in Antarctica, and that with early Tertiary cooling of Australia-Antarctica these groups became restricted to the northern sectors of this still-joined land mass. Since the Cretaceous link between South America and Antarctica was evidently destroyed by the rupture of the Scotia Arc, at the start of the Tertiary and perhaps earlier, marsupials had only one route open to them—into Australia, to which they became restricted. Such a general picture is wholly coherent with the established history of the forest communities known from Antarctica: fossil araucarians, *Podocarpus*-like fossils, and the genus *Acmopyle* are known from Antarctica as late as the Oligocene (Florin, 1963), and fossils possibly referable to *Dacrydium*, as late as the late Cretaceous. It is generally acknowledged that these groups were "forced" northward after the Oligocene; indeed, they are known as Oligocene fossils only from the northern part of the Antarctic Peninsula, and at that time were probably lacking from the entire elevated interior of Antarctica. This wholesale northward shift of faunas and floras after the end of the Cretaceous is documented also by the history of *Agathis*, *Podocarpus* sectio Polypodiopsis, and other conifers which, although not yet known to have occurred on Antarctica, show extinction in southern areas (e.g., parts of New Zealand) from which they were once known.

If the occurrence of Fagaceae in Gondwanalandic areas is explicable on the bases adopted here, then a major block to acceptance of my basic thesis (Schuster, 1972a) is removed. That thesis holds that the presence of intermingled Laurasia-derived and Gondwanaland-derived groups in Australasia, especially New Guinea, does not mean that this was a center of origin—or an ancient migratory route from Asia to Antarctica and beyond—but is to be explained by the fact that in geologically recent times Gondwanalandic Australasian biotas were physically juxtaposed to Laurasia-derived Asian biotas; the short time since this event (perhaps at the start of the Pliocene) has allowed extensive commingling in an area centering on New Guinea, but this mingling rapidly diminishes on both sides of Wallace's Line.

---

from primitive to specialized, from Upper Cretaceous to Pleistocene, is impressive: it clearly suggests origin in North America (they are known from the early Campanian from western North America [Fox, 1971]) and late arrival in Australasia. Fossil marsupials from Australia (Macropodidae, Phascolomidae, Peramelidae, Dasyuridae, Diptrotodontidae, Thylacoleonidae; the last two are known only as fossils) are known chiefly from the late Tertiary and Pleistocene; but the group first appears in the fossil record of Australia in late Oligocene times (about 30–35 million years ago)—by which time six families were found. There is thus no evidence that the marsupials occurred before mid-Tertiary times in Australasia. They presumably had arrived there by the time that continent rifted away from Antarcica, 45–50 million years ago.

I have surveyed this problem in some depth because the ultimate conclusions one is forced to adopt (I use the word "forced" because, puzzling about the ranges of both the marsupials and *Nothofagus* for over a decade, I find them paradoxical in all respects) are rather momentous. First, the Malesian cradle area, on maps of which some phytogeographers have drawn arrows going in every direction all over the world (e.g., Smith, 1967), is an artifact. Some of the organisms cited to demonstrate its antiquity, such as *Nothofagus* and the marsupials, arrived there demonstrably well after the start of the Tertiary (marsupials) or in late Cretaceous times, with arrival in New Guinea not prior to the Pliocene (*Nothofagus*). Second, the entire biota of the general region from New Zealand and Tasmania to New Caledonia, Fiji, and into southeastern Asia—the cradle area accepted by Smith, Takhtajan, and van Steenis—has had an incredibly complex history. There has been infusion of numerous tropical, largely but not wholly lowland, taxa (e.g., *Casuarina, Castanopsis, Lithocarpus, Rhododendron*) from Laurasia since the start of the Miocene—and mostly since the start of the Pliocene—from the west and northwest (southeastern Asia); infusion of Laurasian groups of presumably North American origin (*Nothofagus* and marsupials), which reached Malesia from the south (Antarctica) between late Cretaceous and early Tertiary times; and casual, long-distance immigation.

Some Conclusions

The general conclusion can hardly be avoided—on both geological and paleozoological bases—that until well into the Cretaceous, long after a diversity of angiosperms was on the scene, there persisted a southern land mass, Gondwanaland, throughout which stepwise overland dispersal was possible (figs. 43, 44). Such movement must have been rather easy, except perhaps as limited by climatic constraints, until about 85–100 million years ago, by which time the Indian plate had split away (fig. 26). Dispersal of taxa adapted to nontropical climatic regimes continued longer; it continued to New Zealand and New Caledonia until about 80 million years ago; between Antarctica and South America until about 65 million years ago; between Antarctica and Australia until about 45–50 million years ago.

We thus have a picture, painted in broad strokes, of many millions of years of clearly established overland migratory routes during a time of active angiosperm evolution and dispersal. The apparently very rapid dispersal of angiosperms during Cretaceous times, in both Laurasia and Gondwanaland, reflects the prevalence of migratory routes that had not yet been sundered. The existence of a filter (at the very least) between North and South America during all or most of the Cretaceous, and the narrow distances between Africa and Laurasia, made possible the rapid dispersal of angiosperms through such corridors. Hence, even though early angiosperms largely lacked effective long-distance dispersal mechanisms, and effective animal dispersers had not yet evolved, the angiosperms were able to use stepwise migration efficiently. We need not invoke long-distance, overwater dispersal. The dispersal of conifers and taxads, as determined from the fossil record, suggests that the chief impediment to such stepwise dispersal was climatic and not physical (a broad tropical zone "unfit" for cool- to cold-adapted conifers); a similar impediment surely limited early angiosperm dispersal to some extent, although (judging from mid-Cretaceous pollen records from Australia, Africa, and South America) less than conifer dispersal. With these limitations, we must assume a Cretaceous world where dispersibility, overland, was much easier than one would gather from a steady-state world such as van Steenis (1971) assumes for Cretaceous and later times.

Possibly the rapid evolution of taxa of angiosperms (orders, families, genera, and species) in the time since the end of the Cretaceous reflects the isolation into discrete floras that was physically enforced on the angiosperms after Gondwanaland began its well-documented fission and migration of component parts. This isolation was further exaggerated by widespread late-Cretaceous marine transgression. The period starting 80–90 million years ago, mid-Cretaceous times, thus marks an era of progressively declining gene flow between sundered populations—and an era of increasingly active evolution of endemic floras. Hammond (1971a) put this as a generalization: "A single community of organisms that was divided and separated by the splitting of a continent should follow somewhat different evolutionary sequences."

Evidence from living angiosperms is inadequate to allow sure derivation of an ancestral home for the group. Extant woody ranalean families (34–35 families in 12 orders and 3 subclasses, Magnoliidae, Ranunculidae, and Hamamelididae, of A. C. Smith [1972]), especially the largely relict taxa of the orders Winterales, Magnoliales, Laurales (in part), Illiciales, Trochodendrales, Cercidiphyllales, Eupteleales, and Eucommiales, occur almost equally in both Gondwanalandic and Laurasian provinces—often with one allied family in Gondwanalandic areas, the other in Laurasian areas (fig. 6). It is therefore impossible to derive any specific locus of origin. As noted in an earlier paper (Schuster, 1972a), however, the possibility of early Tertiary infusion of Gondwanalandic taxa into Laurasia (via the Indian plate), and the well-documented infusion of Gondwanalandic taxa into marginal sectors of Laurasia (via the Australasian plate), suggest that considerable enrichment of the Laurasian flora has perhaps taken place. Both infusions occurred after the initial spread of angiosperms, which had been achieved by mid-Cretaceous times. There is only slight basis for placing the origin of angiosperms in the warmer, northern portions of Gondwanaland. Such an assumption is in phase with microfossil data presented by Brenner (in this volume). His earliest "angiosperm province" would lie in latitudes that at the start of the Cretaceous were in the warmer portions of Gondwanaland, which was then just beginning to fissure. It is also in phase with the phylogeny proposed by A. C. Smith (1972), who would recognize as the most primitive angiosperm order the Winterales, with a single family Winteraceae. This group is almost exclusively Gondwanalandic (Schuster, 1972a); only 1 of about 90 species in one of nine genera has crossed Wallace's Line—without, however, having attained the present continental portions of Laurasia (fig. 35).

Extinction (or evolution into widely different modern types) of all very early angiosperm types shown to have existed in early Cretaceous times by Doyle and Hickey (in this volume), makes it impossible, without adequate fossil evidence, to postulate a more clearly defined home for the Angiospermae. The apparent lack of relict angiosperm taxa from Africa—with a few notable exceptions—has been duly emphasized by Takhtajan (1969), who was at a loss to find an explanation for it. The answer seems to reside in two factors: First, Africa was split from the rest of Gondwanaland, except for South America, well before the Cretaceous. Hence, if primitive angiosperms did not originate in Africa, they evidently had limited opportunity to enter it except by long-distance dispersal—a mode in which primitive angiosperms seem singularly deficient (aside from modern derivatives, such as *Actea, Ranunculus,* and *Clematis*). Second, in all Gondwanaland reconstructions, Africa occupies a central position, and until South Atlantic and Indian Ocean evolution it had a pronounced continental character—and thus was not a fit environment for mesophytic taxa. The relatively limited series of primitive taxa that do occur in Africa (e.g., certain Annonaceae, Monimiaceae) apparently got there via South America, after the African–South American plate split off from the remainder of Gondwanaland. Even after that event, presumably by Jurassic times, migration routes via the present-day Antarctic Peninsula and the Scotia Arc were probably effective in long maintaining at least a diffuse connection between South America and East Gondwanaland. Such essentially overland or short-distance migration routes between the various provinces of a long-sundered Gondwanaland are more evident from study of the fossil history of the gymnosperms (Florin, 1963; Schuster, 1972a), of the Hepaticae (Schuster, 1969, 1972b), and the marsupials (Cox, 1973).

The primary conclusions documented in this paper are the following: (1) The history of both the Indian and the Australasian plates is highly complex (figs. 26, 27). (2) Both plates served as vehicles for dissemination of Gondwanalandic taxa into Laurasia. (3) The Australasian plate, probably rather depauperate floristically and depleted biotypically, arrived devoid of tropical-adapted taxa in latitudes with tropical climates, and has received widespread, massive, and still-continuing infusion of taxa of east Asian origin since the start of the Pliocene. (4) Since the Indian plate migrated through several climatic zones, it probably also arrived at the borders of Laurasia rather depleted, and therefore only remnants of the original Gondwanalandic flora are recognizable in India today. Much of the flora has been either so altered in the last 45 million years or so intermingled with Laurasian-derived taxa that the Gondwanalandic origin of much of the area's biota is obscured except for the lower groups (Bryophyta) which show slow evolution rates and can survive in microhabitats. (5) At least limited migration was possible from North America to Australasia, via South America and Antarctica (the "*Nothofagus* track," also employed somewhat later by the marsupials; figs. 40, 41), and some Laurasian taxa reached Australasia by this route prior to the Miocene.

The biotic and geological complexities of both the Indian and Australasian regions, and of southeastern Asia, are so great that one cannot deduce from the present flora, or from the ambiguous and sparse fossil history, anything that would point to a center of origin of the angiosperms in the general region from Australasia to southeastern Asia—the "Yunnan to Queensland" of van Steenis (1971) or the "Assam to Fiji" of Takhtajan (1969). One can conclude that this general region experienced commingling of Gondwanalandic and Laurasian floras and survival of a host of rather primitive taxa; a wide range of climatic extremes persisted, linked with at most very local glaciation during the entire Pleistocene.[46] Climatic and geological events (absence of extensive glaciation; migration of continental

[46] It should be re-emphasized that the northward migration of Australasia—including Tasmania and New Zealand—was exactly at the right time to avoid extensive glaciation and hence extinction of biotas. Feeble glaciation occurred in the North Island of New Zealand and the Kuziosko Plateau of Australia; somewhat more extensive but still very regional glaciation occurred in Tasmania and the South

plates leading to mass rafting of floras and their infusion into other floras) combined to create a situation where complex and intermingled floras result, in which many old and relict groups survive. The geological history of the gymnosperms of that area (emphasized in Schuster, 1972a) clearly shows that southeastern Asia and Australasia represent two centers of survival, which when arbitrarily and artificially considered as one (as Smith [1967, 1970], Takhtajan [1969], and van Steenis [1971] have done) *simulate* a single region so rich in relicts as to also suggest a center of origin; compare figures 6–10.

Since a taxon like *Nothofagus,* whose ability to migrate over water gaps is supposedly almost nonexistent, could disperse from its probable site of origin in North America all the way through South America to Australasia during the Upper Cretaceous, other taxa with greater powers of long-distance dispersal probably migrated more rapidly and showed much less inhibited migratory patterns. Thus complex migration routes must be assumed, even for poorly dispersible taxa ranging from dinosaurs and marsupials to *Nothofagus*. Added to this, the efficacy of migration of many modern types of angiosperms unquestionably increased rapidly in the period after the onset of the Tertiary, when fruit- and seed-eating mammals and birds began to be a factor in dispersal. On that basis we cannot avoid the general conclusion that derivation of any center of origin of the angiosperms from the ranges of present-day taxa is indefensible. If we are ever to find such a center, its recognition must be on the basis of fossil evidence.

The evidently long persistence of Gondwanaland (figs. 43, 44), a supercontinent which included about 60 percent of the land area of the world, is the central fact in the history here portrayed. It is clear that the breakup of this region coincided largely with the Cretaceous and Tertiary evolution and diversification of the angiosperms; it is equally clear that, if Angiospermae originated in Gondwanaland, there were two opportunities for numerous representatives of the group to be rafted northward until contact was made with Laurasia: one in Cretaceous to early Tertiary times (India), one in the Tertiary (Australia). At least until the Middle Cretaceous, filters existed that would allow migration from Africa–South America into both Eurasia and North America (which were then still linked into a supercontinent, Laurasia). These are facts central to all phytogeography. With

Island of New Zealand (see, e.g., Flint, 1947). The migration was almost in phase with the general start of cooling in Tertiary times; as cool zones crept northward, so did the Australasian plate. The local survival in Australia of various cool-adapted genera like *Nothofagus* may be linked with this phenomenon. The entire linkage has had major phytogeographic consequences.

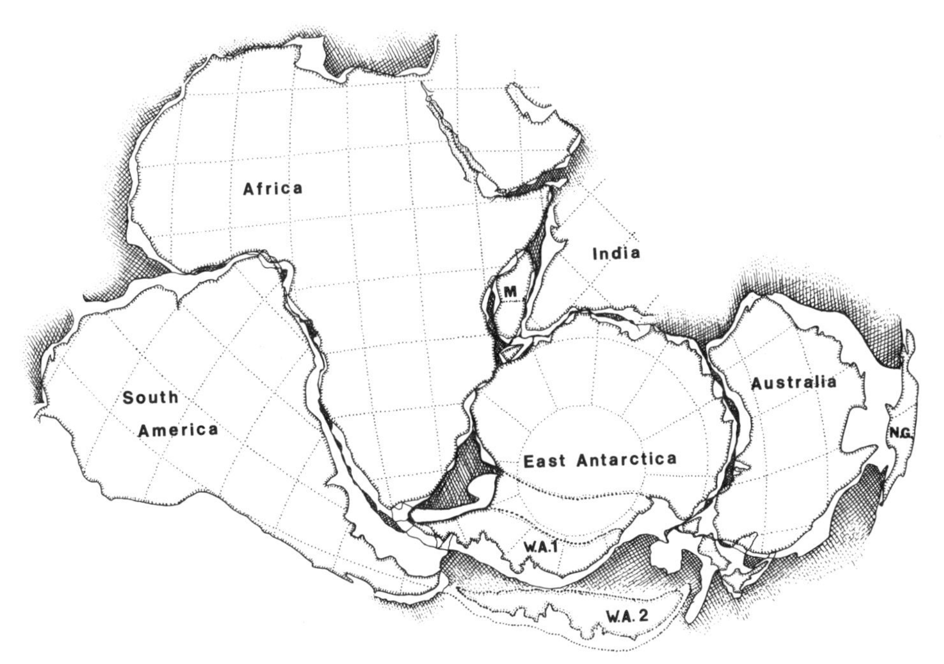

Fig. 43. Orientation of land areas in Gondwanaland. Smith and Hallam's reconstruction has East and West Antarctica (*W.A. 1*) united, and the peninsular portion of West Antarctica lies too far north. If Dalziel and Elliott's (1971) assumption of a rectilinear connection between the cordillera of South America and the mountains of Australasia is correct, then presumably West Antarctica was situated at about *W.A. 2.* In that case, if East Antarctica was oriented as far north as here suggested, then West Antarctica was a separate geological province, perhaps separated by water from East Antarctia—and may have constituted an independent migratory route. *N.G.,* New Guinea. (Modified from Smith and Hallam, 1970.)

the Tertiary impingement of India and Australia against the Asian sector of Laurasia, new migratory routes were created (dashed arrows on fig. 45), which were mostly ruptured either during or prior to the Pleistocene; in the Pacific Basin a series of new migration routes for Asian taxa have been created (dotted arrows, fig. 45), from southeastern Asia into the Pacific and in part as far as Hawaii. The end result is a level of complexity that has been vastly underestimated.

We must reconcile ourselves to a situation where, until at least mid-Cretaceous times, widespread overland migration throughout the Southern Hemisphere was readily possible (as the history of the ratite birds clearly shows [see, e.g., Cracraft, 1973]) and where Cretaceous epicontinental seas, running roughly north-south, dissected much of Laurasia. During this time, north-south migrations between fragmenting Laurasia and Gondwanaland remained commonplace

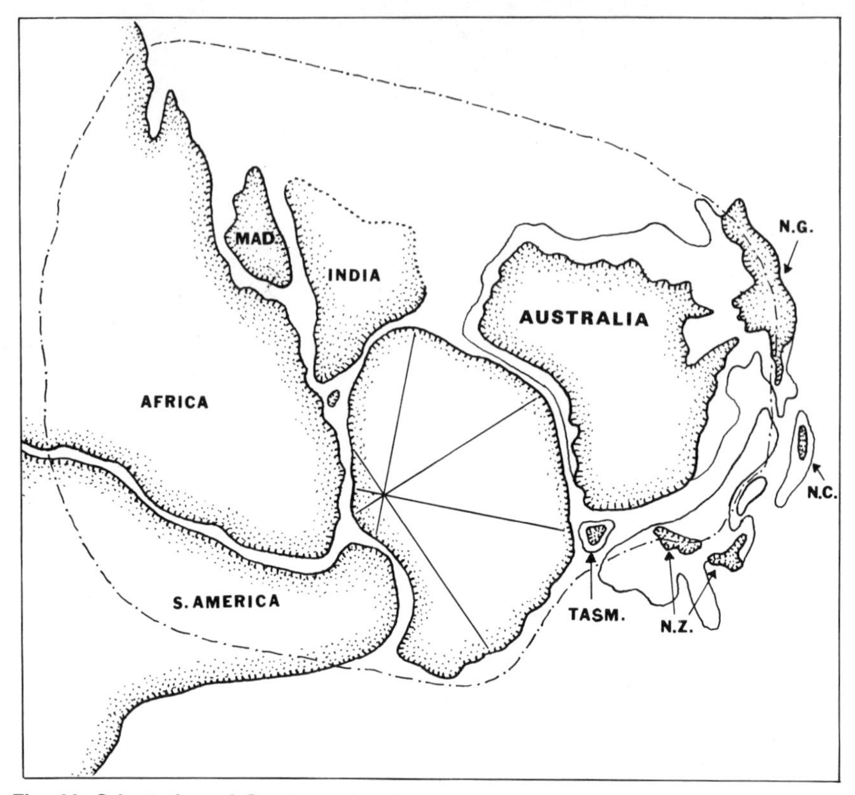

Fig. 44. Orientation of Gondwanalandic land masses at the end of the Paleozoic. (After Schopf, 1970b.)

(via the Europe-Africa and North America–South America bridges or filters). We must equally reconcile ourselves to the following facts: Until the Tertiary, migration between what are now widely spaced continental blocs was possibly easier than east-west migration within Laurasia, owing to the extensive epicontinental seas that dissected it. Until quite late in the Tertiary, the so-called Malesian track did not exist, and probably the complex interrelationships in the biota of Malesia reflect recent transgression. All during the Tertiary, North and South America were isolated. During the Tertiary, with coevolution between mammals and birds, on one hand, and fruiting devices of angiosperms, on the other, long-distance dispersal became progressively more efficient in many groups. Therefore not only migratory routes but migratory modes have shown significant change since Cretaceous times.

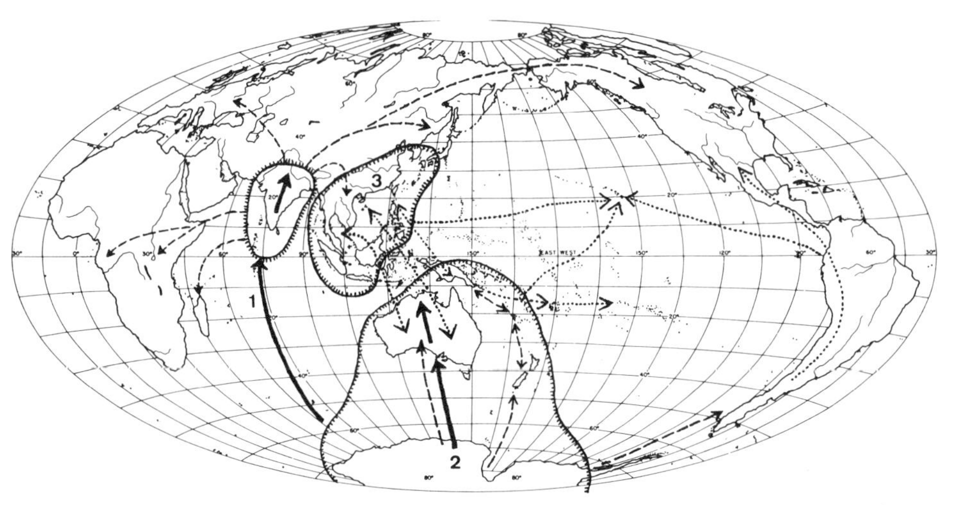

Fig. 45. Generalized schematic portrayal of the complex floristic history of south and southeast Asia. (1) Route of Indian plate migration, with arrows showing subsequent migratory routes (all within the last 40–45 million years). (2) Route of the Australasian plate; New Zealand and New Caledonia are remaining fragments of the eastern edge of the plate, earlier "abraded off" in part by contact with the Pacific plate. (3) Southeast Asian center of survival. Within the last 10–15 million years, routes indicated by dotted arrows became feasible; dashed arrows indicate routes, now no longer open, which were once fully or partially bridged.

References

Audley-Charles, M. G., and Hooijer, D. A. 1973. Relation of Pleistocene migrations of pygmy stegodonts to island arc tectonics in eastern Indonesia. *Nature* 241: 197–98.

Berggren, W. A. 1969. Cenozoic chronostratigraphy, planktonic foraminiferal zonation and the radiometric time scale. *Nature* 224: 1072–75.

Blake, S. T. 1972. *Idiospermum* (Idiospermaceae), a new genus and family for *Calycanthus australiensis*. *Contrib. Queensland Herbarium* 12: 1–37.

Briden, J. C. 1967. Recurrent continental drift of Gondwanaland. *Nature* 215: 1334–39.

Burton, C. K. 1970. The palaeotecnic status of the Malay Peninsula. *Palaeogeogr. Palaeoclimatol. Palaeoecol.* 7: 51–60.

Butler, P. M., and Kielan-Jaworowska, Z. 1973. Is *Deltatherium* a marsupial? *Nature* 245: 105–6.

Carlquist, S. 1965. *Island Life*. Natural Hist. Press, New York.

Chase, C. G. 1971. Tectonic history of the Fiji Plateau. *Bull. Geol. Soc. Amer.* 82: 3087–3109.

Colbert, E. H. 1973. *Wandering Lands and Animals*. Dutton, New York.

Couper, R. A. 1960. Southern Hemisphere Mesozoic and Tertiary Podocar-

paceae and Fagaceae and their palaeogeographic significance. *Proc. Roy. Soc. London, ser. B,* 152: 491–500.
Cox, C. B. 1973. Systematics and plate tectonics in the spread of marsupials. *Paleontology* 12: 113–19.
Cracraft, J. 1973. Continental drift, paleoclimatology, and the evolution and biogeography of birds. *J. Zool. Lond.* 169: 455–545.
Cranwell, L. M. 1963. *Nothofagus:* Living and fossil. In *Pacific Basin Biogeography,* ed. J. L. Gressitt, pp. 387–400. Bishop Mus., Honolulu.
Cranwell, L. M. 1964. Antarctica: Cradle or grave for its *Nothofagus.* In *Ancient Pacific Floras,* ed. L. M. Cranwell, pp. 87–93. Univ. of Hawaii Press, Honolulu.
Croizat, L. 1952. *Manual of Phytogeography.* Junk, The Hague.
Cullen, D. J. 1970. A tectonic analysis of the south-west Pacific. *New Zealand J. Geol. Geophys.* 13: 7–19.
Dalziel, I. W. D., and Elliot, D. H. 1971. Evolution of the Scotia Arc. *Nature* 233: 246–51.
Darlington, P. J., Jr. 1957. *Zoogeography: The Geographical Distribution of Animals.* Wiley, New York.
Darlington, P. J., Jr. 1965. *Biogeography of the Southern End of the World.* Harvard Univ. Press, Cambridge, Mass.
Darwin, C. 1903. Letter to J. D. Hooker, 1881. In *More letters of Charles Darwin,* ed. F. Darwin and A. C. Seward. Appleton, New York.
Davies, H. L., and Smith, I. E. 1971. Geology of eastern Papua. *Bull. Geol. Soc. Amer.* 82: 3299–3312.
Dietz, R. S., and Holden, J. C. 1970. The breakup of Pangaea. *Sci. Amer.* 223: 30–41.
Dietz, R. S., and Holden, J. C. 1971. Pre-Mesozoic oceanic crust in the eastern Indian Ocean (Wharton Basin?). *Nature* 229: 309–12.
Durham, J. W. 1963. Paleogeographic conclusions in light of biological data. In *Pacific Basin Biogeography,* ed. J. L. Gressitt, pp. 355–65. Bishop Mus., Honolulu.
Flint, R. F. 1947. *Glacial Geology and the Pleistocene Epoch.* Wiley, New York.
Florin, R. 1963. The distribution of conifer and taxad genera in time and space. *Acta Hort. Bergiani* 20: 121–312.
Fooden, J. 1972. Breakup of Pangaea and isolation of relict mammals in Australia, South America, and Madagascar. *Science* 175: 894–98.
Fox, R. C. 1971. Marsupial mammals from the early Campanian Milk River Formation, Alberta, Canada. *J. Linn. Soc., Zool.* 50 (suppl.): 145–64.
Freeland, G. L., and Dietz, R. S. 1971. Plate tectonic evolution of Caribbean–Gulf of Mexico region. *Nature* 232: 20–23.
Gamundi, I. J. 1971. Las "Cyttariales" Sudamericanas. *Darwinia* 16: 461–510.
Good, R. D. 1964. *The Geography of the Flowering Plants.* 3d ed. Longmans, London.
Griffiths, J. R. 1971. Reconstruction of the South-west Pacific margin of Gondwanaland. *Nature* 234: 203–7.

Hackman, B. D. 1971. *Rec. Proc. XII Pacific Sci. Congr.*, vol. 1, pp. 1–366.
Hallam, A. 1967. The bearing of certain palaeozoogeographic data on continental drift. *Palaeogr. Palaeoclimatol. Palaeoecol.* 3: 201–41.
Hallam, A. 1972. Continental drift and the fossil record. *Sci. Amer.* 227: 56–66.
Hamlin, G. B. 1972. Hepaticae of New Zealand, Parts I and II. *Rec. Dominion Mus.* 7: 243–366.
Hammond, A. L. 1971a. Plate tectonics: The geophysics of the earth's surface. *Science* 173: 40–41.
Hammond, A. L. 1971b. Plate tectonics II: Mountain building and continental geology. *Science* 173: 133–34.
Hayes, D. E., and Ringis, J. 1973. Seafloor spreading in the Tasman Sea. *Nature* 243: 454–58.
Heirtzler, J. R. 1971. The evolution of the southern oceans. In *Research in the Antarctic,* ed. L. O. Quam. A.A.A.S. Publ. no. 93, pp. 667–84.
Heirtzler, J. R., Veevers, J. V., Bolli, H. M., Carter, A. N., Cook, P. J., Krasheninnikov, V. A., McKnight, B. K., Proto-Decima, F., Renz, G. W., Robinson, P. T., Rocker, K. and Thayer, P. A. 1973. Age of the floor of the eastern Indian Ocean. *Science* 180: 952–54.
Hennig, W. 1966. *The Diptera Fauna of New Zealand as a Problem in Systematics and Zoogeography.* Pacific Insects Monograph 9.
Hermes, J. J. 1968. The Papuan geosyncline and the concept of geosynclines. *Geol. Mijnbouw* 47: 81–97.
Hill, A. W. 1929. Antarctica and problems in geographical distribution. *Proc. 5th Internatl. Bot. Congr.*, vol. 2, p. 1477.
Holloway, J. 1954. Forests and climates in the South Island of New Zealand. *Trans. Roy. Soc. New Zealand* 82: 329–410.
Hughes, N. F., ed. 1973. *Organisms and Continents through Time.* (A joint symposium of the Geological Society, Palaeontological Association, and Systematics Association.) Palaeontological Association Special Paper no. 12, London.
Hurley, P. M. 1971. Possible inclusion of Korea, central and western China, and India in Gondwanaland. Paper presented at the American Geophysical Union Meetings, April, 1971, in Washington, D.C.
Huxley, L. 1918. *Life and Letters of Sir Joseph Dalton Hooker,* vol. 1. J. Murray, London.
Irving, E. 1964. *Paleomagnetism and its Application to Geological and Geophysical Problems.* Wiley, New York.
Jones, J. G. 1971. Australia's Caenozoic drift. *Nature* 230: 237–39.
Karig, D. E. 1970. Ridges and basins of the Tonga-Kermadec Island Arc system. *J. Geophys. Res.* 75: 239–54.
Karig, D. E. 1971. Origin and development of marginal basins in the western Pacific. *J. Geophys. Res.* 76: 2542–61.
Keng, H. 1970. Size and affinities of the flora of the Malay Peninsula. *J. Trop. Geogr.* 31: 43–56.
Kobayasi, Y. 1966. On the genus *Cyttaria* (2). *Trans. Mycol. Soc. Japan* 7: 118–32.

Koponen, T. 1972. Speciation in the Mniaceae. *J. Hattori Bot. Lab.* 35: 142–54.

Kurtén, B. 1969. Continental drift and evolution. *Sci. Amer.* 220: 54–64.

Kurtén, B. 1971. *The Age of Mammals.* Columbia Univ. Press, New York.

Lakhanpal, R. N. 1970. Tertiary floras of India and their bearing on the historical geology of the region. *Taxon* 19: 675–94.

Le Pichon, X., and Heirtzler, J. R. 1968. Magnetic anomalies in the Indian Ocean and sea-floor spreading. *J. Geophys. Res.* 73: 2102–17.

Li, H. L. 1971. Floristic relationships between eastern Asia and eastern North America. Reprinted without change of pagination by the Morris Arboretum, Philadelphia, from *Trans. Amer. Phil. Soc., n.s.,* 42(2): i–ii, 371–429, 56 maps.

McElhinny, M. W. 1967. In *Internatl. Union Geol. Sci. UNESCO Symp. Continental Drift.*

McElhinny, M. W. 1968. Northward drift of India: Examination of recent palaeomagnetic results. *Nature* 217: 342–44.

McElhinny, M. W. 1970. Formation of the Indian Ocean. *Nature* 228: 977–79.

McElhinny, M. W., Giddings, J. W., and Embleton, B. J. J. 1974. Paleomagnetic results and late Precambrian glaciations. *Nature* 248: 557–61.

McIntyre, D. J., and Wilson, G. J. 1966. Preliminary palynology of some antarctic Tertiary erratics. *New Zealand Bot.* 4: 315–21.

McKenzie, D., and Sclater, J. G. 1973. The evolution of the Indian Ocean. *Sci. Amer.* 228: 63–72.

Mallick, D. I. J. 1971. *Rec. Proc. XII Pacific Sci. Congr.,* vol. 1, p. 368.

Pittman, W. C., Herron, E. M., and Heirtzler, J. R. 1968. Magnetic anomalies in the Pacific and sea floor spreading. *J. Geophys. Res.* 73: 2069–85.

Preest, D. S. 1963. A note on the dispersal characteristics of the seeds of the New Zealand podocarps and beeches and their biogeographical significance. In *Pacific Basin Biogeography,* ed. J. L. Gressitt, pp. 415–24. Bishop Mus., Honolulu.

Quantin, P. 1971. *Rec. Proc. XII Pacific Sci. Congr.,* vol. 1, p. 5.

Raven, P., and Axelrod, D. I. 1972. Plate tectonics and Australasian paleobiogeography. *Science* 176: 1379–86.

Ridd, M. I. 1971. South-east Asia as a part of Gondwanaland. *Nature* 234: 531–33.

Saepadmo, E. 1973. Fagaceae. *Flora Malesiana* 7: 265–403.

Sahni, B. 1939. The relationship of the *Glossopteris* flora with the Gondwana glaciation. *Proc. Indian Acad. Sci.* 9B: 1–6.

Sargent, C. S. 1922. *Manual of the Trees of North America (Exclusive of Mexico).* Riverside Press, Cambridge, Mass.

Schodde, R. 1970. Two new suprageneric taxa in the Monimiaceae alliance (Laurales). *Taxon* 19: 324–28.

Schopf, J. M. 1970a. Gondwana paleobotany. *Antarctic J. U.S.* 5: 62–66.

Schopf, J. M. 1970b. Relation of floras of the Southern Hemisphere to continental drift. *Taxon* 19: 657–74.

Schuster, R. M. 1953. Boreal Hepaticae, a manual of the liverworts of Minnesota and adjacent regions. *Amer. Midland Natur.* 49: 257–684.

Schuster, R. M. 1958. Boreal Hepaticae, a manual of the liverworts of Minnesota and adjacent regions. III. Phytogeography. *Amer. Midland Natur.* 59: 257–332.
Schuster, R. M. 1960. Notes on Nearctic Hepaticae. XIX. The relationships of *Blepharostoma, Temnoma* and *Lepicolea,* with description of *Lophochaete* and *Chandonanthus* subg. *Tetralophozia,* subg. n. *J. Hattori Bot. Lab.* 23: 192–210.
Schuster, R. M. 1966. *The Hepaticae and Anthocerotae of North America,* vol. I. Columbia Univ. Press, New York.
Schuster, R. M. 1969. Problems of antipodal distribution in lower land plants. *Taxon* 18: 46–91.
Schuster, R. M. 1972a. Continental movements, "Wallace's Line" and Indomalayan-Australasian dispersal of land plants: Some eclectic concepts. *Bot. Rev.* 38: 3–86.
Schuster, R. M. 1972b. Evolving taxonomic concepts in the Hepaticae, with special reference to circum-Pacific taxa. *J. Hattori Bot. Lab.* 35: 169–201.
Schuster, R. M. 1974. *The Hepaticae and Anthocerotae of North America,* vol. III. Columbia Univ. Press, New York.
Schuster, R. M. 1975. Distributional anomalies of Asiatic and Australasian Hepaticae. *J. Hattori Bot. Lab.,* in press.
Shor, G. G., Jr., Kirk, H. K., and Menard, H. W. 1971. Crustal structure of the Melanesian Area. *J. Geophys. Res.* 76: 2562–86.
Simpson, G. G. 1945. *The Principles of Classification and a Classification of Mammals.* Bulletin of the American Museum of Natural History no. 85.
Simpson, G. G. 1961. Historical zoogeography of Australian mammals. *Evolution* 15:431–46.
Skottsberg, C. 1925. *Juan Fernandez and Hawaii.* Bishop Mus. Bull. no. 16.
Skottsberg, C. 1960. Remarks on the plant geography of the southern cold temperate zone. *Proc. Roy. Soc. London, ser. B.,* 152: 447–57.
Smith, A. C. 1967. The presence of primitive angiosperms in the Amazon Basin and its significance in indicating migrational routes. *Atas Simpos. Biota Amaz.* 4: 37–59.
Smith, A. C. 1969. A reconsideration of the genus *Tasmannia* (Winteraceae). *Taxon* 18: 286–90.
Smith, A. C. 1970. The Pacific as a key to flowering plant history. Harold L. Lyon Arboretum Lecture, Univ. of Hawaii, Honolulu.
Smith, A. C. 1972. An appraisal of the orders and families of primitive extant angiosperms. *J. Indian Bot. Soc.* (Golden Jubilee vol.) 50A: 215–26 [1971].
Smith, A. G., and Hallam, A. 1970. The fit of the southern continents. *Nature* 225: 139–44.
Smith, G. L. 1972. Continental drift and the distribution of Polytrichaceae. *J. Hattori Bot. Lab.* 35: 41–49.
Srivastava, S. K. 1967. Upper Cretaceous palynology: A review. *Bot. Rev.* 33: 260–88.
Sullivan, W. 1974. *Continents in Motion: The New Earth Debate.* McGraw, New York.
Szweykowski, J. 1956. Beiträge zur Geographie der polnischen Lebermoose

II. *Chandonanthus setiformis* (Ehrh.) Howe in der polnischen Tatra. *Acta Soc. Bot. Polon.* 25: 603–13.

Takhtajan, A. 1969. *Flowering Plants, Origin and Dispersal.* Oliver, Edinburgh.

Tarling, D. H. 1971. Gondwanaland, palaeomagnetism and continental drift. *Nature* 229: 17–21.

Tate, G. H. H. 1947. On the anatomy and classification of the Dasyuridae (Marsupialia). *Bull. Amer. Mus. Nat. Hist.* 88: 97–156.

Thenius, E. 1972. Säugetierausbreitung in der Vorzeit. *Umschau* 72: 148–53.

Van Bemmelen, R. W. 1949. Verslag van een petrografisch onderzaek der gesteent collective van het Boven Digoel gebied, enz. *Ingr. Ned.-Indië* 4: 137–45.

Van der Linden, W. J. M. 1971. *Rec. Proc. XII Pacific Sci. Congr.*, vol. 1, p. 390.

Van Steenis, C. G. G. J. 1962. The land-bridge theory in botany, with particular reference to tropical plants. *Blumea* 11: 235–372.

Van Steenis, C. G. G. J. 1971. *Nothofagus,* key genus of plant geography, in time and space, living and fossil ecology and phylogeny. *Blumea* 19: 65–98.

Van Steenis, C. G. G. J., and Van Balgooy, M. M. J., eds. 1966. *Pacific Plant Areas* 2. *Blumea,* suppl. 5.

Veevers, J. J., Jones, J. G., and Talent, J. A. 1971. Indo-Australian stratigraphy and the configuration and dispersal of Gondwanaland. *Nature* 229: 383–88.

Wegener, A. 1924. *The Origin of Continents and Oceans.* 4th rev. ed., trans. J. Biram. Reprint ed., 1966, Dover, New York.

Wellman, P., McElhinny, M. W., and McDougall, I. 1969. On the polar-wandering path for Australia during the Cenozoic. *Geophys. J. Roy. Astron. Soc.* 18: 371–95.

Pollen and Leaves from the Mid-Cretaceous Potomac Group and Their Bearing on Early Angiosperm Evolution

JAMES A. DOYLE, ***Department of Botany University of Michigan, Ann Arbor***

LEO J. HICKEY, ***Division of Paleobotany Smithsonian Institution, Washington, D.C.***

UNDISPUTED MEGAFOSSIL REMAINS of angiosperms, mostly leaves, first become a regular component of the fossil record during the late Lower Cretaceous, approximately 100 million years ago. Although early paleobotanists occasionally remarked on the peculiar morphology of Lower Cretaceous angiosperm leaves (e.g., Lesquereux, 1883, pp. 4–5; Ward, 1888, pp. 129–30; Fontaine, 1889, pp. 291, 347), they generally attempted to identify or at least to compare them with familiar modern families or even genera. These identifications and the high degree of specialization and taxonomic diversification which they imply form the principal basis for the widespread belief that the angiosperms originated and underwent their basic adaptive radiation well before they entered the Cretaceous fossil record (e.g., Seward, 1931; Axelrod, 1952, 1960, 1970; Takhtajan, 1969). Partly due to the assumption of a long latent history for flowering plants, discussions on their origin and phylogeny have tended to be based almost entirely on comparative studies of modern forms.

Various theories have been proposed to explain the "sudden" appearance of supposedly diversified angiosperms in the late Lower Cretaceous. Most of these have been variants of Darwin's suggestion in 1875 that the angiosperms originated and diversified in some isolated "homeland" and simply invaded the areas of the known fossil record, especially Europe and North America, in the Lower Cretaceous (see discussion in Takhtajan, 1969). Early authors (e.g., Heer, Seward, and Berry) suggested that this homeland was the Arctic, a hypothesis rendered implausible by better knowledge of

Arctic fossil floras and modern distributions. More recent authors have favored the Indo-Australian region (Takhtajan, 1969; Smith, 1970), although geological and more detailed distributional evidence suggest that the concentration of living primitive angiosperms in this area is due to the recent juxtaposition of the Asian and Australasian crustal plates by continental drift (see Schuster, 1972; Raven and Axelrod, 1972). Perhaps most sophisticated is the essentially ecological theory of Axelrod (1952, 1960, 1970), who postulates that the angiosperms began their differentiation in the tropical uplands, far from lowland basins of deposition, in the Permo-Triassic. In his most recent formulation, Axelrod (1970) suggests that the angiosperms entered the lowlands in the Lower Cretaceous, rather than much earlier, in response to an increase in climatic equability caused by splitting of the previously continuous Gondwana landmass.

Although many authors (e.g., Stebbins, 1950; Pacltová, 1961; Wolfe, 1972a) have questioned the theoretical basis of older leaf-identification methods, the first direct evidence against the prior-diversification theory came from studies of Mesozoic pollen. First, studies of Triassic, Jurassic, and even early Lower Cretaceous rocks revealed no angiosperm pollen, contrary to what would have been expected even if the flowering plants were diversifying in areas far from depositional basins (Scott, Barghoorn, and Leopold, 1960; Hughes, 1961; Pierce, 1961). Furthermore, workers on mid-Cretaceous pollen assemblages (Pacltová, 1961; Pierce, 1961; Brenner, 1963; Kemp, 1968) noted that the angiosperm pollen types were far less diverse than would have been predicted on the basis of the megafossil identifications. Finally, more detailed studies of mid-Cretaceous sequences from widely separated parts of the world have shown a regular stratigraphic increase in the morphological diversity of the angiosperm pollen flora, such that successively appearing types can be fitted into reasonable evolutionary series, rather than the random appearance of unrelated advanced types, as would be expected with immigration of already differentiated taxa from some other area (Doyle, 1969a; Muller, 1970). The widely recognized usefulness of angiosperm pollen in the stratigraphic correlation of Cretaceous sediments is a practical indication of the consistency of this diversification pattern (Zaklinskaya, 1962a,b; Couper, 1964; Pacltová, 1971).

We undertook the present study because we believed that a reevaluation of the early angiosperm megafossil record might help to resolve the apparent contradiction between pollen and leaf evidence and might produce new insights into the origin and early evolution of the angiosperms. Elaborating on an earlier suggestion (Doyle,

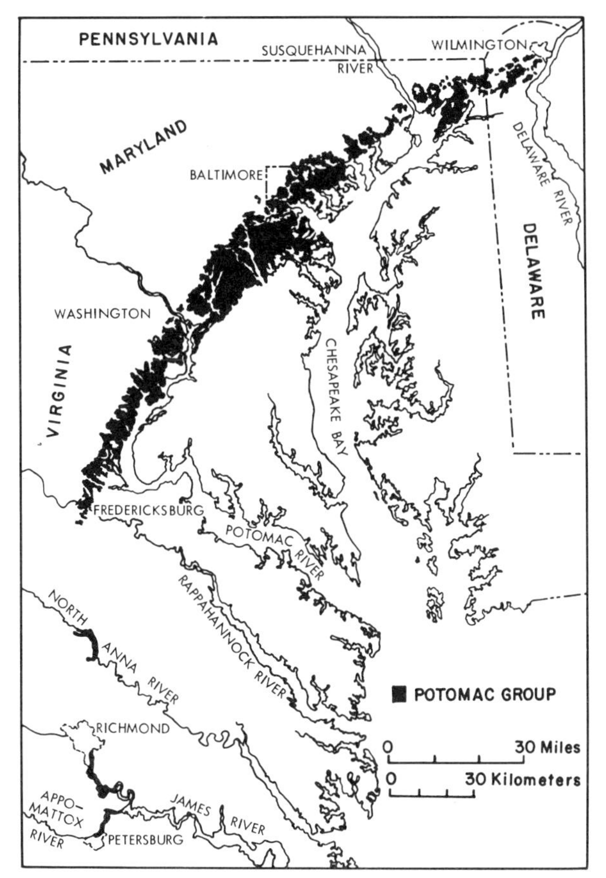

Fig. 1. Outcrop map of the Lower Cretaceous Potomac Group in Virginia, Maryland, and Delaware. Possible extension of the unit into New Jersey is not shown. (Modified after Glaser, 1969.)

1969a, pp. 22–23), we have not attempted systematic identifications with modern taxa but have instead adopted a stratigraphic and comparative-morphological methodology, closely analogous to the approach that is making possible an evolutionary interpretation of the earliest land plants (Banks, 1968, 1972; Chaloner, 1970).

We believed that a strictly morphological analysis of the leaf remains, using Hickey's (1973) classification of leaf architectural characters, might reveal morphological series and complexes that had been obscured by premature attempts at systematic identification. Improved stratigraphic correlations of early angiosperm megafossil localities, made possible by palynology, might reveal temporal, presumably evolutionary, trends in the leaves (just as in the pollen), which had been obscured by the previously vague dating of mid-Cretaceous continental deposits.

Our study has dealt primarily with the Potomac Group of the Atlantic Coastal Plain of the United States (fig. 1), which has yielded one of the richest and most intensively studied early angiosperm megafossil floras, previously described by Fontaine (1889), Ward (1895, 1905), and Berry (1911a, 1916) and frequently cited as evidence for the high diversity of Lower Cretaceous angiosperms. However, we also make comparisons with Lower Cretaceous floras from other geographic areas (some of which we have studied firsthand), such as Kansas, the Black Hills, western Canada, Portugal, Kazakhstan, and Siberia, for correlation purposes and to demonstrate that the regularities seen in the Potomac sequence are not simply a local effect (see Doyle [1969a] and Muller [1970] for a similar treatment of the pollen record).

Geological Background

The Potomac Group, which represents the oldest exposed unit of the middle Atlantic Coastal Plain, consists of a seaward-dipping wedge of fluvial-deltaic sediments exposed in a belt up to 30 km wide extending from the Nottoway, Appomattox, and James rivers south of Richmond, Virginia, up through Fredericksburg, Virginia, along the Potomac River, through Washington, D.C., Baltimore, northeastern Maryland and Delaware, and an uncertain distance into southern New Jersey (fig. 1; see also Hansen, 1969; Owens and Sohl, 1969; Wolfe and Pakiser, 1971). Plant megafossils have been collected from localities scattered along this belt, from the James River to northeastern Maryland. In its type area between Washington and Baltimore, the Potomac Group has traditionally been divided into three lithostratigraphic units or formations (following Clark and Bibbins, 1897). The Patuxent Formation (predominantly cross-bedded sand and gravel) lies at the base of this sequence, followed by Arundel Clay (presumably swamp-deposited, dark, organic-rich, and siderite-bearing clay), and the Patapsco Formation (mostly sand and variegated, often red, clay). For many years the predominantly sandy beds which overlie typical Patapsco strata in Maryland and Delaware were incorrectly identified (McGee, 1888; Clark, 1897; Berry, 1916) with the younger and lithologically distinct Raritan Formation, the oldest unit in the northern New Jersey Coastal Plain, and excluded from the Potomac Group. These "Maryland Raritan" beds have since been returned to the Potomac Group (Weaver et al., 1968) and mostly recently have been treated simply as part of the

Patapsco Formation (Wolfe and Pakiser, 1971). Both the Potomac and the Raritan are disconformably overlain by the much younger Magothy Formation (Santonian-Campanian [Doyle, 1969a; Wolfe and Pakiser, 1971]). Outside the Baltimore-Washington area the Arundel Clay is not recognizable; there is no other consistent way to subdivide the Potomac sequence lithologically, and it is therefore referred to simply as the Potomac Formation (Jordan, 1962, 1968; Owens, 1969).

The continental nature of the outcrop sequence below the Woodbridge Clay Member of the New Jersey Raritan Formation (which is independently dated by marine mollusks as mid-Cenomanian [Doyle, 1969a; Sohl, cited in Wolfe and Pakiser, 1971]) precludes direct correlations with the marine faunal biostratigraphic zones upon which the standard European stage sequence is based (table 1), and has hampered correlations within the Potomac Group. Except for rare and stratigraphically undiagnostic vertebrate and freshwater invertebrate remains, paleontological evidence for correlation consisted until recently of plant megafossils. Because of

Table 1. Subdivisions of the Cretaceous, with the absolute ages of their boundaries in millions of years. Age data from E. G. Kauffman, unpublished data.

System	*Series*	*Stage*		*Age of base in years* ($\times 10^6$)
Tertiary				63
Cretaceous	Upper	Maestrichtian		70
		Campanian		82.5
		Santonian		84.5
		Coniacian		87.4
		Turonian		90.5
		Cenomanian		93.5
	Lower	Albian		100.5
		Aptian		108.5
		Barremian	"Neocomian"	
		Hauterivian	"Neocomian"	
		Valanginian	"Neocomian"	
		Berriasian	"Neocomian"	136
Jurassic				

the sporadic distribution of lithologies suitable for plants and the difficulties in demonstrating superposition, both common problems in continental sequences, only three broad floristic units could be recognized (Lower Potomac, Patapsco, and "Raritan" floras), and some localities were in fact assigned to the wrong unit (e.g., Baltimore, discussed later). In the past 15 years, however, palynological studies have markedly clarified the situation, thanks to the relative ubiquity of pollen and spores in both outcrop and subsurface samples and the availability of extensive, vertically controlled well sequences (Groot and Penny, 1960; Groot, Penny, and Groot, 1961; Brenner, 1963, 1967; Doyle, 1969a,b, 1970, 1973; Wolfe and Pakiser, 1971). In addition, the existence of comparable pollen and spore assemblages in faunally dated nearshore marine rocks of other areas has made possible more secure correlations with the European stages, particularly the Aptian, Albian, and Cenomanian (Brenner, 1963; Norris, 1967; Hedlund and Norris, 1968; Doyle, 1969a, 1970; Kemp, 1970; Pacltová, 1971; Wolfe and Pakiser, 1971).

In his pioneering study on the pollen and spores of the Potomac Group of Maryland, Brenner (1963) recognized two informal biostratigraphic units or palynological zones: zone I corresponding to the Patuxent and Arundel Formations, and zone II (subdivided into subzones II-A and II-B) corresponding to the Patapsco Formation. This scheme has been extended to younger beds by Doyle (1969b, 1973) and Doyle and Hickey (1972). Subzone II-C of zone II and zone III together correspond to the "Maryland Raritan" (Patapsco-Raritan transition zone of Doyle [1969a]), while zone IV corresponds to the lower two members of the New Jersey Raritan Formation (Farrington Sand and Woodbridge Clay). Subsequently, detailed studies of closely spaced core samples from two wells drilled through the Potomac Formation near Delaware City, Delaware (studied preliminarily by Brenner [1967]) have revealed gradual changes (particularly in the angiosperm element) which allow still finer subdivision and correlations within these units (Doyle, 1970, and unpublished data). The principal modifications of Brenner's interpretation of the stratigraphy of these wells are the assignment of the upper part of his "Patapsco" (actually equivalent to the lower part of the "Maryland Raritan") to subzone II-C, and of the beds which he correlated with the New Jersey Raritan Formation to zone III (fig. 2).

The pollen sequence (discussed in more detail below) may be summarized briefly as follows (see left side of fig. 28). In all but the top of zone I, as well as in presumably somewhat older (Barremian)

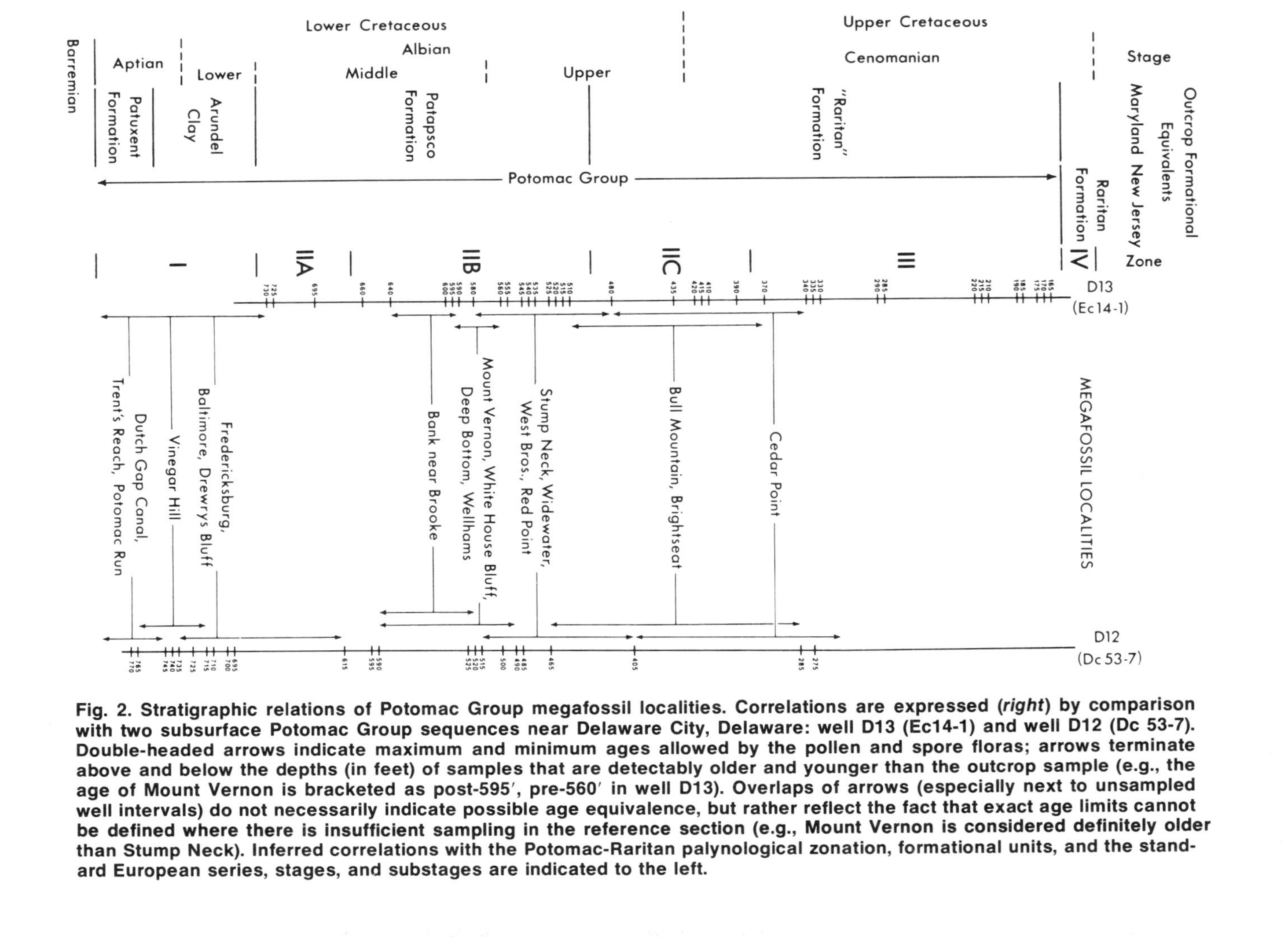

Fig. 2. Stratigraphic relations of Potomac Group megafossil localities. Correlations are expressed (*right*) by comparison with two subsurface Potomac Group sequences near Delaware City, Delaware: well D13 (Ec14-1) and well D12 (Dc 53-7). Double-headed arrows indicate maximum and minimum ages allowed by the pollen and spore floras; arrows terminate above and below the depths (in feet) of samples that are detectably older and younger than the outcrop sample (e.g., the age of Mount Vernon is bracketed as post-595′, pre-560′ in well D13). Overlaps of arrows (especially next to unsampled well intervals) do not necessarily indicate possible age equivalence, but rather reflect the fact that exact age limits cannot be defined where there is insufficient sampling in the reference section (e.g., Mount Vernon is considered definitely older than Stump Neck). Inferred correlations with the Potomac-Raritan palynological zonation, formational units, and the standard European series, stages, and substages are indicated to the left.

rocks elsewhere (Couper, 1958; Kemp, 1968), the only angiosperm pollen consists of highly subordinate tectate-columellate monosulcates, with a single polar germination furrow, or sulcus, of the *Clavatipollenites-Retimonocolpites-Liliacidites* complex. More distinctively dicotyledonous tricolpate pollen, with three longitudinal furrows, or colpi, enters in the upper part of zone I, as in the Lower or Middle Albian of many other areas, but slightly earlier (Aptian) in the African–South American tropics (Brenner, in this volume). In subzones II-A and II-B, both tricolpates and monosulcates become progressively more diverse; especially in the upper part of subzone II-B, as in the Upper Albian of other areas (Pacltová, 1971), many of the tricolpates develop rudimentary thin areas or ora in the centers of their colpi (Doyle, 1969a). In subzone II-C and zone III, these "tricolporoidates" (many of them small, smooth-walled, and triangular) diversify in size, shape, and sculpture. By the top of zone III, many of them can be described as truly tricolporate. Finally, in zone IV, as in the mid-Cenomanian of Europe (Pacltová and Mazancová, 1966; Médus and Pons, 1967), the first triangular triporates (with round pores rather than elongate colpi) of the Normapolles complex appear.

Since the length of time involved in the early angiosperm record is often not appreciated, it is useful to note that the Cretaceous is one of the longest geological periods, extending some 73 million years, from approximately 136 to 63 million years ago. Extrapolating from available radiometric data on the ages of the Aptian, Albian, and Cenomanian, the Potomac–lower Raritan interval probably represents about 18 million years (table 1).

Our first step in integrating the pollen and leaf sequences was to make palynological preparations of pieces of matrix from the classic Potomac megafossil collections of the Smithsonian Institution, as well as from several new localities (enumerated in the next section). Palynological correlations with the Delaware City reference section, summarized in figure 2 and discussed in detail below, are expressed as double-headed arrows indicating maximum and minimum allowable ages in terms of sample depths in the well sequence.

The megafossil localities fall at five main stratigraphic levels, with an unfortunate gap representing subzone II-A and the lowest part of subzone II-B. Because of the greater transportability of pollen than leaves, there are more conspicuous environmentally controlled (facies) differences among megafossil assemblages from one presumed horizon than among pollen assemblages. In some cases (e.g., Fredericksburg and Baltimore), these differences were incorrectly ascribed by one or another of the early workers to differences in age. Although each locality yields only a partial representation of the total leaf flora, palynological correlations between facies (made possible by the greater overlap in elements, and the better vertical control from wells of the ranges of facies-restricted forms) have allowed us to reconstruct what appears to be a fairly complete picture of the range of leaf morphology and ecological adaptations of the angiosperm flora for at least three horizons (upper zone I and middle and upper subzone II-B). In some cases, intensive collecting has revealed "overlaps" in the leaf floras as well, which help confirm the palynological correlations. While the greater facies restriction of leaves means that they are less reliable for stratigraphic applications and demonstration of evolutionary series, it should be noted that this also makes them potentially more valuable as indicators of the ecological associations and habitat preferences of early angiosperms.

In the following sections, the pollen and megafossil floras from each horizon are discussed and compared. In our pollen identifications we have frequently recognized two degrees of comparison for forms differing slightly (*cf.*) and more markedly (*aff.*) from described species. These graded comparisons and the bracket ("reference point") correlation conventions are roughly comparable to the more elaborate system of Hughes and Moody-Stuart (1969). As these authors have stressed, such methods allow more precise correlations than the method of identification of zones, especially in highly gradational sequences where biological changes within zones are often greater than those between them.

Leaves are assigned to the earliest valid species with which they can be identified. Numerous synonymies which we have developed to deal with the overly proliferated number of species have been disregarded here, pending a later systematic survey of the taxonomic status of the Potomac leaf flora. Because the morphologic limits of these leaf species are more poorly defined than are those of the pollen species, we have not adopted the *cf.* and *aff.* conventions used for the pollen. Generic misidentifications—e.g., involving the assignment of extinct forms to modern genera, or, in the case of the pollen, to invalid form genera—are indicated by quotation marks around the generic name.

Finally, although we discuss only leaves and pollen, problematic angiosperm inflorescences and other reproductive structures often occur in association with these remains, and should eventually be subjected to the same sort of analysis.

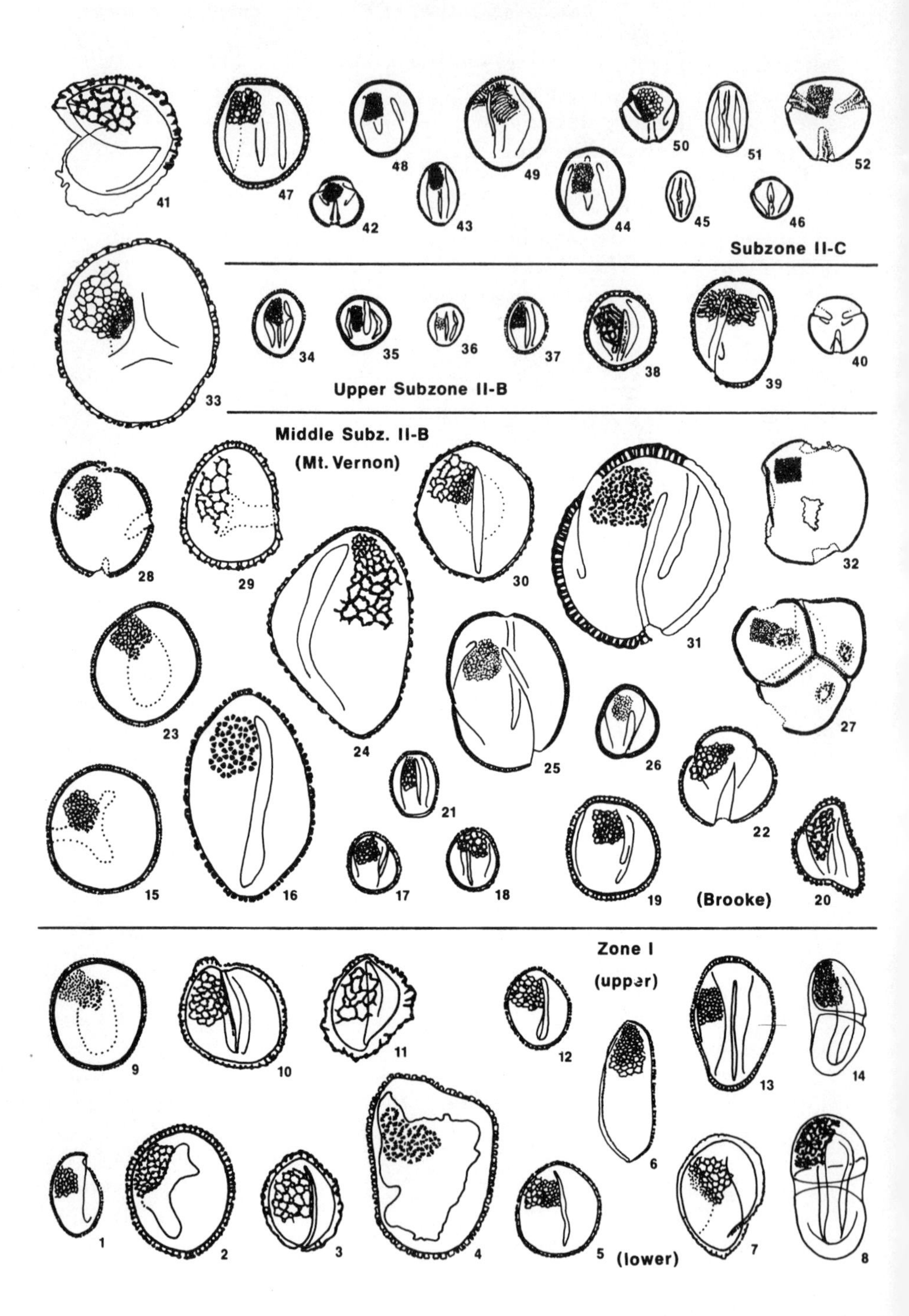
Subzone II-C
Upper Subzone II-B
Middle Subz. II-B
(Mt. Vernon)
(Brooke)
Zone I
(upper)
(lower)

Potomac Pollen and Leaf Sequence

ZONE I

The six principal lower-Potomac megafossil localities that we investigated palynologically fall into two groups of three, which can be correlated with the lower and upper parts of zone I in Delaware well D12 (pre-745′ and post-725′ respectively), while a seventh locality (Vinegar Hill) appears to be intermediate in age (fig. 2). The most secure correlations with the standard stage sequence can be made by comparison with the British marine Aptian–Lower Albian sequence described by Kemp (1968, 1970).

In the three putatively older localities—Fish Hut above Dutch

Fig. 3. Angiosperm pollen types from Potomac Group megafossil localities (except 8 and 14). Schematic drawings of pollen grains, arranged roughly according to the presumed relative stratigraphic positions of their source localities, emphasize diagnostic characters of size (all × 300), shape, apertures, and exine structure as seen in optical section and surface view (insets). See text for discussion of morphological and stratigraphic significance, and for citation of localities and authors of names of taxa. (1) *Clavatipollenites* cf. *hughesii* (71-8). (2) *Clavatipollenites* sp. A (71-14). (3) *Retimonocolpites* sp. A (71-15). (4) Aff. new genus A (71-15). (5) *Liliacidites* sp. C (= *Retimonocolpites* sp. of Doyle, 1973) (71-15). (6) *Liliacidites* sp. A (= *Retimonocolpites* sp. C of Doyle, 1973) (71-8). (7) *Liliacidites* sp. B (= *Retimonocolpites* sp. E of Doyle, 1973) (71-8). (8) Aff. *Decussosporites* (not an angiosperm) (71-15). (9) *Clavatipollenites* aff. *tenellis* (71-6). (10) *Retimonocolpites* cf. *dividuus* (71-21). (11) Cf. "*Peromonolites*" *peroreticulatus* (65-X). (12) Aff. "*Clavatipollenites*" *minutus* (71-6). (13) Aff. "*Tricolpopollenites*" *crassimurus* (71-6). (14) *Decussosporites microreticulatus* (not an angiosperm) (71-6). (15) Cf. *Asteropollis asteroides* (71-13). (16) New genus A (71-13). (17) Cf. "*Tricolpopollenites*" *micromunus* (71-13). (18) "*Retitricolpites*" *fragosus* (71-13). (19) Aff. "*Retitricolpites*" *magnificus* (71-13). (20) Cf. "*Retitricolpites*" *vermimurus* (71-13). (21) Cf. "*Retitricolpites*" *prosimilis* (71-5). (22) "*Retitricolpites*" *georgensis* (71-5). (23) *Clavatipollenites* cf. *tenellis* (71-5). (24) *Liliacidites* sp. D (cf. *Retimonocolpites* sp. B. of Doyle, 1973) (71-5). (25) Cf. "*Tricolpopollenites*" *crassimurus* (71-5). (26) *Tricolpites* cf. *albiensis* or *T.* cf. *sagax* (71-5). (27) Aff. *Ajatipollis* (71-4). (28) *Stephanocolpites fredericksburgensis* (71-4). (29) Aff. "*Peromonolites*" *reticulatus* (71-4). (30) *Liliacidites* aff. sp. E (aff. *Retimonocolpites* sp. D of Doyle, 1973) (71-4). (31) "*Retitricolpites*" *geranioides* (71-4). (32) *Penetetrapites mollis* (71-5). (33) *Liliacidites* sp. E (cf. *Retimonocolpites* sp. D of Doyle, 1973) (69-37). (34) Aff. "*Tricolpopollenites*" *micromunus* (71-9). (35) *Tricolpites* cf. *albiensis* (71-9). (36) "*Tricolpopollenites*" *minutus* (71-18). (37) "*Retitricolpites*" *prosimilis* (71-9). (38) "*Retitricolpites*" *vermimurus* (71-9). (39) Aff. "*Retitricolpites*" *paraneus* (71-18). (40) Aff. "*Tricolporopollenites*" *triangulus* (71-9). (41) Cf. "*Liliacidites*" *textus* (71-2). (42) Aff. "*Tricolpopollenites*" *micromunus* (71-2). (43) "*Retitricolpites*" *prosimilis* (71-1). (44) *Tricolpites* cf. *sagax* (71-3). (45) *Tricolporoidites* cf. *subtilis* (71-1). (46) Cf. "*Tricolporopollenites*" *triangulus* (71-3). (47) Aff. "*Retitricolpites*" *magnificus* (71-3). (48) *Tricolpites* aff. *nemejci* (71-3). (49) "*Retitricolpites*" *paraneus* (71-3). (50) *Tricolporoidites* aff. *bohemicus* (71-3). (51) Aff. "*Tricolporopollenites*" *distinctus* (71-3). (52) *Tricolporoidites* sp. A (71-3).

Gap Canal (71-15), Potomac Run (71-14), and Trent's Reach (71-8)—all from the undifferentiated Potomac Group ("Patuxent") of Virginia, the angiosperm element is exclusively monosulcate and, as throughout zone I, highly subordinate to pteridophyte spores and gymnosperm pollen (Brenner, 1963; Doyle, 1969a, 1973; Wolfe, Doyle, and Page, 1975). However, several partially intergrading morphologic complexes assignable to the form genera *Clavatipollenites* Couper (1958), *Retimonocolpites* Pierce (1961), and *Liliacidites* Couper (1953) are present (fig. 3: 1–7). Monosulcate pollen is found not only in modern monocots and magnoliid dicots, but also in various living and fossil gymnosperms (e.g., Ginkgoales, Cycadales, Bennettitales, Pentoxylales, and the seed-fern family Peltaspermaceae). As first noted by Couper (1958) and recently reaffirmed by Van Campo (1971), the feature which indicates that these three Lower Cretaceous genera are angiospermous is their tectate-columellate exine structure, with an array of well-defined columellae (bacula) connecting nexine and tectum, rather than the superficially similar "reticulum" formed by a spongy or alveolar layer in gymnosperms.[1]

Some monosulcates in the lower zone I outcrop samples (as in D12-770′ and 765′) are closely comparable to *C. hughesii* Couper in being small, elliptical, and finely columellate (fig. 3: 1). These are associated with generally similar but larger, thicker-walled, more coarsely pilate forms, often with an irregular-shaped sulcus (fig. 3: 2, *Clavatipollenites* sp. A, possibly intergrading with *"Peromonolites" reticulatus* Brenner); small, coarsely reticulate forms (fig. 3: 3, *Retimonocolpites* sp. A); and a complex of rarer forms (fig. 3: 5–7, *Liliacidites* spp. A, B, C) with a tendency for differentiation of sculpture into open-reticulate and more finely reticulate to tectate on different parts of the same grain, a character which is of interest in being restricted to monocots today (Doyle, 1973). Perhaps most remarkable in their morphological differentiation from *Clavatipollenites* are rare large grains (fig. 3: 4, aff. new genus A) with a single aperture covering almost half the grain and well-defined "crotonoid" or "stellate" sculpture (Arkhangel'skii, 1966), a pattern commonly associated with certain advanced, tricolporate-derived

[1] Using this criterion, the oldest angiospermous pollen yet described is *Clavatipollenites hughesii* Couper (1958; see also Kemp, 1968) from the (upper?) Barremian of England. Scanning electron microscopy of comparable zone I forms shows that the exine is tectate-perforate rather than clavate (Wolfe, Doyle, and Page, 1975). Judging from the published photographs, the older (Jurassic) grains assigned to *Clavatipollenites* by Pocock (1962), Schulz (1967), and Tralau (1968) and accepted by Muller (1970), have alveolar rather than columellate exine structure and, as was recognized by Pocock (1970) and Tralau (1968), they are more likely cycadophytic than angiospermous.

dicot families, such as Euphorbiaceae, Buxaceae, and Thymeleaceae, but also found in a few monocots (e.g., *Lilium:* Erdtman, 1952; Krutzsch, 1970a; Muller, 1970). The three lower localities are further bracketed by the absence of the peculiar gymnosperm *Decussosporites microreticulatus* Brenner (fig. 3: 14), which appears at D72-745′ and is common in Arundel and younger beds, and by the presence of more clearly bisaccate forms (fig. 3: 8, aff. *Decussosporites*), as in D12-770′, 765′, and several outcrop Patuxent localities.

The three megafossil localities which correlate with the upper part of zone I (post-725′ in well D12, pre-730′ in well D13) are Fontaine's (1889) Fredericksburg locality (65-X, 71-21) and a new locality at Drewrys Bluff (71-111) in the undifferentiated Potomac Group of Virginia, and the Baltimore, Maryland, locality (71-6) which was assigned by Berry (1911a) to the Patapsco Formation, but which on lithostratigraphic as well as palynological grounds should be referred to the Arundel Clay (as noted by Fontaine, in Ward, 1905). These localities have a more diversified angiosperm assemblage, including several forms first seen in the 715′–695′ interval in well D12: (1) monosulcates resembling *Liliacidites* sp. C in sculpture but closer to *Clavatipollenites hughesii* in size (fig. 3: 12, aff. "*C.*" *minutus* Brenner); (2) larger, more spheroidal *Clavatipollenites* with often indistinct apertures (fig. 3: 9, *C* aff. *tenellis* Phillips and Felix 1972, a predominately zone II species); (3) medium-sized, open-reticulate monosulcates with a reduced number of columellae and a tendency for the reticulum to detach and the sulcus to fold inward (fig. 3: 10, *Retimonocolpites* cf. *dividuus* Pierce); and (4) rare specimens of at least one species of medium-sized, prolate, reticulate tricolpates (fig. 3: 13, aff. "*Tricolpopollenites*" *crassimurus* Groot and Penny, a predominantly zone II species). The last two complexes, considered by Brenner (1963) to be restricted to Patapsco (zone II) and younger beds, have been found subsequently at several Arundel and Arundel-equivalent localities (Wolfe, Doyle, and Page, 1975); considering the absence of zone II "index" spores, these are believed to represent genuine range extensions. The three localities also yield both *Decussosporites microreticulatus* (fig. 3: 14) and small, very coarsely reticulate monosulcates with columellae rare or absent (fig. 3: 11, cf. "*Peromonolites*" *peroreticulatus* Brenner). The inferred intermediate age of the Vinegar Hill, Maryland, locality (71-11), always referred to the Arundel Clay, is based on the presence of these last two species and the absence of the four latest zone I complexes cited above.

The presence of *Clavatipollenites* (and, less definitely, of "ephedroid" gymnosperm pollen) indicates that zone I is no older than

Barremian (cf. Brenner, 1963; Couper, 1964; Doyle, 1969a; Kemp, 1970). In fact, the presence even in basal zone I of *Parvisaccites rugulatus,* which appears at the base of the dated marine Aptian of England (Kemp, 1970), and the diversity of angiospermous monosulcates other than *C. hughesii* (though perhaps in part an effect of more favorable facies) suggest it is Aptian or younger. The fact that *Decussosporites microreticulatus* extends well down into the British Aptian (Kemp, 1970) suggests that some of zone I, especially the lower part, is pre-Albian, but the British data also suggest that its upper boundary falls within the Albian, probably close to the Lower–Middle Albian boundary. Both small, finely reticulate to tectate tricolpates (such as *"Tricolpopollenites" micromunus* Groot and Penny and *Tricolpites albiensis* Kemp) and the zone II index spore *Apiculatisporis babsae* Brenner, neither seen in D12-715′ through 695′ but both characteristic of D13-730′ and other basal subzone II-A samples, first enter at the base of the British Middle Albian (Kemp, 1970). Consistent with a Lower Albian age for uppermost zone I is Kemp's (1968, 1970) report of loosely reticulate monosulcates comparable to *Retimonocolpites dividuus* (as *Clavatipollenites rotundus*) and medium-sized reticulate tricolpates (*Tricolpites* sp. 2) from the Lower Albian, although the age of the locality yielding the latter is disputed by Hughes (personal communication, 1970).[2]

As with the pollen, megafossil angiosperm remains are highly subordinate through zone I, in relation to both the total flora and the small number of beds in which they occur. Only at Fredericksburg and Baltimore, in the upper portion of zone I, are the angiosperms represented by more than a few fragmentary specimens. Invariably the beds in which they are found are moderately coarse, grading from fine to medium sand, as at Fredericksburg, to very fine sand at the remaining localities. Where observable, the bedding of these strata is inclined from the horizontal; thus they are presumed to represent near-channel and levee deposits. In contrast, only conifer, cycadopsid, and fern remains are found in adjacent, finer-grained,

---

[2] Our resulting dating of the first Potomac tricolpates as Lower Albian is a contradiction, though not a serious one, to the conclusion of Brenner (in this volume) that tricolpates do not occur in Laurasia until the Middle Albian. We would suggest that this contradiction arises partly from the fact that the ages of first occurrences of tricolpates in most of the areas cited by Brenner are poorly controlled downward. For example, the British section, though well dated, is marine and hence depauperate in pollen (see Kemp, 1970) while Lower Albian pollen floras below the Oklahoma Middle Albian flora of Hedlund and Norris (1968) have not been described. In fact, Hedlund and Norris's data cannot be used to date the first appearance of tricolpates in the Potomac sequence since their flora correlates securely with Potomac horizons well above the base of zone II (middle subzone II-B).

parallel-bedded or laminated units whose deposition is inferred to have taken place in lower-energy, more stable sites such as swamps and flood basins.

All of the angiosperm leaves found in zone I are simple, belong to a limited number of types (increasing upward in the zone from approximately 5 to 9 or 10), and fall into the elliptic, ovate, or obovate shape classes. Their most striking characteristic, however, is the general disorganization of their venation, expressed in such features as the highly irregular size and shape of intercostal areas (areas between secondary veins), the irregularly ramifying courses and poor differentiation of the tertiary and higher vein orders, and the frequently poor demarcation of petiole from blade.

All of these are features used by Hickey (1971) to denote "first rank" leaves which, he concluded, on the basis of comparative morphological analysis of modern leaves, represent the primitive conditions for dicots. Such disorganization plus brochidodromous arching of the secondaries well within the margin and the presence of a primary vein composed in its proximal portion of apparently discrete strands (both also observed by Wolfe [1972b]), combine to give zone I leaves an "archaic" appearance first noted by Fontaine (1889) and Ward (1888) but generally ignored until recently, partly as a result of Berry's (1911a) unfounded comparison of the Fredericksburg leaves with Gnetales (see Wolfe, 1972b; cited also in Doyle, 1969a; Wolfe, Doyle, and Page, 1975). These characters, especially in combination, are practically restricted to basically monosulcate magnolialian families such as Winteraceae (see Wolfe, 1972b) and Canellaceae today. However, rather than considering the first-rank condition a magnolialian character per se, we would prefer to think of it as a grade of evolution through which passed other lines in addition to those leading to the modern Magnoliales.

Typical examples of entire-margined leaves with highly disorganized venation, brochidodromous secondaries, and poor differentiation of blade and petiole are the narrow obovate *Rogersia* Fontaine (fig. 4) (Fredericksburg and Dutch Gap), the possibly intergrading larger, elliptic *Ficophyllum* Fontaine (figs. 5, 6) (Dutch Gap, Potomac Run, and Fredericksburg), and *Celastrophyllum latifolium* Fontaine (fig. 7), a broad elliptic form from Baltimore.

Proteaephyllum reniforme Fontaine (fig. 8) (Dutch Gap and Fredericksburg) is a reniform leaf with a distinct midrib nearly equaled in strength by the crowded lower secondary veins which radiate into the lamina in nearly palmate fashion and branch to form several orders of brochidodromous loops well within the margin. Among zone I leaves it represents the most extreme example of breakdown

Fig. 4. ***Rogersia angustifolia*** **Fontaine (USNM 192339), from Fredericksburg, Virginia (zone I) (× 1).**

of the midrib (see Wolfe, Doyle, and Page, 1975). This and other features suggest a relationship to the possibly herbaceous ovate-peltate, actinodromous complex which becomes abundant higher in the section.

Although zone I leaves are predominantly entire, marginal teeth occur in two forms: *Quercophyllum tenuinerve* Fontaine (fig. 28: 1) from Fredericksburg and new collections from Dutch Gap and Drewrys Bluff, and *Proteaephyllum dentatum* Fontaine (fig. 9) from Baltimore. These teeth are irregularly spaced glandular processes of the doubly convex type (A-1 of Hickey [1973]), a shape very rare among living angiosperms (as in Illiciales); glandular teeth of this shape with such highly irregular spacing and vascular morphology seem, in fact, to be completely unknown.

The only lobate zone I leaf type, which is common at Baltimore, was named *Vitiphyllum* (fig. 10) by Fontaine (1889) because of a supposed relationship to the modern Vitaceae (Berry, 1911a; Axelrod, 1970). However, when examined in detail, this leaf shows the same disorganized venation characterizing the other zone I leaves. In addition, the lobes of foliar tissue simply conform to the branches of

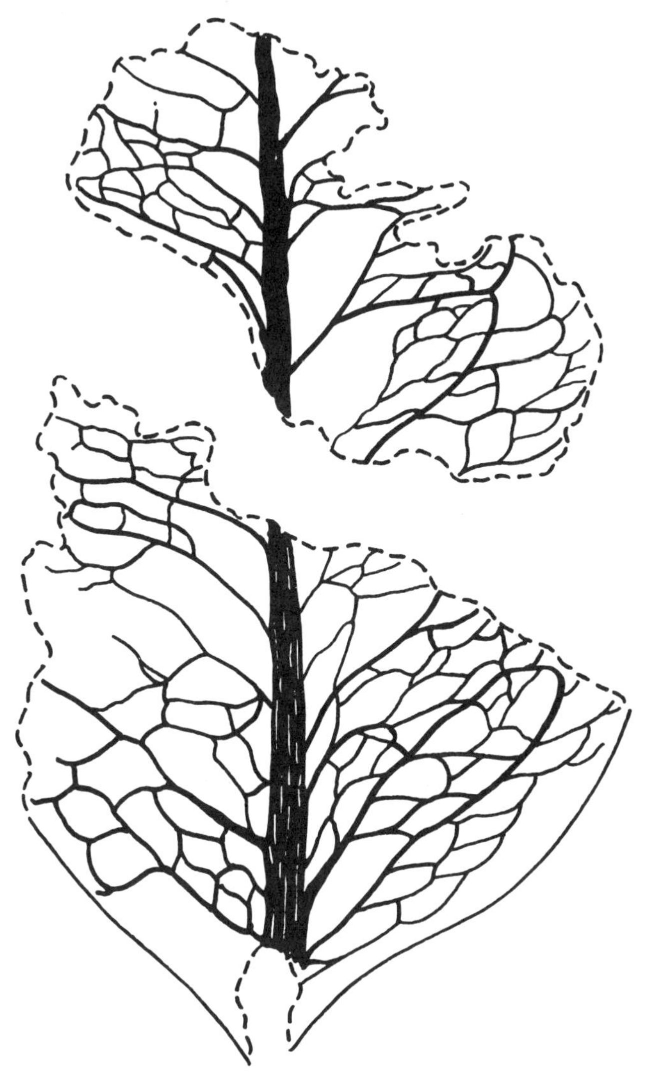

Fig. 5. Broken leaf of *Ficophyllum* from Fredericksburg, Virginia (zone I). Apical portion originally described and illustrated as *Ficophyllum crassinerve* Fontaine (1889, p. 292, pl. 157, fig. 4; USNM 192347a) and the lower portion as *Proteaephyllum ellipticum* Fontaine (1889, p. 285, pl. 142, fig. 1, la; USNM 192347b) (× 1).

the primary venation, with the intervening sinuses completely unbraced to resist stresses.

Another striking characteristic of zone I leaves is their small size; most of them are less than 5 cm long (also noted for Albian leaves

Fig. 6. Detail of the marginal and intercostal venation of a fragment of a large, broad leaf of *Ficophyllum* (originally identified as *Ficus virginiensis* Fontaine, USNM 192353) from Fredericksburg, Virginia (zone I) (× 2).

Fig. 7. Leaf fragment belonging to the *Celastrophyllum latifolium* Fontaine complex from Baltimore, Maryland (zone I). Originally described and illustrated as *Celastrophyllum obovatum* Fontaine (in Ward, 1905, p. 560, pl. 117, fig. 2, USNM 31814) (× 1).

found in the U.S.S.R. [Vakhrameev, 1952; Samylina, 1960, 1968]). Leaves of *Ficophyllum,* some of which exceed 20 cm, are a marked exception; interestingly, they show a corresponding increase in the regularity of their secondary and intercostal venation (fig. 6).

Acaciaephyllum Fontaine is the leaf type which differs most radically from the others of zone I. The best specimen of this form (fig. 11) consists of a bent (hence probably herbaceous) axis about 5 cm

Fig. 8. *Proteaephyllum reniforme* Fontaine (USNM 3915) from Fredericksburg, Virginia (zone I) (× 1).

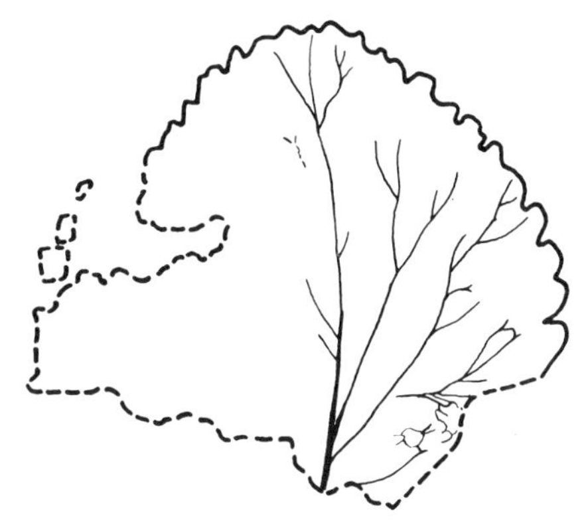

Fig. 9. *Proteaephyllum dentatum* Fontaine (USNM 31820) from Baltimore, Maryland (zone I) (× 1).

Fig. 10. *Vitiphyllum multifidum* Fontaine (USNM 31824) from Baltimore, Maryland (zone I). Note the highly irregular venation and lack of venational bracing of the sinuses (× 1).

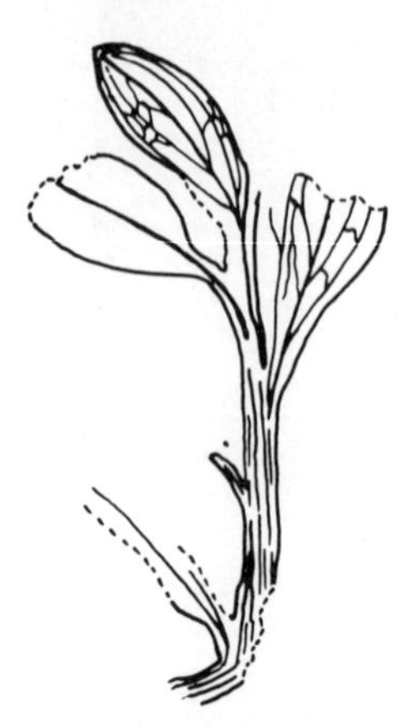

Fig. 11. ***Acaciaephyllum spatulatum*** **Fontaine (USNM 175802A), bent axis with attached obovate, acrodromous leaves from Dutch Gap Canal near Richmond, Virginia (zone I) (× 1).**

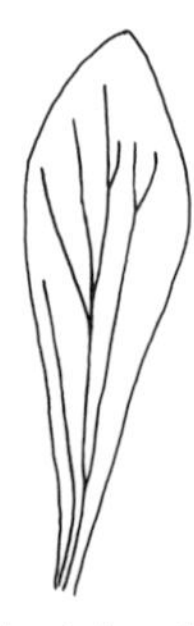

Fig. 12. ***Acaciaephyllum spatulatum*** **Fontaine (USNM 205256) from Dutch Gap Canal near Richmond, Virginia (zone I) (× 1).**

long, bearing several narrow obovate leaves with long sheathing bases. Although highly variable (compare figs. 11 and 12), the secondary veins of the majority of these leaves form concentric arches which converge and appear to fuse toward the leaf apex. This pattern, the fine higher-order crossveins, and the greatly elongate leaf bases with essentially parallel venation suggest monocotyledonous affinities (Doyle, 1973), which accord with the occurrence of monosulcate pollen with monocotyledonous sculpture features in the same beds. However, despite their uniqueness in the flora these leaves show the same type of irregular vein behavior and poor differentiation of vein orders as their contemporaries, as well as great plasticity of size and shape. *Plantaginopsis* Fontaine (in Ward, 1905), an irregularly parallel-veined, elongate leaf from Baltimore (fig. 28: o), is possibly another member of this complex but almost no material of it remains available for study.

Though the existence of monosulcate angiosperm pollen in Barremian-Aptian rocks elsewhere suggests that megafossils should exist, angiosperm leaves of zone I age are practically unknown. The few that have been reported, including a minute, entire-margined, pinnately veined leaf with disorganized venation from presumed Neocomian sediments of Transbaikalian Siberia (Vakhrameev, 1973) and a crenulate-margined reniform from the Aptian of Portugal, known as *Hydrocotylophyllum lusitanicum* Teixeira, agree well with the range of morphology seen in zone I.

MIDDLE SUBZONE II-B

The next group of localities includes Bank near Brooke (71-13), Deep Bottom (71-74), Mount Vernon (71-4), and White House Bluff (71-5) in Virginia, and Wellhams (71-12) in Maryland. The pollen floras from these localities contain a greater diversity of angiosperm pollen types than in subzone II-A or lower subzone II-B, including both additional angiospermous monosulcates and tricolpates that vary widely in size and sculpture (fig. 3: 15–32). The relative proportions (and to a lesser extent the specific composition) of elements vary markedly from sample to sample, the most extreme differences in the angiosperm flora being between Brooke (with unusually abundant tricolpates) and Mount Vernon and White House Bluff (with unusually abundant monosulcates). Detailed comparisons with the Delaware wells indicate that most of these differences are of environmental (facies) rather than stratigraphic (time) significance: the only detectable age difference is that Brooke (and presumably the palynologically barren but lithostratigraphically related Aquia Creek locality) seems to be slightly older than the other localities (post-D13-640′ and pre-D13-590′ or D12-520′, versus post-D13-590′, pre-D13-560′ or D12-490′). The flora of the higher of these two intervals is especially important in dating the Potomac sequence, since it can be correlated closely with the flora described by Hedlund and Norris (1968) from the Fredericksburgian sequence of Oklahoma, whose nearby Texas equivalents are dated by ammonites as late Middle Albian (Young, 1966).

Of the morphological complexes occurring at both Brooke and Mount Vernon, several are very long-ranging, extending to the base of zone II or even into zone I. These include *Clavatipollenites* aff. *hughesii*, the larger *C.* cf. *tenellis* (fig. 3: 23) and very small, nearly smooth to finely reticulate, and medium-sized, tectate-reticulate tricolpates (fig. 3: 26, 17, 25; cf. holotype of *Tricolpites albiensis* Kemp or *T.* cf. *sagax* Norris, cf. "*Tricolpopollenites*" *micromunus* Groot and Penny, and cf. "*Tricolpopollenites*" *crassimurus* Groot and Penny). Other types appear first at 660′ or 640′ in well D13, or 590′

in D12: (1) coarsely stellate or crotonoid monosulcates (fig. 3: 16, new genus A); (2) medium-sized tricolpates with coarser-reticulate sculpture in the mesocolpia (fig. 3: 22, "*Retitricolpites*" *georgensis* Brenner); (3) apparent tri-, tetra-, and pentachotomosulcate and polycolpoidate relatives of *Clavatipollenites tenellis* (fig. 3: 15, 28; cf. *Asteropollis asteroides* and *Stephanocolpites fredericksburgensis* Hedlund and Norris); (4) permanent tetrads of grains with three irregular apertures (fig. 3: 27, aff. *Ajatipollis* Krutzsch [1970b]). Of these, only aff. *Ajatipollis* is not known from Oklahoma; new genus A is not reported by Hedlund and Norris, but it occurs in their type slides deposited at the Smithsonian Institution. Finally, several elements are found only above D13-640′ (from D13-600′ onward): (1) rugulo-reticulate tricolpates (fig. 3: 20, "*Retitricolpites*" *vermimurus* Brenner); (2) small reticulate tricolpates with coarser sculpture at the poles or in the mesocolpia (fig. 3: 18, 21; "*Retitricolpites*" *fragosus* Hedlund and Norris, and cf. "*Retitricolpites*" *prosimilis* Norris); (3) medium-sized, fairly thick-walled tricolpates (fig. 3: 19, aff. "*Retitricolpites*" *magnificus* Habib [1970], in part = "*Retitricolpites*" *virgeus* sensu Brenner [1963]). Of these, all but "*R.*" *prosimilis* are reported from Oklahoma.

Among the presumably facies-related peculiarities of the Mount Vernon–White House Bluff assemblage are the presence or greater abundance of fairly long-ranging species of *Clavatipollenites* (including *C.* cf. *tenellis*, fig. 3: 23; and the *Asteropollis-Stephanocolpites* complex, fig. 3: 15, 28), *Retimonocolpites* (including forms related to "*Peromonolites*" *reticulatus* Brenner, fig. 3: 29), and of stratigraphically more restricted monocotyledonoid monosulcates (fig. 3: 24, *Liliacidites* sp. D, cf. *Retimonocolpites* sp. B of Doyle [1973], coarsely reticulate forms first seen at D13-595′). However, these samples, and to a certain extent Deep Bottom and Wellhams, contain additional forms which enter between 600′ and 580′ in well D13, or at 525′ or 520′ in well D12. These include such typical Oklahoma Fredericksburgian forms as: striato-reticulate tricolpates (fig. 3: 39, aff. "*Retitricolpites*" *paraneus* Norris sensu Hedlund and Norris); large, very coarsely columellate tricolpates (fig. 3: 31, "*Retitricolpites*" *geranioides* sensu Brenner); and the peculiar, thin-walled, irregularly tricolpoidate *Penetetrapites mollis* Hedlund and Norris (fig. 3: 32, often with a thin area at one pole). In addition, tricolporoidate variants (especially of the *Tricolpites albiensis* and "*Tricolpopollenites*" *micromunus* complexes), though present at Brooke (and even down to D13-660′), are more common at Mount Vernon, White House Bluff and related localities.

An upper age limit for the Mount Vernon–Brooke and Oklahoma

floras (pre-D13-560′, or pre-D12-490′) is provided by the presence of the *Asteropollis-Stephanocolpites* complex (largely restricted to horizons below D13-560′ and D12-490′), and the absence of the forms restricted to D13-560′ and younger rocks cited later (though monosulcate forms apparently related to the trichotomosulcate *Liliacidites* sp. E do occur, fig. 3: 30), and the rarity of very small (8–12 μ) tricolp(oroid)ates (especially typical *"Tricolpopollenites" minutus* Brenner, see fig. 3: 36).

Features of the angiosperm leaf flora in subzone II-B closely parallel those of the pollen. In contrast to their highly subordinate position in zone I, angiosperm leaves are locally abundant in subzone II-B for the first time, consist of an increased number and variety of morphological types (approximately 36 for the whole subzone), and are distributed in a wider variety of lithofacies. Leaf architecture in subzone II-B is characterized by increases in the rigidity of organization of vein patterns and in regularity of vein behavior in a number of lines.

Two important angiosperm leaf types which first appear in the middle subzone II-B localities are palmately veined peltate and lobate-cordate-ovate leaves, which intergrade with each other in shape, venation, and marginal characters (e.g., shallow, broad crenulations) (fig. 28: s, t). Their shape and the tendency of their primary veins to branch dichotomously and to form several orders of brochidodromous loops well within the margin suggest derivation from the reniform complex of zone I, of which several representatives survive to this level. All three complexes are most common at the same localities (Mount Vernon, White House Bluff) noted earlier for the unusual abundance of monosulcate angiosperm pollen.

Truly peltate leaves are represented by *Menispermites virginiensis* Fontaine (figs. 13, 14) (Brooke, Mount Vernon) which ranges up to 8 cm in diameter with approximately 10 primary veins radiating from the funnel-shaped leaf base. A rudimentary regularity in the primary venation manifests itself in the tendency to begin regular dichotomous branching about midway to the broadly crenulate margin.

"Populus" potomacensis Ward (fig. 15) (White House Bluff, Mount Vernon), the most typical member of the lobate-cordate complex, is characterized by its ovate shape and a base ranging from shallowly lobate to so deeply cordate that it appears peltate in imperfect specimens. The primary veins, which radiate actinodromously or acrodromously from the base of the leaf, show a high degree of regularity in their courses. An additional ovate form known as *Menispermites tenuinervis* Fontaine has a truncate to obtuse base,

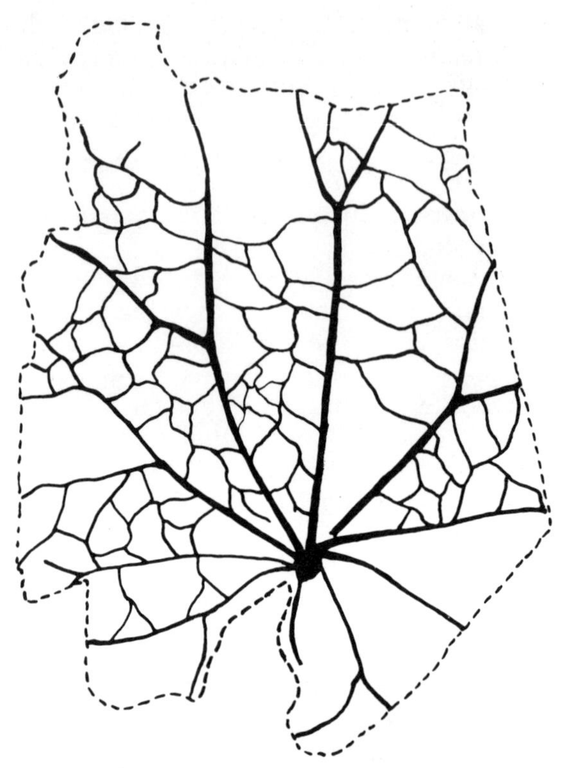

Fig. 13. *Menispermites virginiensis* Fontaine (USNM 3248) from Brooke, Virginia (middle subzone II-B). A fragment of a large peltate leaf showing the highest degree of vein regularity reached in this form (× 1).

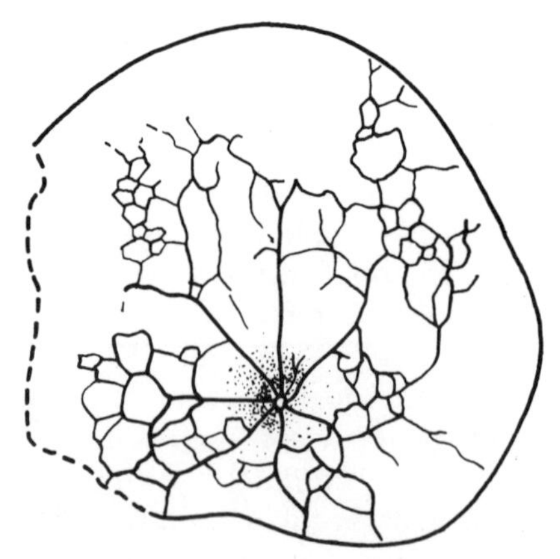

Fig. 14. *Menispermites virginiensis* Fontaine (USNM 5392) from Mount Vernon, Virginia (middle subzone II-B). A small peltate leaf with an apparently infundibuliform base and somewhat less regular venation (× 1).

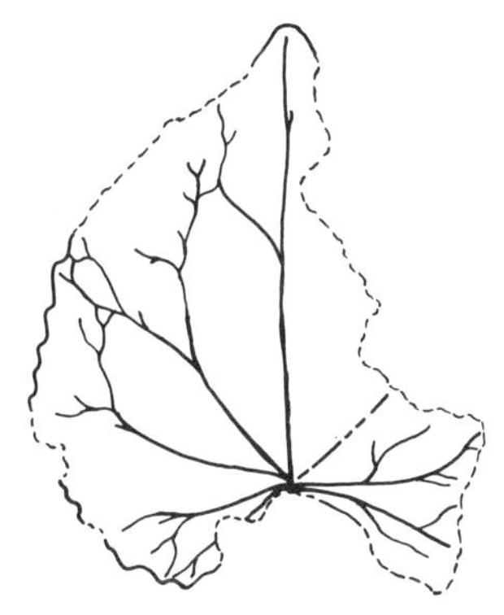

Fig. 15. *"Populus" potomacensis* Ward (USNM 205257) from Mount Vernon, Virginia (middle subzone II-B) (× 1).

indicating its possible transitional relationship to the peltate or reniform complexes.

The most common subzone II-B reniform leaf, *Populophyllum reniforme* Fontaine (fig. 16) (Brooke and Aquia Creek), has a base which varies from shallowly lobate to deeply cordate with only a narrow slit separating the lobes. Hence, this form too is morphologically intermediate between the ovate-cordate and peltate complexes. Cuticle studies (Mersky, unpublished data) reveal a very different epidermal structure from that of associated pinnatifid leaves (*Sapindopsis*, see below); in fact, its features are suggestive of modern monocots and certain herbaceous dicots. Finally, a rarer reniform type from Brooke and Mount Vernon recalls *Proteaephyllum reniforme* from zone I (with which it was identified by Fontaine [1889] and Ward [1895]), but its venation is completely flabellate, with no trace of a pinnate medial primary vein.

An aquatic or semiaquatic habitat for the peltate and, to some extent, the lobate and reniform complexes is suggested both by functional-morphological analogy (not necessarily homology!) with modern aquatic forms and by the sedimentary associations. The characteristic dichotomizing and brochidodromous looping of the primary venation well within the margin are more analogous to the venation of modern aquatics such as Nymphaeales and *Nymphoides* than to the near-marginal camptodromous venation of the peltate-lobate leaves of lianas such as the Menispermaceae, as are the presence of long, broad, and apparently lax petioles in the first two complexes, and of funnel-shaped leaf bases in *Menispermites virginiensis*. In addition, they usually occur in clay and silt with parallel lamination (suggesting quiet water deposition) and often in well-

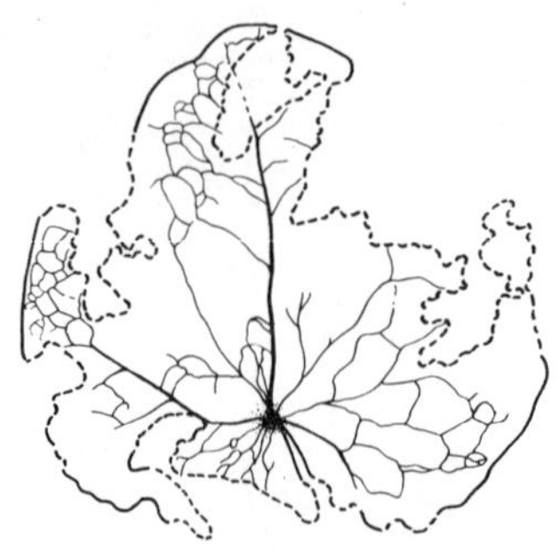

Fig. 16. ***Populophyllum reniforme*** **Fontaine (USNM 201913) from Aquia Creek, Virginia (middle subzone II-B). A very deeply cordate leaf with an apparently infundibuliform leaf base (× 1).**

spaced clusters, suggesting growth in place. Still another possible aquatic adaptation is seen in a fragmentary, presumably actinodromous, leaf known as *Aristolochiaephyllum cellulare* Fontaine, which displays an elaborate pattern of several orders of brochidodromous veins near the margin which are so greatly thickened that they project several millimeters from the plane of the leaf. The result is an alveolar structure strikingly similar to the hexagonal air cells found on the undersides of the leaves of modern *Victoria regina* and *Euryale*. From a comparative morphologic viewpoint, the presence of peltate and cordate-lobate aquatics with many leaf architectural features (though not all the details) seen in modern Nymphaeales is consistent with the diversity of the pollen flora at this level, since Nymphaeales have basically monosulcate or (in *Nelumbo*) tricolpate pollen.

Actinodromous fragments described as *Aristolochiaephyllum crassinerve* Fontaine (Brooke, Mount Vernon, Quantico?) indicate the existence of larger members of this complex with superficially better-organized secondary and tertiary venation, approaching third rank. In detail, however, the higher-order venation, though percurrent, is more irregular in course than in typical third-rank leaves, and the thick veins are somewhat spongy-looking and poorly demarcated. The sandy matrix of this form constitutes a significant exception to the usual sedimentary association of the peltates and lobates with fine-grained sediments. Perhaps its more rigidly organized venation was an adaptation to a higher-energy aquatic environment.

Simple leaves with disorganized first- or second-rank venation persist and diversify in the middle subzone II-B flora. These include entire-margined, brochidodromous forms such as *Celastrophyllum brookense* Fontaine, *Ficophyllum eucalyptoides* Fontaine, a new elliptic species from Brooke, and a new *Rogersia* from Fort Foote (pa-

Fig. 17. ***Sapindopsis magnifolia*** **Fontaine (USNM 3369) from Brooke, Virginia (middle subzone II-B). A pinnatifid leaf with a narrow wing of foliar tissue on the rachis (× 1).**

lynologically barren but lithostratigraphically close to Mount Vernon). Nonentire forms all have doubly convex (A-1) glandular serrations ranging in size from small in *Celastrophyllum saliciforme* Ward (Mount Vernon), to moderate in the semicraspedodromous *C. acutidens* Fontaine (see fig. 28: u) (Brooke, Mount Vernon, Fort Foote) and in *C. hunteri* Ward (White House Bluff), to large in the craspedodromous leaves of *Quercophyllum grossidentatum* Fontaine (Brooke). A new crenate form with first-rank hyphodromous venation

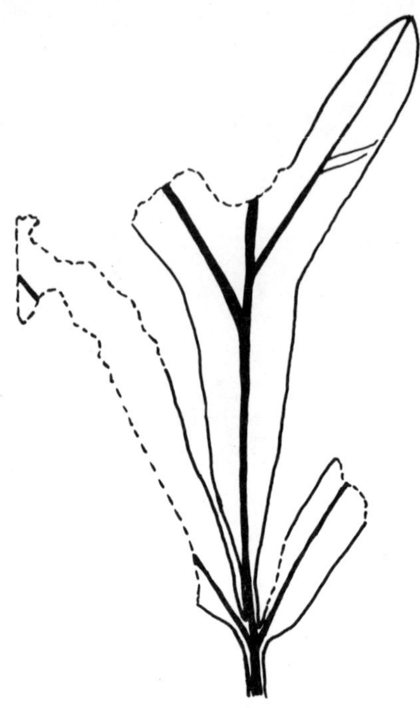

Fig. 18. ***Sapindopsis magnifolia*** **Fontaine (USNM 201914) from Brooke, Virginia (middle subzone II-B). A leaf fragment with the foliar wing becoming very broad apically. Originally described and illustrated as** ***S. variabilis*** **Fontaine (1889, p. 298, pl. 154, fig. 2) (× 1).**

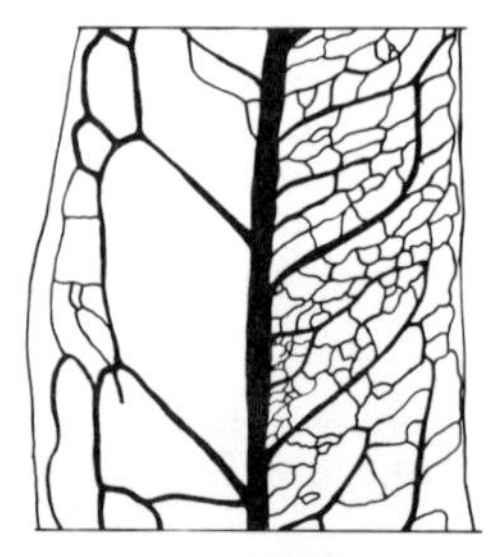

Fig. 19. ***Sapindopsis magnifolia*** **Fontaine (USNM 201915) from Brooke, Virginia (middle subzone II-B). Detail of venation (× 2).**

and leaves arranged oppositely on axes occurs at Brooke and Aquia Creek.

The most striking innovation in angiosperm leaves in the middle of subzone II-B is the development of both pinnately and palmately lobed forms. The pinnatifid leaves belong to *Sapindopsis magnifolia* Fontaine (figs. 17–19) (including *S. variabilis* Fontaine) and occur in great abundance at Brooke (noted earlier for the unusual abundance of tricolpate angiosperm pollen) in the lower few centimeters of a clay just above an abrupt contact with a sand. They are also found at Aquia Creek and in silty beds above the peltate-reniform bed at White House Bluff. In this type the lamina of each lobe is decurrent into a wing of foliar tissue running along the rachis. The pattern of lobation is highly variable, with the wing frequently widening apically and the terminal lobe occasionally bifurcating. The venation of this form is relatively irregular (ranging from high first to low second rank), with closely spaced secondaries that tend to form brochidodromous loops. The highly unusual outline of these leaves is quite unlike that of most modern pinnatifid blades as well as modern winged compound leaves of the Sapindaceae (with which they were originally compared) and the Cunoniaceae, where the leaflet lamina is constricted at the petiole.

A comparable trend toward palmate lobation manifests itself at Brooke in a complex including *Araliaephyllum obtusilobum* Fontaine (fig. 20), of which some variant members resemble the upper lobes of certain *Sapindopsis* leaves. This palmately lobed leaf has two (rarely one) lateral lobes and primary veins that originate at slightly different levels (palinactinodromous, as in modern *Platanus*) rather than at the same point (actinodromous). Their secondaries are slightly irregular in spacing and course but show a tendency to form marginal loops. At the leaf base the tertiaries tend to be irregularly percurrent between the primaries, but distally they become thin, irregular, and poorly defined. The second rank or at most low third rank of the palmately lobed leaves in this part of subzone II-B is in marked contrast to the "platanoids" of succeeding beds. From its first appearance, the palmately lobed group is restricted to coarse, generally crossed-bedded sediments such as silt and very fine-grained sand.

Finally, an ovate, sagittate, apparently campylodromous leaf called *Alismaphyllum Victor-Masoni* (Ward) Berry (White House Bluff) (fig. 28: r) must be noted as a possible poorly preserved monocot, though in the absence of higher-order venation a relationship to the ovate-cordate complex cannot be ruled out.

Angiosperm leaf assemblages comparable to the Brooke–Mount Vernon flora are known from essentially contemporaneous strata in other parts of the world. Peltate, lobate-cordate, palmately lobed, and low-rank serrate and cordate simple leaves make up the Middle Albian angiosperm floras of Kazakhstan, eastern Siberia, and Portugal (Teixeira, 1948; Vakhrameev, 1952; Samylina, 1960, 1968; Krassilov, 1967) with the addition of pinnatifid leaves in the Middle Albian of

Fig. 20. A member of the *Araliaephyllum obtusilobum* Fontaine complex from Brooke, Virginia (middle subzone II-B), originally identified as *A. aceroides* by Fontaine (USNM 201916). Note the weakness and irregularity of the second- and third-order veins (× 1).

western Canada ("lower Blairmore" [Bell, 1956]) and in the upper Lakota Formation (sensu Waage, 1959) of western South Dakota and northeastern Wyoming (Ward, 1899).

UPPER SUBZONE II-B

Palynology indicates that two small "Patapsco" megafossil collections of Berry (1911a) from Widewater (71-9) and Stump Neck (71-18), on the Virginia and Maryland banks of the Potomac River respectively, are significantly younger than the other Patapsco localities: i.e., post-580′, pre-480′ in well D13, corresponding to the upper part of subzone II-B as defined here. We have enlarged the known megafossil flora from this interval by additional collecting at Stump Neck and at two other Patapsco localities further north in Maryland, West Brothers (65-2a, = station 29 of Brenner[1963]) and Red Point (69-37). Correlations with floras from other areas, though unfortunately somewhat inexact, are consistent with the (early) Upper Albian age inferred from the more precise correlations of the under- and overlying beds.

Most pollen elements remain the same as in the middle subzone II-B samples, although there is a significant increase in the proportion of more characteristically Upper Cretaceous *Araucaria-*, *Phyllocladus-*, *Sequoia-*, and *Sciadopitys*-like conifer pollen (see Brenner, 1963), and of finely reticulate tricolporoidates (fig. 3: 34–36, aff. "*Tricolpopollenites*" *micromunus; Tricolpites* cf. *albiensis; "Tricolpopollenites" minutus*, etc.), as in the Upper Albian of many areas (Pacltová, 1971). In addition, very small, finely reticulate tricolp(oroid)ates (fig. 3: 37, 36; "*Retitricolpites*" *prosimilis*, and especially "*Tricolpopollenites*" *minutus*), and rugulo- and striato-reticulate tricolpates (fig. 3: 38, 39; "*Retitricolpites*" *vermimurus* and aff. "*R.*" *paraneus* [Doyle, 1969a, fig. 2: j,k]), though present in older samples, are more abundant here, as in D13-560′ through 510′. A post-D13-560′ or post-D12-500′ age is more positively indicated by the presence of monocotyledonoid trichotomosulcates (fig. 3: 33, *Liliacidites* sp. E, =*Retimonocolpites* sp. D of Doyle [1973]), and the first rare and still almost spheroidal members of the smooth-walled triangular tricolpor(oid)ate complex which becomes abundant in zone III and younger rocks (fig. 3: 40, aff. "*Tricolporopollenites*" *triangulus* Groot, Penny and Groot). Finally, upper age limits are provided by the presence of "*Retitricolpites*" *vermimurus* (not found above D13-510′ or D12-465′) and the absence of subzone II-C index forms cited in the next section.

Simple leaves are still an important component of the upper subzone II-B assemblage and show some advances over those in lower strata. An oblong, closely serrate, pinnately hyphodromous leaf from Red Point has straight-convex–sided teeth (B-1), making it the first simple leaf to show any but the convex-convex (A-1) type. Several new peltate forms occur at the base of a clay plug (infilled, cut-off stream meander) at the Quantico locality (71-54), which is barren palynologically but is inferred on lithostratigraphic grounds, and owing to the presence of typical pinnately compound leaves, to correlate with upper II-B. These fossils of peltate forms include fragments of a new large, entire-margined, actinodromous leaf with percurrent secondary but irregular higher-order venation; a new medium-sized (4–6 cm), entire, thick leaf with actinodromous pri-

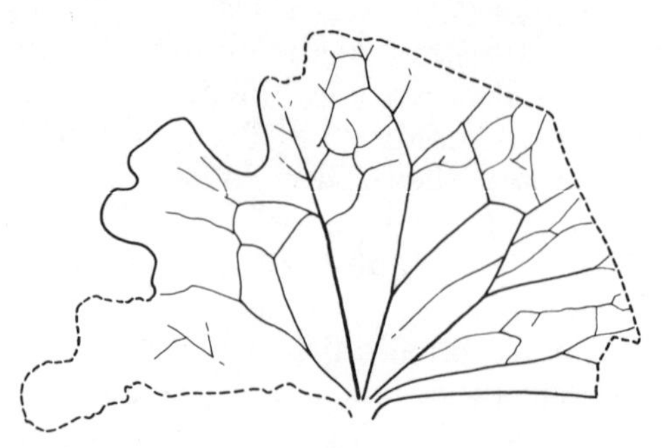

Fig. 21. A lobed, actinodromously veined leaf (USNM 201917) from Quantico, Virginia (correlated with upper subzone II-B), similar to some members of the *Populophyllum reniforme* Fontaine complex (× 1).

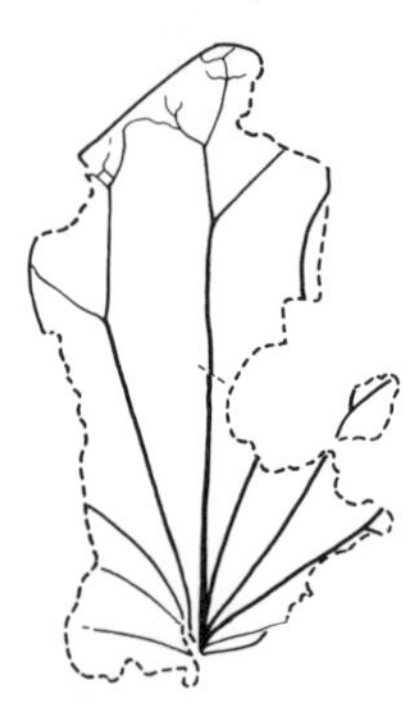

Fig. 22. *Menispermites potomacensis* Berry (USNM 201918A) from near Widewater, Virginia (correlated with upper subzone II-B) (× 1).

mary veins branching toward the margin and a granular texture; a new small, crenate, actinodromous leaf lacking the basal sinus of the other two; and an irregularly lobate, actinodromous leaf with irregularly ramifying secondary and higher-order venation whose architecture is similar to that of the *Populophyllum reniforme* complex (fig. 21). *Menispermites potomacensis* Berry (fig. 22), which occurs at Stump Neck, Widewater, Red Point, and possibly Quantico, is another possible derivative of the cordate-reniform complex. Its architectural features include an ovate, sometimes shallowly lobate, shape and approximately six acrodromous primary veins fanning out from the decurrent base.

The most notable advance in the upper portion of subzone II-B, however, is the appearance of a complex of truly pinnately compound leaves (figs. 23, 24) (some mistakenly identified by Berry [1911a] with the Brooke species *Sapindopsis magnifolia*), with each leaflet attached to the rachis by a true petiolule and with a tendency

Fig. 23. *Sapindopsis* with compound leaves from upper subzone II-B. The illustration is a modification of Berry's (1911a) pl. 86, fig. 1, based on numerous incomplete specimens from Red Point and West Brothers clay pit in Maryland (× 1).

Fig. 24. *Sapindopsis* n. sp. (USNM 201919) from Red Point, Maryland (upper subzone II-B). Detail of venation (× 2).

toward more regular second- and third-order venation. The bases of the leaflets vary from slightly to highly asymmetrical (suggesting the bizarre Upper Cretaceous genus *Fontainea*), while the tertiary veins show a tendency to align themselves perpendicular to the midvein. Gross leaf and leaflet shape; the course, close spacing, and thinness of the secondary veins; similarities in the intercostal vein pattern; and cuticular evidence (Mersky, 1973, and unpublished data) are consistent with the supposition that these compound leaves arose from the pinnatifid complex seen at Brooke, which in fact persists into upper II-B as a rare element at Red Point.

This apparently vigorously evolving complex shows much of the same variability in gross form as does Brooke *Sapindopsis* (including bifurcation of the terminal leaflet) and includes types both with and without laminar glands and marginal teeth. Large leaves (the length of one leaflet exceeding 15 cm) with entire margins, no laminar glands, and brochidodromous to eucamptodromous secondary veins are abundant at Red Point. An even more variable complex with small, laminar resin dots is dominant at West Brothers clay pit, where it occurs in a silty clay bed whose lowest portion contains fragments of charcoal and reworked silt clasts together with a virtual mat of horizontal lignitized branches and small logs. Some members of this complex appear to have palmately lobed apical leaflets; others show incipient to strong development of marginal teeth. These vary from the A-1 type when small to concave-convex (C-1) when larger. The secondary veins of this form branch below the tooth, sending one branch into it and the other into the area of the sinus or to the superjacent secondary. The tooth shape and marginal venation are similar to those found in primitive members of the Rosidae, such as the Cunoniaceae, although a distinct spine or process at the apex of the serration is unlike modern members of the subclass. A small, toothed, low-second-rank, pinnately compound leaf from Quantico may represent stilll another member of the complex, given its very small size and the absence of laminar glands. Toothed pinnates at this level in the Potomac Group are consistent with the presence of similar forms in the Upper Albian Cheyenne Sandstone of Kansas (Berry, 1922), the Fall River Formation in the Black Hills (Hickey, personal observations, 1973; Waage, personal communication, 1973), and correlative sediments in Kazakhstan (*Anacardites neuburgae* Vakhrameev [1952]).

The palmately lobed complex, which appeared in middle subzone II-B and becomes dominant in subzone II-C, is also present in coarser lithofacies (cross-bedded and laminated silt and fine sand) of upper subzone II-B at Stump Neck and Widewater. The best-known

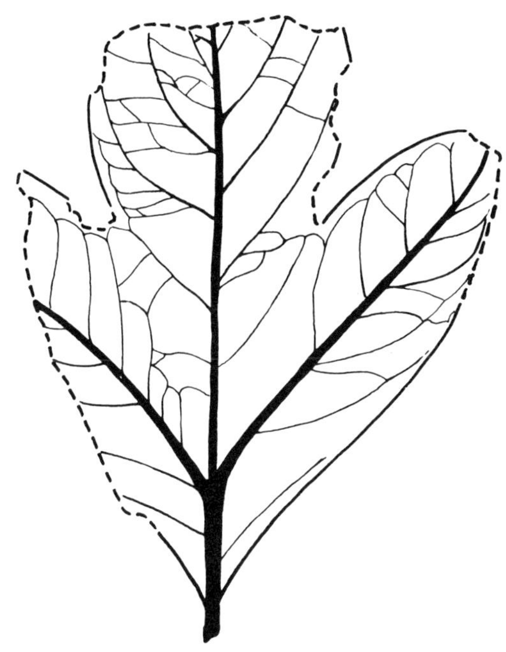

Fig. 25. *"Sassafras" potomacensis* Berry (USNM 201920) from Stump Neck, Maryland (upper subzone II-B). This specimen originally described and figured as *"S." parvifolium* Fontaine by Berry (1911a, p. 486, pl. 94, fig. 2). Note the moderate increase in the strength and regularity of the second- and third-order venation over that in *Araliaephyllum obtusilobum* (× 1).

member of this complex, *"Sassafras" potomacensis* Berry (fig. 25) (including *"Sassafras" parvifolium* Fontaine sensu Berry, and *Araliaephyllum crassinerve* sensu Berry) can range up to 10–20 cm long, with three to seven lobes and as many suprabasal palinactinodromous primary veins. Its secondary veins are fairly regular but the tertiaries are thin, closely spaced, and irregularly and obliquely percurrent across the intercostal areas. This form, which incidentally bears the earliest-known angiosperm leaf galls, shows that a moderate increase in vein regularity was taking place independently among palmately lobed forms.

SUBZONE II-C–?ZONE III

The three final pollen localities, Bull Mountain (71-1), Brightseat (71-2), and Cedar Point (71-3), are all from the predominantly sandy "Raritan" beds which overlie the typical Patapsco Formation in Maryland but are older than the type New Jersey Raritan Formation (see Doyle, 1969a; Wolfe and Pakiser, 1971; and above). The Bull

Mountain and Brightseat localities are believed to fall in the interval above 510′ and below 370′ in Delaware well D13, which was identified by Brenner (1967) as part of subzone II-B, but which is here designated as a new subzone, II-C. Cedar Point appears to be slightly younger (certainly post-D12-480′), and may fall either in the upper part of subzone II-C or in the basal part of zone III (below 340′ in D13, but possibly as high as 275′ in D12). Close correlations with pollen floras from marine-dated Upper Albian rocks in England (Kemp, 1968, 1970) and the Upper Albian–?basal Cenomanian lower Colorado Group of the western Canadian plains (Norris, 1967) indicate a lower age limit within the Upper Albian, although the fact that floras comparable to those from immediately younger beds (Pierce, 1961; Hedlund, 1966; Pacltová, 1971) lie above the base of the Cenomanian would also allow a basal Cenomanian age, especially for Cedar Point.

Many of the monosulcate and tricolp(oroid)ate angiosperms are the same as in the preceding group of localities, including several forms common to the Canadian Upper Albian, e.g., finely structured, tectate tricolpates (fig. 3: 44, *Tricolpites* cf. *sagax;* fig. 3: 43, "*Retitricolpites*" *prosimilis*), and larger reticulate tricolpates (fig. 3: 47, aff. "*Retitricolpites*" *magnificus*, = "*Retitricolpites*" *vulgaris* sensu Norris). However, "*Retitricolpites*" *vermimurus* is absent, and the most common angiosperms are a complex of very small, reticulate tricolporoidates, which are closely comparable to the dated British Upper Albian forms illustrated by Kemp (1968) as *Tricolpites albiensis* (fig. 3: 42). More indicative are several complexes known only above D13-510′ and D12-465′: (1) medium-sized, very coarsely reticulate monosulcates (fig. 3: 41, cf. "*Liliacidites*" *textus* Norris); (2) very small, smooth, thin-walled tricolporoidates (fig. 3: 45, *Tricolporoidites* cf. *subtilis* Pacltová, = "*Psilatricolpites*" *parvulus* sensu Norris[?]); and (3) two species of the rugulate, bisaccate conifer *Rugubivesiculites* (*R. reductus* Pierce, and *R. rugosus* Pierce). Both species of *Rugubivesiculites* enter somewhat above the base of the Canadian Upper Albian sequence (Viking Formation), while "*L.*" *textus* and "*P.*" *parvulus* enter still higher (upper shale unit). In addition, Cedar Point and possibly Bull Mountain yield forms seen only above D13-480′: (1) slightly thicker-walled, smooth tricolporoidates (fig. 3: 51, aff. "*Tricolporopollenites*" *distinctus* Groot and Penny); (2) small reticulate tricolporoidates with coarser mesocolpial sculpture (fig. 3: 50, *Tricolporoidites* aff. *bohemicus* Pacltová); and (3) prolate tricolpates with relatively tall, thin columellae (fig. 3: 48, *Tricolpites* aff. *nemejci* Pacltová), while Cedar Point yields finely striate tricolpates (fig. 3: 49, typical "*Retitricolpites*" *paraneus*) and, more interesting from an evolutionary point of view, both small and medium-sized triangular tricolporoidates, the latter with broad, ill-defined colpus margins (fig. 3: 46, 52; cf. "*Tricolporopollenites*" *triangulus* and *Tricolporoidites* sp. A).

Finally, the age of all these samples is bracketed above by the absence of many forms typical of all but the base of zone III, such as larger and thicker-walled triangular and prolate tricolpor(oid)ates, "*Retitricolpites*" *vulgaris* Pierce, and typical *Tricolpites nemejci,* all elements found in such Lower Cenomanian deposits as the lower and middle Peruc Formation of Czechoslovakia (below the appearance of triporate Normapolles [Pacltová, 1971]), the Woodbine Formation of Oklahoma (Hedlund, 1966), and the "Dakota" Formation of Minnesota (Pierce, 1961).

The angiosperm leaf flora of subzone II-C is dominated by palmately lobed and apparently related, simple, pinnately veined "platanoid" leaves, possibly because of the predominance in the Maryland "Raritan" of fluvial sand, with which this complex is typically associated. Many of the same species, as well as a greater variety of simple (including peltate) and pinnately compound types, are typical of the more extensively collected Dakota flora of Kansas (Lesquereux, 1892) and mid-Cretaceous beds in Kazakhstan (Vakhrameev, 1952). Both floras were formerly considered Upper Cretaceous, but both are now known on marine evidence to extend down into the Upper Albian (Vakhrameev, 1952; Bayne, Franks, and Ives, 1971), an excellent agreement with the palynological dating of subzone II-C.

With the exception of rare survivors of "*Sassafras*" *potomacensis* (with its characteristic galls), most subzone II-C "platanoids" exhibit strong, rigidly organized tertiary veins, perpendicularly traversing the broad intercostal areas, subparallel quaternary veins between the tertiaries, and nondecurrent junction of the lateral and medial primaries in a thickened, padlike area. The most widely distributed type is *Araliopsoides cretacea* (Newberry) Berry (fig. 26), found at White Point, Bull Mountain, East Washington Heights, and numerous other localities. This palmately trilobed leaf ranges up to 15 cm long and has an expanded petiole base, suggesting either the development of a regular abscission mechanism or the presence of an axillary bud, both of which might be related to the deciduous habit. Other important subzone II-C platanoids include peltate-based, palmately lobed, actinodromous leaves known as *Aspidiophyllum* Lesquereux (Shannon Hill and East Washington Heights) and two spe-

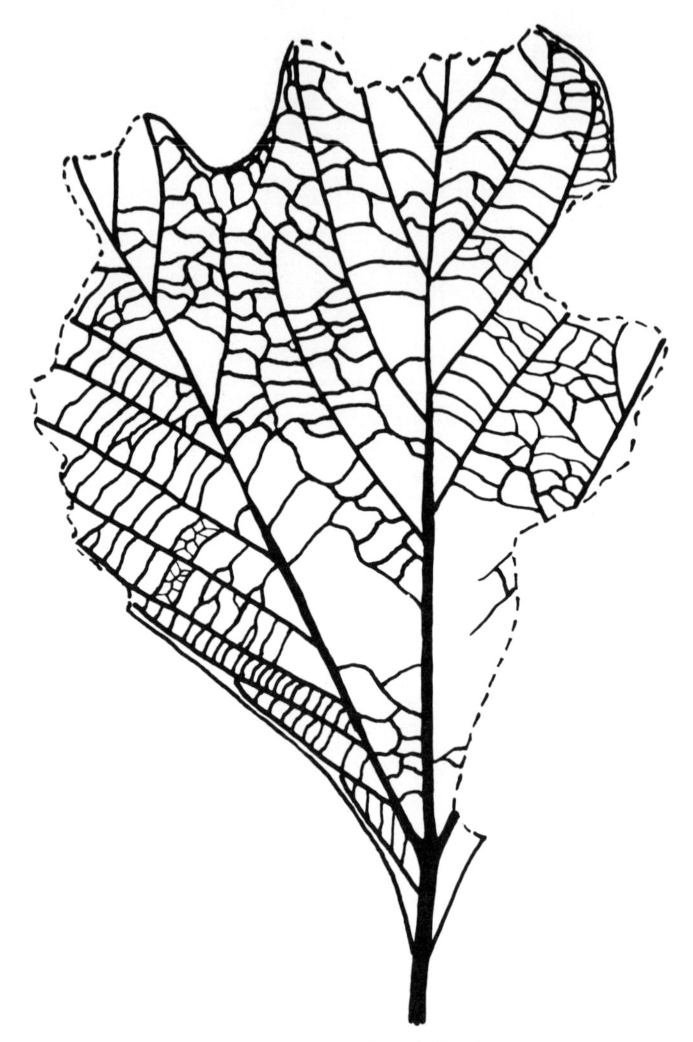

Fig. 26. *Araliopsoides cretacea* (Newberry) Berry (USNM 201921A) from Bull Mountain, Maryland (subzone II-C–?zone III) ($\times$ 1).

cies of *Protophyllum* Lesquereux (fig. 27) (Cedar Point, Shannon Hill, and East Washington Heights), whose higher-order venation suggests it is a secondarily unlobed member of the same complex.

The rare simple, nonpalmate (or palmate-derived) leaves found thus far in the Maryland "Raritan" all have entire margins and second- or lower-rank venation. A small assemblage of leaves from the Hylton Pits of New Jersey (Berry, 1911b), not studied palynologically by us but probably the source of a subzone II-C or zone III

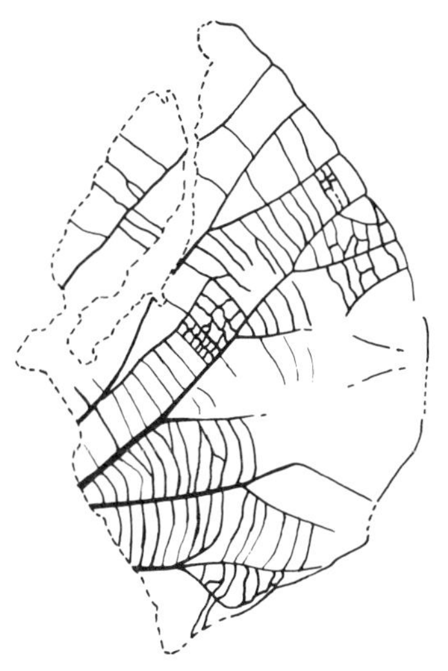

Fig. 27. Fragment of a leaf of *Protophyllum multinerve* Lesquereux (USNM 201922) from Cedar Point, Maryland (subzone II-C–?zone III) (× 1).

pollen flora (locality 11063) described by Wolfe and Pakiser (1971), may provide additional evidence on the flora of the finer-grained facies of this interval. It consists largely of a number of apparently simple, pinnately veined, second-rank leaves which Berry assigned to such genera as *Laurus*, *Eucalyptus*, and *Andromeda*, as well as a low-rank emarginate leaf known as *Liriodendropsis* Hollick, a type which becomes considerably more common in the Upper Cretaceous. This flora thus provides something of a transition from the Potomac flora to that of the succeeding New Jersey Raritan Formation, which is rich in non-"platanoid" forms (Newberry, 1895; Hollick, 1906; Berry, 1911b). These data, as well as those from certain other Upper Albian–Cenomanian floras (e.g., Nazaré in Portugal, dominated by low-rank, simple leaves [Saporta, 1894]), suggest that the rank of simple leaves lagged behind that of contemporaneous pinnately compound and palmate types.

To summarize the Potomac Group megafossil sequence (see fig. 28), in zone I the angiosperm leaves are rare, simple, mostly entire, relatively undiverse in shape, and with highly irregular "first-rank" pinnate venation frequently accompanied by such characters as poor differentiation of blade from petiole and multistranded midribs. The one lobate form has unbraced sinuses, and those with nonentire margins all have doubly convex (A-1) teeth. By the middle of subzone II-B, angiosperm leaves have become abundant and include new lobate-cordate, peltate, and both pinnately and palmately lobed

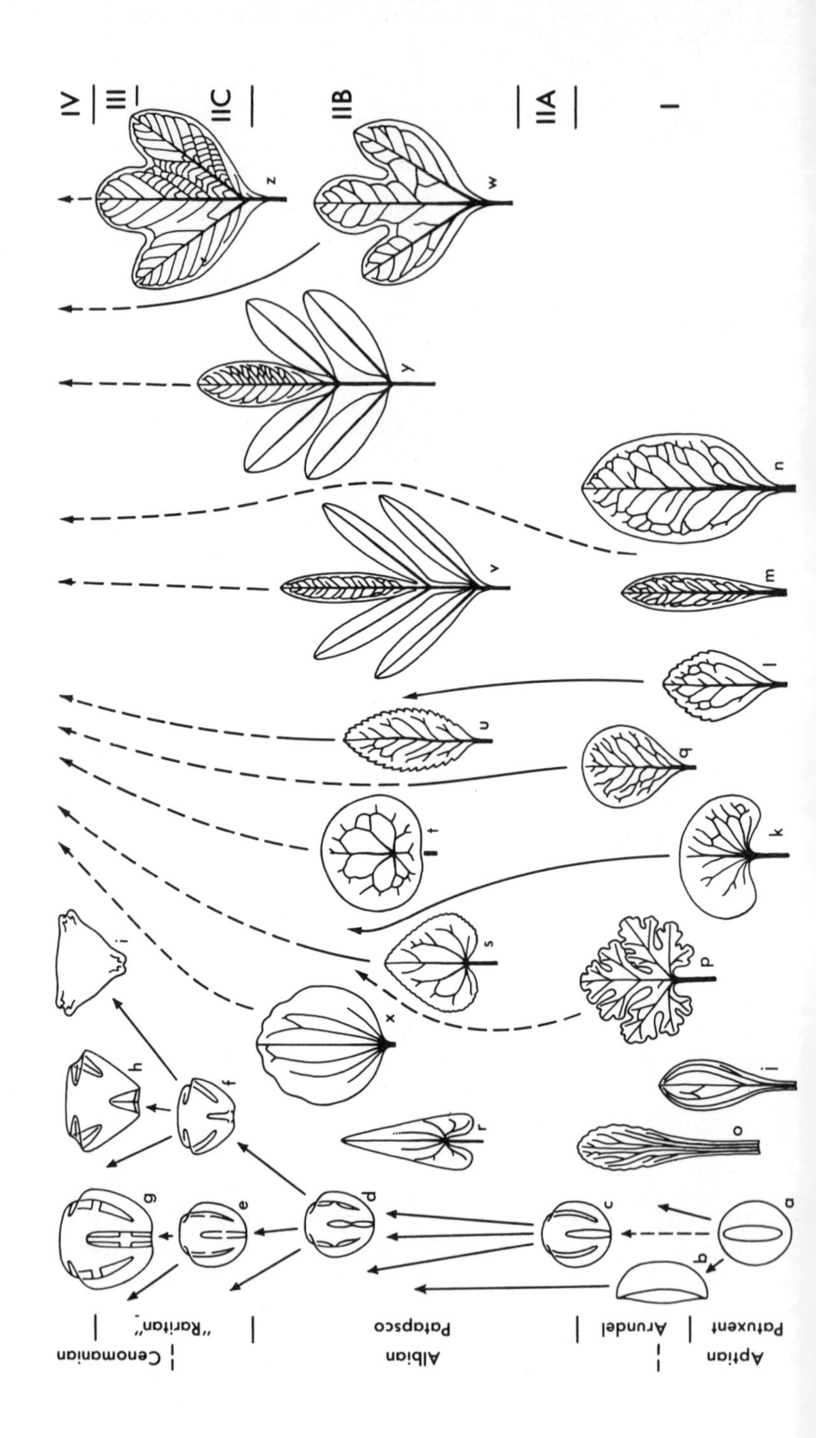
Aptian
Albian
Cenomanian
Patuxent
Arundel
Patapsco
"Raritan"
I
IIA
IIB
IIC
III
IV

Fig. 28. Summary of Potomac leaf and pollen sequence. Principal leaf and pollen types are plotted against Atlantic Coastal Plain lithologic units and their presumed equivalents in the standard stage sequence (*left*) and the Potomac-Raritan pollen zonation (*right*). In the case of the pollen, arrows indicate both evolutionary transformations and stratigraphic range extensions. The dashed arrow indicates a transformation for which there is only indirect fossil evidence. In the case of the leaves, solid arrows indicate upward extensions of morphological complexes within the Potomac sequence; dashed arrows, range extensions inferred from other areas.

Pollen types indicated: (a) Generalized tectate-columellate angiospermous monosulcates (*Clavatipollenites, Retimonocolpites* spp.). (b) Reticulate monocotyledonoid monosulcates (*Liliacidites* spp.). (c) Generally tectate-reticulate tricolpates (*Tricolpites* spp.). (d) Generally tectate-reticulate tricolporoidates (*Tricolpites, Tricolporoidites* spp.). (e) Small, generally smooth-walled prolate tricolporoidates (*Tricolporoidites* spp.). (f) Small, generally smooth-walled, oblate-triangular tricolporoidates (*Tricolporoidites, Perucipollis* spp.). (g) Larger, often more highly sculptured, prolate tricolpor(oid)ates (*Tricolporoidites, Tricolporopollenites* spp.). (h) Larger, often more highly sculptured, oblate-triangular tricolpor(oid)ates (*Tricolporoidites, Tricolporopollenites* spp.). (i) Primitive members of the triangular triporate Normapolles complex (*Complexiopollis, Atlantopollis* spp.).

Leaf types indicated: (j) Acrodromous, narrowly obovate, monocotyledonoid (*Acaciaephyllum*). (k) First rank, pinnately veined, reniform (*Proteaephyllum reniforme*). (l) First rank, serrate (*Quercophyllum*). (m) First rank, narrowly obovate (*Rogersia*). (n) First rank, broadly elliptical (*Ficophyllum*). (o) Parallelodromous, elongate (*Plantaginopsis*). (p) Lobate reniform (*Vitiphyllum*). (q) First rank, obovate (*Celastrophyllum*). (r) Campylodromous, sagittate (*Alismaphyllum*). (s) Actinodromous ovate-cordate-lobate (*"Populus" potomacensis, Populophyllum reniforme*). (t) Actinodromous, peltate (*Nelumbites*). (u) Pinnately veined, serrate (*Celastrophyllum*). (v) Second rank, pinnatifid (*Sapindopsis*). (w) Second rank (palin)actinodromous, palmately lobed (*Araliaephyllum*). (x) Acrodromous, lobate elliptical (*Menispermites potomacensis*). (y) Third rank, pinnately compound, sometimes serrate (*Sapindopsis* spp.?). (z) Third rank (palin)actinodromous, palmately lobed (*Araliopsoides, "Sassafras"*, etc.).

shape classes. Most of the angiosperm leaves found in this subzone show a moderate increase in the organization of their venation, with the secondaries showing greater regularity in course and behavior. In the upper portion of subzone II-B, pinnately compound leaves appear; both these and the palmately lobed forms have somewhat more regular tertiary venation than their assumed precursors. Straight-convex (B-1) and concave-convex (C-1) teeth appear in the simple and pinnately compound leaf complexes, respectively. The palmately lobed leaves and their derivatives in subzone II-C and basal zone III show increased diversity in form and the development of a rigid vein network extending to the fourth order. Floras of this interval also demonstrate the appearance of several simple forms transitional to later floras that retain less organized venation than their palmate associates.

Consistency of the Leaf and Pollen Records

Because leaves are more sporadically distributed and more sensitive to local environmental effects, they are more difficult to analyze for evolutionary transformations than is pollen, which often provides relatively complete series of intermediates. However, a number of evolutionary trends are suggested when the principal Potomac angiosperm leaf types are plotted against the pollen sequence (fig. 28). Solid lines indicate known stratigraphic ranges in the Potomac sequence, while dashed lines indicate evidence of range extensions in other geographic areas, such as the Middle Albian of Kazakhstan or the "Dakota" of Kansas. Direct evolutionary derivations of one leaf type from another are not indicated, even where strongly suspected; however, leaves belonging to the same morphological complex are loosely grouped.

In the pollen record (fig. 28, left), there is a major trend in aperture condition from monosulcate to tricolpate, tricolporoidate, tricolporate, and finally triporate. The exact path of the first transformation is uncertain (actual intermediates are unknown), though the record is consistent with its having occurred only once; later steps clearly occurred by parallel evolution in many lines (Doyle, 1970, and unpublished data; Wolfe, Doyle, and Page, 1975). Superimposed on this progressive trend are various divergent (and often reversed) trends in size, shape, and sculpture, leading among the tricolpates and their derivatives from small or medium-sized to very small or large; from prolate to oblate and eventually markedly triangular; from finely reticulate to tectate-psilate, more coarsely reticulate, or striato-recticulate.

The leaf record shows a closely analogous pattern. There is a general trend in several morphological complexes toward organization and regularization of successively higher vein orders. Initially all vein orders were poorly differentiated from each other and irregular in their course and manner of branching and anastomosing; this pattern was associated with decurrency of the secondaries, highly irregular shape of intercostal areas, and poor differentiation of petiole and blade. In the next stage the secondary veins are well differentiated and relatively uniform in spacing, course, and behavior (thus with regularly shaped intercostal areas) but with tertiary and higher-order veins irregularly ramifying and anastomosing. In the final stage the tertiary veins are also well differentiated and regular in course; specifically, they are percurrent, that is, running directly from one secondary to the next. This pattern corresponds closely to the "leaf ranking" trend of Hickey (1971), constituting direct paleontological evidence for its validity (see Hickey and Doyle, 1972), although modern groups show several other modes of regularization of tertiaries besides percurrency (Hickey, unpublished data).

There are also divergent trends involving shape and major venation pattern. One trend, possibly beginning with somewhat irregularly shaped, pinnately veined reniform leaves with a tendency for clustering of the lower secondaries, leads to several palmately veined forms: actinodromous ovate leaves with either lobate or peltate bases, completely flabellately veined reniform leaves, and possibly acrodromous elliptical leaves. Another trend, presumably beginning with pinnately veined simple leaves (there being no evidence for anything but simple leaves in zone I), results in pinnately and palmately lobed leaves (the latter with peculiar palinactinodromous or "pseudopalmate" major venation) and, among the pinnately lobed members of this possibly natural complex, to truly pinnately compound leaves. The still more basic divergence into the dicotyledonous pinnate-"reticulate" and monocotyledonous, closed acrodromous-"parallel" venation patterns had apparently already occurred before the beginning of the Potomac record, thus paralleling evidence from the pollen record (Doyle, 1973).

Consideration of the modern systematic distribution of leaf and pollen characters seen associated at successive Potomac horizons also suggests that leaf and pollen evolution were more or less concomitant, and places limits on the age of certain modern higher taxa.

First, all zone I angiosperm leaves show highly disorganized first-rank venation, and all but a few angiosperm pollen grains from upper zone I are monosulcate. In the modern flora, both characters are concentrated in the order Magnoliales (*sensu stricto* of Takhtajan

[1969]), though not necessarily associated in the same families. However, it is important to realize that since both these characters appear to be primitive and since primitive characters in fossils do not necessarily indicate the existence of the modern phylogenetic group (clade) in which they are retained (Hennig, 1966), they should not be taken as evidence of Magnoliales per se or of magnolialian characters in other organs, but rather as evidence of some of the properties of a primitive evolutionary grade through which many lines may have passed. The record itself indicates that many features found in particular modern Magnoliales (e.g., monoporate pollen shed in permanent tetrads in Winteraceae, or vessels and third- or, rarely, low fourth-rank leaves in Magnoliaceae) are more specialized than what is seen in zone I. Though some zone I leaf types are retained in the Magnoliales (Wolfe, 1972b), similar types probably occurred in the ancestors of tricolpate dicots, whereas others (such as *Acaciaephyllum* and the reniforms) suggest other modern monosulcate groups (monocots and Nymphaeales), and some (such as *Vitiphyllum*) are unlike anything now living.

Second, the association in lower zone I of monosulcate grains with reticulate sculpture differentiated into coarser and finer areas and leaves with apically closed acrodromous ("parallel") major venation and finer crossveins is consistent with the restriction of both sets of characters to the monocots today, besides indicating that the basic split between monocots and dicots occurred very early in the radiation of angiosperms (Doyle, 1973).

Third, the existence of a complex of ovate-cordate and more definitely aquatic peltate leaves in subzone II-B, which is dominated by monosulcate and tricolpate pollen, and of possible ancestral types (the reniforms) in zone I, is consistent with the presence of similar (but more advanced) leaves in the modern aquatic order Nymphaeales (*sensu lato*), which have basically monosulcate or, in the case of *Nelumbo*, tricolpate pollen.

Fourth, the appearance and dominance (in some facies) of palmately lobed, palinactinodromous platanoid leaves in latest subzone II-B and subzone II-C is consistent with the dominance of tricolpate and tricolporoidate pollen at this level, since similar (though more advanced) leaves are found in the relict genus *Platanus*, which has tricolpate pollen.

Fifth, leaves transitional to the pinnately compound type appear only in middle subzone II-B, and truly pinnately compound leaves in upper subzone II-B, where tricolpates and the first tricolporoidates are diversifying; pinnately compound leaves today are restricted to tricolpate and tricolpate-derived dicots, and mostly to a series of related tricolporate orders of the subclass Rosidae (Cronquist, 1968; Takhtajan, 1969), such as Rosales and Sapindales (Hickey and Wolfe, 1975). In fact, the venation of the serrations in correlative Kansas (Cheyenne) and Kazakhstan compound leaves is of a type characteristic of "lower" Rosidae.

The pattern of progressive morphological specialization and diversification observed in both pollen and leaf records suggests strongly the occupation of successively higher and more divergent adaptive zones (or ways of life), that is, adaptive radiation. In the following section, we relate the trends to general evolutionary theory on the origin and radiation of higher adaptive types and to specific hypotheses on the origin and evolution of the angiosperms. In order to do so, we propose ecological-adaptive bases for the trends and summarize evidence from lithologic associations and modern functional analogies which bear on them. In the case of the pollen, we suggest that the progressive aperture trend should be considered in terms of new selective pressures associated with germination on a stigma, one of the main functional correlatives of angiospermy (a hypothesis hinted at by Hughes [1961], and elsewhere elaborated by Doyle [1970], Walker [1971], Smart and Hughes [1972], and Sporne [1972]), and the divergent trends to different modes of pollination ecology. We consider the main "leaf ranking" trend in terms of the two presumed main functions of venation, namely, as a fluid distribution system and a mechanical support framework. Trends in shape and major venation are considered in terms of changes in habit and light-gathering strategies (Horn, 1971). If these or analogous explanations are verified by experimental or mathematical analysis of form, they will constitute additional arguments for the consistency of the pollen and leaf records and their interpretation as reflecting the adaptive radiation of the angiosperms (Muller, 1970). Finally, we discuss how our speculative model helps explain many aspects of one of the most distinctive structures of the angiosperms, the dicotyledonous leaf.

A Model for the Early Adaptive Evolution of the Angiosperms

One of the most important tasks of the modern synthetic theory of evolution is to explain the macroevolutionary changes from one major adaptive type to another (usually equivalent to the origin of higher taxa) in terms of the same process of selection and accumulation of random minor mutations that are believed to be responsible for microevolutionary changes at and below the species level (Mayr, 1963; Schaeffer and Hecht, 1965; Stebbins, 1970a, and in this volume). In his discussion of this problem, Bock (1965) reemphasizes the improbability that the complex structures which characterize a new adaptive zone arose in a single-phase shift from one zone to another; the selective pressures of a new adaptive zone (e.g., aerial life in birds, terrestrial life in amphibians) cannot operate on half-formed structures, and the concept of saltational evolution is unacceptable in the light of modern knowledge of developmental genetics. Rather, Bock argues that the shift from one zone to another occurs in a *transitional adaptive zone* where, under new selective pressures, structures which had arisen under different selective pressures in the ancestral adaptive zone (seen in retrospect as preadaptations) take on new functions. Selection for mutations representing adjustment or perfection of these new functions (postadaptations) may in turn result in preadaptation for the next step. In this way, what are seen in retrospect as the key innovations for occupation of a new adaptive zone arise stepwise under pressures of an intermediate way of life; e.g., lungs and stout fins in fishes living in fresh-water environments subjected to fragmentation during drought, or wings and feathers in arboreal reptiles. Intermediate forms would be expected to show a mosaic of characters of the ancestral and descendant groups. Once this process has culminated in attainment of a radically new function, a new major adaptive zone is reached, a whole new set of selective pressures comes into play, and a group undergoes a major adaptive radiation resulting in the continued integration of still other adaptive features with those acquired earlier.

Although many authors have discussed the possible role of insects in the origin of the flower (e.g., Faegri and Van der Pijl, 1966; Takhtajan, 1969) and of higher vertebrates in dispersal (Corner, 1949, 1964; Van der Pijl, 1972), the most comprehensive attempt to identify the transitional adaptive zone in which some gymnospermous ancestor was transformed into an angiosperm is that of Stebbins (1965, 1970b, and in this volume). Stebbins points out that most of the features which are common to all angiosperms and which set them apart from gymnosperms are connected with reproduction rather than vegetative structure. These characters constitute a syndrome involving shortening of the life cycle and greater efficiency of reproduction than in gymnosperms.

Rather than attempt to define this concept of reproductive efficiency any more rigorously than does Stebbins, we may give several examples. First, insect pollination, made possible by aggregation of sterile and fertile appendages into a flower and resultant juxtaposition of micro- and megasporophylls, permits far more efficient transfer of pollen, reducing the total amount of pollen which need be produced, as well as introducing the potential for new isolating mechanisms, speciation, and subsequent adaptive divergence (Takhtajan, 1969). In addition, the flower is an ephemeral structure whose delicate parts are discarded with a minimum wastage of material if fertilization does not occur. Carpel closure, besides being advantageous in protecting the developing ovules from desiccation and predatory pollinators, allows pollination and fertilization to occur while the ovules are still in an essentially primordial state, with highly reduced megagametophytes; unfertilized ovules abort with less waste than in gymnosperms, where fertilization does not occur until the seeds are nearly mature. Finally, the mechanism of double fertilization ensures that the embryo-nourishing tissue (endosperm) is not elaborated until fertilization has occurred, in contrast to the gymnosperms where the nourishing tissue (the megagametophyte) is already elaborated (cf. Strasburger, 1900, cited in Maheshwari, 1950; Foster and Gifford, 1959). Many of these features are cited by Takhtajan (1969, and in this volume) and Corner (1964) as examples of neoteny, since they represent shifts in the relative timing of developmental events so that reproductive (adult) processes occur when structures are still in a morphologically juvenile stage.

In his discussion of the probable growth habit of the first flowering plants, Stebbins (1965) argues that pressures for rapid and efficient reproduction are relatively weak in stable forest environments, where vegetative vigor and habitat adaptations are more important; in fact, it is in certain forest situations that gymnosperms (conifers) still compete most successfully with angiosperms. Conversely, reproductive efficiency is at a premium in unstable, disturbed conditions. Stebbins therefore suggests that the angiosperms arose in disturbed environments, specifically semiarid regions with well-drained soil and periodic drought, analogous to the environments where the reproductively most specialized forms today (e.g., Gramineae, Labiatae, Compositae) are most diversified. Additional evidence comes from an analysis of carpel closure in terms of cycles of transference of function in structures associated with the megasporangium (Stebbins, 1970b, and in this volume). The outer integument (assumed by Gaussen [1946] and Stebbins [1970b] to be homologous with the reflexed cupule of Mesozoic seed ferns), which presumably originated as a protective structure, became adapted to dispersal (e.g., the fleshy seed coat of *Magnolia*), and the protective function was taken over by the carpel (megasporophyll?). Stebbins emphasizes that such cycles are not universal: they are found pri-

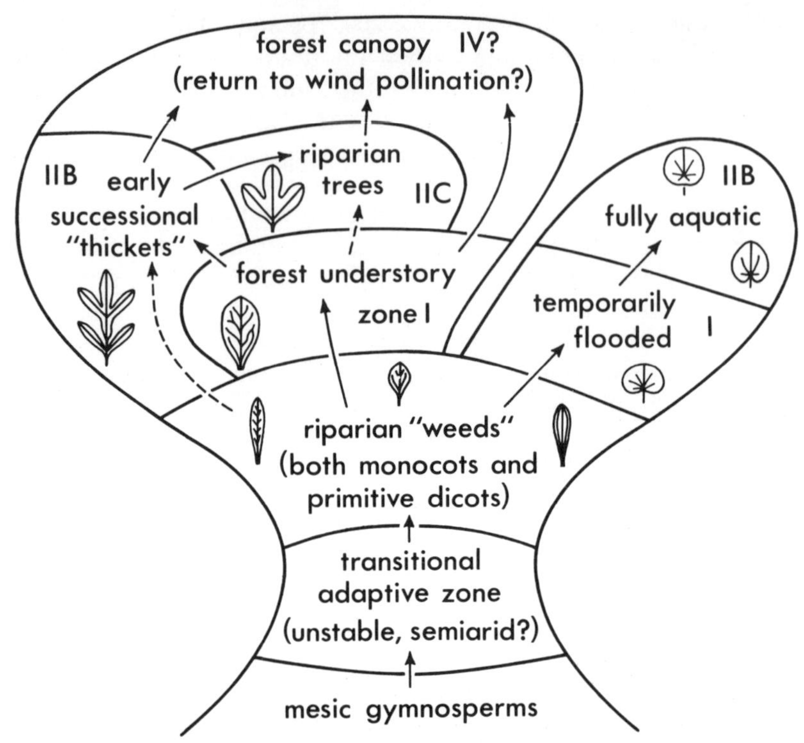

Fig. 29. Proposed model for early ecological-adaptive evolution of the angiosperms. It is postulated that angiosperms arose from mesic gymnospermous ancestors (seed ferns?) under pressures for efficient reproduction in an unstable, semiarid, transitional adaptive zone (see Stebbins, 1965, and in this volume), and that their primary adaptive radiation occurred when they reinvaded mesic habitats after perfection of the angiospermous reproductive system and xeromorphic-related vegetative features. The course of ecological evolution from "riparian weeds" onward is partially based on functional interpretation of observed trends and sedimentological associations of Potomac Group leaves and pollen. Note the early divergence into "forest" and "aquatic" trends. Solid arrows indicate inferred derivations; dashed arrows, alternatives considered less probable. Roman numerals (I, IIB, IIC, and IV) refer to Atlantic Coastal Plain pollen zones by which particular adaptive strategies are inferred to have been achieved.

marily in semixeric groups, where there are conflicting selective pressures for large seeds (required for effective seedling establishment in dry soils), efficient dispersal (ideally by animals, which can seek out favorable habitats), and protection of the large seeds during their longer maturation period.

Many aspects of our model for the early ecologic-adaptive evolution of the angiosperms (fig. 29) are consistent with Stebbins' disturbed semiarid transitional zone. We suggest that, by the Aptian, under the selective pressures of such an environment the angiosperms had already acquired the reproductive features which unite them, and that the Potomac record corresponds to the first stages of their radiation into a new adaptive zone characterized by unprecedented reproductive efficiency and intense coevolution with the animal world.

The pollen record suggests that two of the most important reproductive adaptations had already evolved. Reticulate exine structure, seen in lower zone I *Retimonocolpites, Liliacidites*, and even *Clavatipollenites*, may be taken as evidence of insect pollination (Faegri and Van der Pijl, 1966). Presumably sculpture allows pollen grains to stick to one another and to pollinating insects. We would interpret the main aperture trend from monosulcate to tricolpate and tricolporate, which begins almost immediately above the base of the Potomac sequence, in terms of progressively finer adjustment to the selective pressures involved in germination on a stigma (Hughes, 1961; Doyle, 1970; Walker, 1971; Smart and Hughes, 1972; Sporne, 1972). Whereas there is no advantage for more than one pollen aperture in the moist pollen chamber or micropyle of a gymnosperm, grains with more than one spot for germination should have a definite advantage on the exposed surface of a stigma, where much of the time a monosulcate grain would land with its single aperture facing out into the air. The subsequent origin of the tricolporate condition may be interpreted as a postadaptational division of labor between colpi, which allow the grain to compensate volumetrically for changes in humidity, and ora, which are specialized for germination (Wodehouse, 1935; Henrickson, 1967). Muller (1969) points out that the typically angiospermous differentiation of the exine into nexine, tectum, and columellae, seen already in Barremian *Clavatipollenites*, is far better adapted to absorb shear stresses involved in expansion and contraction accompanying size changes than is a single homogeneous layer. It is tempting to suggest that the elaboration of the exine in angiosperms is also a response to the greater stresses placed on it during germination on a stigma.

Special physical-environmental conditions, such as partial xerophytism, may have provided impetus for the origin of angiosperm reproductive features and possibly some vegetative ones (see discussion below, on the origin of the dicot leaf). Once these features were perfected, however, their unprecedented reproductive efficiency, vegetative flexibility (Takhtajan, 1969), and symbiotic ties with the animal world would have preadapted the angiosperms to a variety of habitats in other climatic regimes. Although associated coevolutionary diversification with respect to pollination ecology and seed dispersal must have continued (reflected in the diversification in pollen size, shape, and sculpture?), we suggest that much of the radiation was into a variety of adaptive subzones characterized by diverse vegetative specializations. This hypothesis would help to explain why, although the patterns of Potomac leaf and pollen evolution are both consistent with the concept of a Cretaceous adaptive radiation and with the systematic distribution of leaf and pollen characters in the modern flora, upper Potomac leaves cover, at least in their primary features, a larger portion of the total morphological spectrum seen in modern leaves than does the pollen (Vakhrameev, 1973). It would also explain the fact that monosulcate pollen and putatively primitive floral features are found today among groups with such divergent habit as woody magnolialian dicots, aquatic nymphaealian dicots, and monocots.

In mesic flood-plain environments such as the site of Potomac deposition, the postulated semixeromorphic, weedy early angiosperms would be most likely to colonize disturbed habitats, the most extreme of which are stream margins subject to periodic fluctuations in water table, flooding, and rapid erosion and sedimentation, rather than immediately invading the stable conifer- and fern-dominated forest vegetation (hence riparian "weeds," in fig. 29). This explanation is consistent with the sporadic distribution of zone I angiosperm pollen species, and with the fact that zone I leaves are largely restricted to coarser-grained sediment (cross-bedded sand or sandy clay and silt), which is interpreted as stream-levee, point-bar, and channel deposits—just where one might expect stream-margin plants to be preserved. These leaves are rare or absent in finer-grained deposits from gymnosperm- and fern-dominated flood basins and back swamps.

Under the more traditional postulate of an ancestral forest-tree habit for the angiosperms, it is difficult to explain in selectionist terms the observed early appearance of monocots and herbaceous dicots (see Stebbins, 1965). However, weedy riparian shrubs would already be under selective pressure for rhizomatous habit and, at the fluctuating aquatic margins of their habitat, for such adaptations as cordate or peltate leaves capable of swiveling at the top of the petiole, aerenchyma, and stomata on both leaf surfaces. Although a similar leaf shape is found in mesic herbs and lianas, the lobate-cordate-peltates of subzone II-B have "herbaceous" cuticle features (Mersky, unpublished data) and leaf architectural features (primaries dichotomizing and looping well within the margin, funnel-shaped blade bases, long fleshy petioles, etc.) more suggestive of modern aquatic groups such as Nymphaeales than lianous groups such as Menispermaceae, and their lithological associations suggest pond habitats ("fully aquatic," in fig. 29). The older (zone I) reniform and reniform-lobate members of the complex might well still occupy the transitional, periodically flooded niche indicated in figure 29.

With their postulated weedy habit, primitive angiosperms might most readily begin their invasion of forest environments as understory shrubs rather than as canopy trees. We would interpret the large, simple, low-rank leaves of the zone I *Ficophyllum* complex as the first forms with specific forest understory adaptations (fig. 29). Horn (1971) has argued that "monolayer" light-gathering strategies, often involving large, undissected leaves arranged in a single plane, are favored in the understory, where it is advantageous for each leaf to intercept as much as possible of the available low-intensity light. Only very rarely do plants with leaves below mid-second rank enter the canopy; in addition, those with large, simple, low-rank leaves (e.g., Winteraceae) are mostly shurbs or small trees often growing in rainforest understory conditions (Takhtajan, 1969; Carlquist, personal communication, 1973). Understory conditions are consistent with the scarcity of *Ficophyllum* even where it is most abundant, in the rich gymnosperm-fern assemblage at Fredericksburg.

According to Horn's model, evolution from understory to canopy conditions would be accompanied by selective pressure for a shift from a monolayer to a multilayer light-gathering strategy. Since full sunlight is several times more intense than is necessary for optimum photosynthetic efficiency, it is advantageous for a canopy plant to filter a large proportion of incident light down through several partially shaded, lower leaf layers. Since the optimum distance of placement for a lower layer is a function of the minimum rather than the maximum diameter of the leaf, Horn cites lobation and dissection as one of the principal strategies for partial shading. We would suggest that a shift from understory to canopy is represented by the pinnatifid to pinnately compound (*Sapindopsis*) and palmately lobed (platanoid) complexes of middle and upper subzone II-B. It may be significant that it is in these complexes that the major trend for increase in regularity of successive vein orders is most pronounced. We suggest that this ranking trend is more closely related to the support function of venation than to its conductive function (Hickey and Doyle, 1972), since secondary reversions in rank appear to be most common among xeric and arctic-alpine leaves today (Hickey, 1971), which must be under at least as much water stress as mesic leaves but under less mechanical stress by virtue of their smaller size and coriaceous texture. This suggestion should be weighed, however, against the fact that many features of third- and fourth-rank leaves show remarkable resemblances to optimal transport networks (Sen, 1971). In particular, the observed tendencies for strong and regular brochidodromous looping of the secondaries near the margin (rather than irregular ramification and anastomosis of their branches), percurrent tertiaries, and venational bracing of the sinuses between teeth and lobes might be interpreted as adaptations against ripping in canopy conditions. The fact that *Sapindopsis* is the first angiosperm leaf type to dominate the flora completely at any locality (Brooke) is consistent with the idea that it formed dense stands. If the lobate *Sapindopsis* and platanoid complexes evolved from understory shrubs, they are more likely to have first formed thickets in locally disturbed situations ("early successional thickets," in fig. 29) than to have directly replaced the conifers. Although we have no objective evidence on the stature of *Sapindopsis,* the early successional hypothesis is supported by the fact that the dense mat of *Sapindopsis* leaves at Brooke occurs at an abrupt lithologic contact between a fluvial sand and a floodplain siltstone unit, and that the pinnately compound leaves at the West Brothers pit occur in a silty unit with intraformational silt clasts just above a charcoal-rich (forest fire?) bed.

The palmately lobed platanoids that are so abundant in subzone II-C and latest Albian rocks of other areas are the first forms with a leaf morphology similar to any familiar modern trees. Their striking association with either channel-sand or stream-levee deposits suggests, however, that even they still occupied unstable, specifically stream-margin, environments much like those of modern *Platanus* ("riparian trees," in fig. 29). Laterally equivalent clayey beds are dominated by different angiosperms, or even ferns and gymnosperms (Rushforth, 1971). We suggest that definitive replacement of the conifers by angiosperms as forest-canopy dominants, made possible by the accumulation of sufficient vegetative advantages (e.g., vessels, leaf morphology) to outrival the conifers, did not occur until well into the Upper Cretaceous, possibly in zone IV. This reconstruction is supported by the subordinate relationship of angiosperms to conifers in Upper Albian (Norris, 1967) and even Cenomanian pollen floras, which in fact led Pierce (1961) to suggest that during deposition of the "Dakota" Formation of Minnesota (mid-Cenomanian?), conifers were still climax forest dominants, whereas dicots largely occupied disturbed habitats. A return to wind pollination would become advantageous only as angiosperms began to form dense (ideally deciduous) forest stands (Whitehead, 1969). Since smooth and porate grains are characteristic of wind-pollinated angiosperm groups today, we interpret the appearance of triporate Normapolles in zone IV as marking the sacrifice by angiospermous trees of one of their now superfluous reproductive adaptations.

The fact that no primitive angiosperms now occupy the semixeric or stream-margin habitats of the postulated transitional adaptive zone is actually predicted by general theory (Bock, 1965); they would be outrivaled by forms produced in the subsequent radiation which have not only the features that arose in the transition zone but also superior postadaptations (see Stebbins, in this volume). Probably the oldest adaptive zones which have been occupied continuously by lines differentiated in the primary radiation are the forest understory (still occupied by the Winteraceae and other Magnoliales), temporarily flooded areas (*Barclaya* and some other Nymphaeales, and monocots), and early successional niches (Saxifragales, Rosales, some Sapindales and Rutales).

A variety of approaches might be taken to test a model such as we have presented. Further elucidation of the environmental significance of the sedimentary associations of early angiosperms and the functional significance of leaf and pollen characters by modern ecological analogy and physical-mathematical analysis might test our hypotheses on the adaptive significance of observed trends. Studies of the first sedimentary environments to yield angiosperms and the patterns of early migration of angiosperms might help identify the transitional adaptive zone. Especially consistent with a model of evolution in semiarid conditions followed by radiation into other environments is the fact that tricolpate pollen, though representing only one group of angiosperms, first appears in the (Barremian-) Aptian of South America, Africa, and Israel in association with physical evidence of aridity (Jardiné and Magloire, 1965; Müller, 1966; Brenner, in this volume), but remains relatively undiverse there until the same time that tricolpates appear and begin to differentiate rapidly in Laurasia (Lower and Middle Albian). Finally, an analysis of the interrelationships of functional and ontogenetic factors in modern leaves might allow certain features of the early angiosperm fossils to be recognized as "holdovers" that arose as adaptations to conditions in the transitional adaptive zone. In the next section, as an example of such an analysis, we argue that the course of ecological evolution which we have postulated explains many aspects of the earliest angiosperm leaves and their subsequent evolution and, in fact, provides a novel explanation for the origin of the dicot leaf.

A Model for Dicot Leaf Evolution

The reticulate venation of the typical dicot leaf, already established in earliest Potomac times, is characterized by a hierarchy of progressively finer vein orders and by extensive anastomosis within and between vein orders (except usually for the finest order, of freely ending veinlets). Because the dicot leaf is such a familiar structure, it is easy to overlook the fact that this pattern is practically unique among vascular plants, the closest analogs being seen in the aberrant gymnospermous genus *Gnetum* (Gnetales) and a few ferns (especially Dipteridaceae, Cheiropleuriaceae, and some other members of the "polypodioid" complex). It contrasts even with the situation in most of those ferns and gymnosperms that are often characterized as having reticulate venation (e.g., *Glossopteris*, *Sagenopteris*), where the reticulum is made up of elongate areoles delimited by veins of essentially one order.

The marked difference between the venation of the dicot leaf and the leaves of ferns and gymnosperms was recognized by Němejc (1956, pp. 73–74), who interpreted it as evidence against derivation of the dicot leaf from a compound, fernlike leaf with dichotomous pinnule venation by a simple process of webbing. Němejc suggested that a more radical reorganization of leaf morphology was required, namely arresting of leaf ontogeny at an early stage (neoteny), perhaps associated with xeromorphic reduction of a compound frond, and subsequent evolution in new directions.

Our model, summarized schematically in figure 30, resembles that of Němejc in postulating a stage of reduction associated with xerophytism, followed by a stage of *secondary reexpansion,* but it also incorporates certain correlations between ontogenetic processes and mature leaf architecture, many of which were first pointed out to us by W. H. Wagner, Jr. Taken together, the great differences in the pattern of leaf development in angiosperms and other vascular plants and the disorganized venation of zone I leaves suggest some kind of fundamental reorganization of ontogenetic patterns and mature form not long before the angiosperms entered the record, and we shall try to demonstrate that the secondary-expansion model provides a specific and parsimonious explanation for both sets of phenomena. Many aspects of this model are subject to testing and refinement by comparative studies on the relationships between developmental processes and venation in modern plants.

Ontogenetic studies (Pray, 1960, 1962) indicate that two phases of meristematic activity are involved in the growth of fern leaves with open dichotomous venation (e.g., *Nephrolepis* and *Regnellidium,* the latter complicated by presence of a marginal vein). The frond axis (petiole, rachis) is produced by activity of an apical meristem, accompanied by intercalary growth and elongation. The pinnules are produced by rows of marginal meristematic cells which arise on either side of the embryonic rachis. The entire dichotomous venation pattern is initiated during marginal growth, with the veins differentiat-

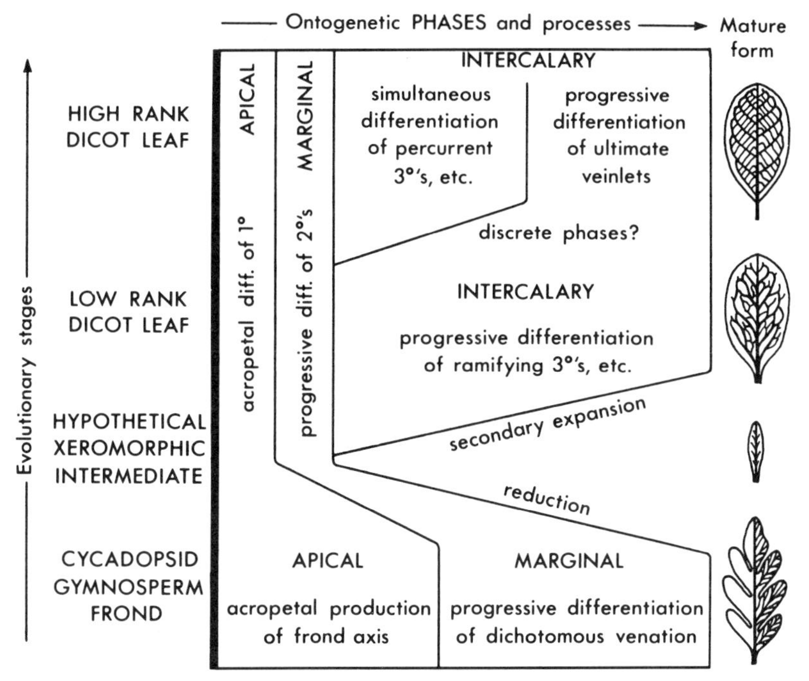

Fig. 30. Model for the origin of the dicot leaf and the subsequent trend for increased regularity of successively higher vein orders ("ranking") in terms of ontogenetic and phylogenetic processes. Four postulated evolutionary stages are arranged vertically. Successive phases of meristematic activity, their relative duration, and the process of vein differentiation known or assumed to be operative in each case are indicated from left to right. Transformation of a pinnately compound, cycadopsid gymnosperm (seed-fern?) leaf, with dichotomous pinnule venation and with a growth pattern assumed to be analogous to that seen in ferns, into a dicot leaf, with several orders of venation and relatively brief phases of apical and marginal meristematic activity is postulated to have occurred as a result of evolutionary reduction associated with xerophytism, followed by a secondary reexpansion of leaf area mediated by a new developmental process, intercalary growth. The trend observed in the Potomac sequence for increase in "rank" (especially for percurrency of tertiary and higher vein orders) is interpreted as having resulted from the gradual addition of developmental processes, specifically "simultaneous" rather than "progressive" differentiation of tertiary and higher vein orders, and perhaps discrete phases of vein differentiation.

ing acropetally behind the outward-growing marginal meristem from cell lineages established by differing division behavior of marginal cells.

In contrast, typical simple, pinnately veined dicot leaves go through three overlapping phases of meristematic activity, each associated with a different pattern of vein differentiation (Esau, 1953;

Pray, 1955, 1963; Slade, 1957). In the *apical* phase, apical meristematic activity produces a peg-shaped leaf primordium, in which procambium destined to become the future midrib differentiates acropetally. When the primordium is still usually less than a millimeter long, apical growth is followed by a relatively brief *marginal* phase, during which two marginal meristems produce an embryonic lamina. Unlike the situation in ferns, only the pinnately arranged secondary veins are initiated during this phase. Like the dichotomizing veins of ferns, the secondaries differentiate progressively (acropetally) outward from the midrib, following the direction of marginal growth; however, no close relation between veins and marginally produced cell lineages has been reported. Most of the leaf area and all the finer venation are produced during the following *intercalary* phase as a result of the activity of a plate meristem, representing meristematic activity spread over the whole blade. Tertiary and higher vein orders are differentiated successively in the meristematically expanding areas of leaf tissue between the secondary veins.

Pray (1960, 1962) and Slade (1957) suggest that the difference between open dichotomous and reticulate venation is causally related to the relative predominance of marginal versus intercalary growth. However, according to Hara (1964), the simple reticulate venation of the fern *Onoclea* (a basically dichotomous system with veins connected by anastomoses of the same order) is also initiated during strictly marginal growth. Hence, we suggest that the most basic morphological result of the predominance of intercalary growth in dicots is not the presence of anastomoses but rather the existence of a hierarchy of successively finer vein orders, each differentiating in the expanding areas between already differentiated veins.

Pray (1955, 1963) and Slade (1957) report that unlike the secondary veins, tertiary and higher-order veins (except for the ultimate veinlets) usually appear to differentiate "simultaneously" rather than "progressively" across areas blocked out by existing veins; that is, all the cells making up a row destined to become a future vein differentiate simultaneously into procambium rather than differentiating in a particular sequence. However, Pray (1955, 1963) seems to have demonstrated that the highest order of freely ending veinlets does differentiate progressively (rather than by breakage as suggested by Slade [1957]). It is interesting in connection with the trend observed in the fossil record that all the leaves reported to show simultaneous differentiation of tertiary and higher-order venation are third rank, with regularly percurrent tertiaries. In contrast, *Aucuba* (Cornaceae), the only second-rank dicot (with an irregularly ramifying intercostal network) whose venation has been investigated ontogenetically

(Pray, 1963), shows progressive differentiation of tertiary and higher-order veins. Hence, we suggest that the observed trend from ramifying to percurrent tertiary and higher-order venation reflects a shift from progressive differentiation of all vein orders to simultaneous differentiation of tertiary and finer veins (except for ultimate veinlets). With less basis, except the gross correlation between order and time of differentiation, we also suggest that the poor definition of vein orders in low-rank leaves (as in those of the lower Potomac sequence and modern Winteraceae) reflects continuous differentiation of finer and finer veins, whereas in high-rank leaves with well-defined vein orders, the veins differentiate in discrete phases.

Our model for the origin of the dicot leaf and the subsequent ranking trend (fig. 30) begins with a compound leaf (in view of the other evidence for a cycadopsid rather than a coniferopsid gymnosperm ancestor [see Bailey, 1949; Cronquist, 1968; Takhtajan, 1969; Bierhorst, 1971]), presumably with a pattern of growth analogous to that seen in ferns. Next we suggest that such a leaf underwent a process of reduction, as an adaptation to the semiarid transition zone we have already postulated for other reasons, in which it lost most of its capacity for apical and marginal growth. The result was a small, simple leaf with a highly reduced blade and only vestiges of the previous laminar venation. The reduced nature of this prototype would place restrictions on the subsequent course of evolution if the plants were subjected to selective pressures for secondary expansion of leaf area, as would be expected during a secondary invasion of mesic habitats such as we have postulated in Potomac times. With loss of the potential for producing a pinnately compound structure, the leaf might well be forced to increase its area by a different developmental process, such as intercalary growth. In addition, under our assumptions on the relationship between ontogeny and venation, selection would favor intercalary rather than marginal reexpansion because of its effect of producing a more efficient fluid distribution network: expansion of a simple leaf by marginal growth might be predicted to produce a *Taeniopteris*-type leaf, with a midrib and only one order of fine, relatively inefficient secondary veins to conduct fluids between the midrib and the blade, whereas expansion by intercalary growth would produce a leaf with large secondary "trunk bundles" permitting rapid diffusion to and from the margin, and a hierarchical system of finer veins supplying the intercostal areas between them.

This secondary-expansion hypothesis provides an explanation for the leaf ranking trend. In the initial stages of phylogenetic reexpansion of the blade, one might expect that control of vein differentiation would be rudimentary and that tertiary and higher-order veins, like the secondaries, would differentiate and ramify progressively into the new areas produced by plate meristematic activity. The result would be an irregularly ramifying and anastomosing system of poorly defined vein orders, just as in lower Potomac leaves. Gradually, under selective pressures for structural strength, mutations would accumulate for developmental mechanisms producing a regular, rigid array of structural elements, particularly for simultaneous differentiation of tertiary and higher-order veins (resulting in percurrency), and perhaps discrete phasing of vein differentiation. This process would result in a gradual stratigraphic trend for increase in leaf rank, as is seen in the Potomac sequence.

An alternative hypothesis of a gradual shift from apical and marginal to intercalary growth, without an intermediate reduced stage, would less readily explain the disorganized venation of early angiosperms. Some remnant of the typically highly organized, fernlike venation would be expected to survive during such a gradual process. Possible examples are the Permian (seed-fern?) genus *Gigantopteris*, with several orders of rigidly organized but nonangiospermous reticulate venation, or the fern family Dipteridaceae, with rigidly organized, angiospermlike higher-order venation but dichotomously organized compound to bilobate leaves.

Various indirect evidence is consistent with the idea that the dicot leaf represents the secondarily expanded blade of a simple-leafed, xeromorphic ancestor (a phyllode). At least two analogies from unrelated groups suggest that reduction and secondary expansion might result in leaf architectural features like those seen in early angiosperms. One is the gymnospermous genus *Gnetum*, with remarkably dicotlike, low-rank, simple leaves. Since the evidence from anatomy, embryology, and reproductive morphology indicates that *Gnetum* is related to, but more advanced than, the xeromorphic genera *Ephedra* and *Welwitschia*, and that all three genera share a coniferopsid (hence simple-leafed) rather than cycadopsid ancestry (see Bierhorst, 1971), the dicotlike leaves of *Gnetum* are best interpreted in terms of secondary expansion of a simple needle- or scalelike leaf associated with invasion of rain-forest habitats. In addition, Wagner (1964) has suggested that the simple, spatulate leaves of *Ophioglossum*, with several orders of irregular, somewhat dicotlike reticulate venation, arose by a similar process of reduction and secondary expansion from fernlike, compound, dichotomously veined leaves of the type found in the related genus *Botrychium*. In fact, an apparent reduction series to a simple, almost linear leaf can be seen within *Botrychium*. The small size of many Albian leaves (also attributed by

Vakhrameev [1952] to a xeromorphic ancestry), the poor differentiation into petiole and blade in lower Potomac angiosperms, and the spatulate shape of such forms as *Rogersia* are also suggestive.

Finally, a leaf reduction trend can be inferred among Mesozoic seed ferns, from bi- or tripinnate (Peltaspermaceae and some Corystospermaceae), to simply pinnate (other Corystospermaceae) and palmately compound (*Sagenopteris*, Caytoniaceae), perhaps culminating with the simple *Glossopteris*-like leaves described by Delevoryas and his colleague (Delevoryas, 1969; Delevoryas and Gould, 1971) from the Jurassic of Mexico. An analogous but earlier trend may have been involved in the origin of the simple *Taeniopteris*-type leaves of the earliest Permian cycads (Mamay, 1969), perhaps derived from a compound-leafed neuropterid (medullosan seed fern). These examples may have a more direct bearing on the plausibility of our model for origin of the dicot leaf, since they are from groups which, on other grounds, have the most reasonable potential as angiosperm ancestors.

The possibility that the origin of angiosperms involved xeromorphic forms with reduced leaves helps explain the nonrecognition of angiosperm precursors (Stebbins, 1965) and suggests new directions in the search for angiosperm ancestors. We further suggest that other morphological features which set the angiosperms apart from most of the gymnosperms might also be reinterpreted as holdovers from a xeromorphic transitional phase. For example, the scalariform pitting of tracheids in modern vesselless angiosperms (rare in the secondary xylem of gymnosperms except Bennettitales and a few reduced Cycadales [Takhtajan, 1969]) might be explained as a paedomorphic prolongation of the metaxylem pitting pattern into the secondary xylem (see Takhtajan, 1969, and in this volume), as a result of the reduction of secondary growth in the semiarid transition zone followed by its reaugmentation during reinvasion of mesic habitats (cf. cases of paedomorphosis in secondarily woody forms cited by Carlquist [1962]). The elaboration of protective buds and axillary branching, both features unusual among cycadopsid gymnosperms, might also be originally xeromorphic adaptations which, of course, allowed evolution in new directions after reinvasion of mesic habitats. Finally, the possibility should not be overlooked that many of the condensations and paedomorphic features in floral structure cited above represent pleiotropic transfers from the vegetative to the reproductive sphere of new ontogenetic mechanisms first selected as xerophytic adaptations and subsequently for their role in protection and reproductive efficiency. Such transfer of the process of intercalary expansion might be invoked in explaining the origin of the angiosperm carpel from the rachis of a corystospermaceous or caytoniaceous megasporophyll with pinnately borne reflexed cupules (= anatropous, bitegmic ovules, according to Gaussen [1946] and Stebbins [1970b, and in this volume]).

Conclusion

The application of an integrated morphological and stratigraphic methodology demonstrates the consistency of the Potomac pollen and leaf records and allows both to be interpreted in terms of a Cretaceous primary adaptive radiation of the angiosperms. Of course, it would be unwarranted to consider lower Potomac forms to be the first angiosperms, or eastern North America to be their long-sought "homeland"; some prior evolution must have occurred to produce the observed number of zone I pollen and leaf types, and, in fact, monosulcate angiosperm pollen exists at least a stage earlier (Barremian of England). However, considering the limited range of morphological differentiation in both the pollen and leaf records, there is no need to postulate any extensive prior diversification (e.g., as suggested by Axelrod, 1952, 1960, 1970), and it is possible that the origin of angiosperms was much closer to their first appearance than has been thought (see also Hughes, 1961).

More important than their bearing on the rather sterile controversy on the age of the angiosperms, our observations indicate that the fossil record is now of major significance as evidence for the solution of Darwin's "abominable mystery" of their origin. First, the Cretaceous record drastically reduces the number of theories of angiosperm phylogeny (and derivative taxonomic systems) that can be seriously entertained (Doyle, 1969a, 1973; Muller, 1970; Wolfe, Doyle, and Page, 1975). By demonstrating the direction of evolution in pollen and leaf characters, the record rules out theories which consider the "Amentiferae" ("higher Hamamelididae," generally with advanced triporate pollen and third- or fourth-rank leaves) to be relatively primitive, while favoring theories which assign this position to the Magnoliales (some of which retain primitive monosulcate pollen and first-rank leaves). On the other hand, the occurrence from the beginning of the Potomac record of monocotyledonoid and herbaceous dicotyledonoid forms cautions against assuming that the first angiosperms had exclusively "woody magnoliid" vegetative characters (see Doyle, 1973). Finally, instead of being fully modernized and lacking clues to their mode of origin, early angiosperm leaves exhibit sedimentary associations and striking morphological features which, when analyzed in terms of modern ecological and ontogenetic analogies, provide novel evidence on the origin and ancestors of the group.

As directions for future research on the evolutionary significance of the Cretaceous angiosperm record, and as tests of the hypotheses and generalizations we have presented, we urge intensive studies of the following:

1. Critical parts of the Potomac Group (e.g., zone I and lower zone II), and of correlative (Aptian-Albian) and immediately older (Neocomian) sequences in other geographic areas (e.g., African–South American tropics, eastern Asia, the Wealden of England, Portugal, and the U.S.S.R.), to document origins and evolutionary transformations of leaf and pollen types, and directions of geographic and ecological diffusion.
2. Additional characters of fossil pollen and leaves (e.g., using scanning electron microscopy and cuticle analysis), and of other organs (wood, fruits, and seeds).
3. Sedimentary associations, taphonomy (preservation), and paleoecology of early angiosperms.
4. Comparative functional morphology, ecology, development, and taphonomy of modern pollen and leaves, of both angiosperms and other vascular plants.

Acknowledgments

We thank W. H. Wagner, Jr., for discussion of the ontogenetic aspects of leaf morphology, Andrew Stephenson and Judy Rankin for discussion of the adaptive and ecological aspects of the early angiosperm record, and Marcie Mersky for critical comments and the use of unpublished cuticle data. J. A. Doyle's part of this research was begun as a Smithsonian Postdoctoral Research Associate, subsequently supported by University of Michigan Rackham grant no. 360557, and completed as an associate of the Laboratoire de Palynologie, C.N.R.S., Montpellier, France. L. J. Hickey's studies were supported, in part, by Smithsonian Research Foundation grants no. 427215 and no. 430019.

References

Arkhangel'skii, D. B. 1966. Zvezdchataya skul'ptura ekziny pyl'tsevykh zeren [Stellate sculpture of the exine of pollen grains]. In *Znacheniye palinologicheskogo analiza dlya stratigrafii i paleofloristiki* [The significance of palynological analysis for stratigraphy and paleofloristics], ed. M. I. Neyshadt, pp. 22–26. Nauka, Moscow.

Axelrod, D. I. 1952. A theory of angiosperm evolution. *Evolution* 6: 29–60.
Axelrod, D. I. 1960. The evolution of flowering plants. In *The Evolution of Life*, ed. S. Tax, pp. 227–305. Univ. of Chicago Press, Chicago.
Axelrod, D. I. 1970. Mesozoic paleogeography and early angiosperm history. *Bot. Rev.* 36: 277–319.
Bailey, I. W. 1949. Origin of the angiosperms: Need for a broadened outlook. *J. Arnold Arbor.* 30: 64–70.
Banks, H. P. 1968. The early history of land plants. In *Evolution and Environment*, ed. E. T. Drake, pp. 73–107. Yale Univ. Press, New Haven.
Banks, H. P. 1972. The stratigraphic occurrence of early land plants. *Palaeontology* 15: 365–77.
Bayne, C. K., Franks, P. C., and Ives, W., Jr. 1971. Geology and groundwater resources of Ellsworth County, Central Kansas. *Bull. Kansas Geol. Surv.* 201: 1–84.
Bell, W. A. 1956. *Lower Cretaceous Floras of Western Canada.* Canada Geological Survey Memoir no. 285. Ottawa.
Berry, E. W. 1911a. Systematic paleontology, Lower Cretaceous: Fossil plants. In *Lower Cretaceous,* ed. W. B. Clark, pp. 214–508. Maryland Geological Survey, Baltimore.
Berry, E. W. 1911b. *The Flora of the Raritan Formation. Bull. Geol. Surv. New Jersey* 3. Trenton.
Berry, E. W. 1916. Systematic paleontology, Upper Cretaceous: Fossil plants. In *Upper Cretaceous,* ed. W. B. Clark, pp. 757–901. Maryland Geological Survey, Baltimore.
Berry, E. W. 1922. The flora of the Cheyenne Sandstone of Kansas. *U.S. Geol. Surv. Profess. Paper* 127-I: 199–225.
Bierhorst, D. W. 1971. *Morphology of Vascular Plants.* Macmillan, New York.
Bock, W. J. 1965. The role of adaptive mechanisms in the origin of higher levels of organization. *Syst. Zool.* 14: 272–87.
Brenner, G. J. 1963. *The Spores and Pollen of the Potomac Group of Maryland.* Bulletin of the Maryland Department of Geology, Mines and Water Resources 27. Baltimore.
Brenner, G. J. 1967. Early angiosperm pollen differentiation in the Albian to Cenomanian deposits of Delaware (U.S.A.). *Rev. Palaeobot. Palynol.* 1: 219–27.
Carlquist, S. 1962. A theory of paedomorphosis in dicotyledonous woods. *Phytomorphology* 12: 30–45.
Chaloner, W. G. 1970. The rise of the first land plants. *Biol. Rev. Cambridge Phil. Soc.* 45: 353–77.
Clark, W. B. 1897. Outline of present knowledge of the physical features of Maryland. In *Maryland Geological Survey*, vol. 1, pp. 139–228. Baltimore.
Clark, W. B., and Bibbins, A. B. 1897. The stratigraphy of the Potomac Group in Maryland. *J. Geol.* 5: 479–506.
Corner, E. J. H. 1949. The Durian theory or the origin of the modern tree. *Ann. Bot. (London), n.s.* 13: 367–414.

Corner, E. J. H. 1964. *The Life of Plants*. World Pub., New York.
Couper, R. A. 1953. Upper Mesozoic and Cainozoic spores and pollen grains from New Zealand. *Bull. New Zealand Geol. Surv. Paleontol.* 22: 1–77.
Couper, R. A. 1958. British Mesozoic microspores and pollen grains. *Palaeontographica* 103B: 75–179.
Couper, R. A. 1964. Spore-pollen correlation of the Cretaceous rocks of the Northern and Southern Hemispheres. In *Palynology in Oil Exploration*, ed. A. T. Cross, pp. 131–42. Society of Economic Paleontologists and Mineralogists Special Publication 11. Tulsa, Oklahoma.
Cronquist, A. 1968. *The Evolution and Classification of Flowering Plants*. Houghton, Boston.
Delevoryas, T. 1969. Glossopterid leaves from the Middle Jurassic of Oaxaca, Mexico. *Science* 165: 895–96.
Delevoryas, T., and Gould, R. E. 1971. An unusual fossil fructification from the Jurassic of Oaxaca, Mexico. *Amer. J. Bot.* 58: 616–20.
Doyle, J. A. 1969a. Cretaceous angiosperm pollen of the Atlantic Coastal Plain and its evolutionary significance. *J. Arnold Arbor.* 50: 1–35.
Doyle, J. A. 1969b. Angiosperm pollen evolution and biostratigraphy of the basal Cretaceous formations of Maryland, Delaware, and New Jersey. *Geol. Soc. America Abstracts with Programs for 1969* pt 7: 51. (Abstr.)
Doyle, J. A. 1970. Evolutionary and stratigraphic studies on Cretaceous angiosperm pollen. Ph.D. dissertation, Harvard University.
Doyle, J. A. 1973. Fossil evidence on early evolution of the monocotyledons. *Quart. Rev. Biol.* 48: 399–413.
Doyle, J. A., and Hickey, L. J. 1972. Coordinated evolution in Potomac Group angiosperm pollen and leaves. *Amer. J. Bot.* 59: 660. (Abstr.)
Erdtman, G. 1952. *Pollen Morphology and Plant Taxonomy. Part I. Angiosperms.* Chronica Botanica, Waltham, Mass.
Esau, K. 1953. *Plant Anatomy*. Wiley, New York.
Faegri, K., and Van der Pijl, L. 1966. *The Principles of Pollination Ecology*. Pergamon, Oxford.
Fontaine, W. M. 1889. *The Potomac or Younger Mesozoic Flora*. U.S. Geological Survey Monograph 15. Washington, D.C.
Foster, A. S., and Gifford, E. M. 1959. *Comparative Morphology of Vascular Plants*. Freeman, San Francisco.
Gaussen, H. 1946. *Les gymnospermes actuelles et fossiles*. Pt. III. Travaux du Laboratoire Forestier, Toulouse.
Glaser, J. D. 1969. *Petrology and Origin of Potomac and Magothy (Cretaceous) Sediments, Middle Atlantic Coastal Plain. Maryland Geol. Surv. Rept. Investigations* 11. Baltimore.
Groot, J. J., and Penny, J. S. 1960. Plant microfossils and age of nonmarine Cretaceous sediments of Maryland and Delaware. *Micropaleontology* 6: 225–36.
Groot, J. J., Penny, J. S., and Groot, C. R. 1961. Plant microfossils and age of the Raritan, Tuscaloosa, and Magothy Formations of the eastern United States. *Palaeontographica* 108B: 121–40.

Habib, D. 1970. Middle Cretaceous palynomorph assemblages from clays near the Horizon Beta deep-sea outcrop. *Micropaleontology* 16: 345–79.

Hansen, H. J. 1969. Depositional environments of subsurface Potomac Group in southern Maryland. *Bull. Amer. Assoc. Petrol. Geol.* 53: 1923–37.

Hara, N. 1964. Ontogeny of the reticulate venation in the pinna of *Onoclea sensibilis*. *Bot. Mag. (Tokyo)* 77: 381–87.

Hedlund, R. W. 1966. *Palynology of the Red Branch Member of the Woodbine Formation (Cenomanian), Bryan County, Oklahoma. Bull. Oklahoma Geol. Surv.* 112. Norman.

Hedlund, R. W., and Norris, G. 1968. Spores and pollen grains from Fredericksburgian (Albian) strata, Marshall County, Oklahoma. *Pollen Spores* 10: 129–59.

Hennig, W. 1966. *Phylogenetic Systematics*. Univ. of Illinois Press, Urbana.

Henrickson, J. 1967. Pollen morphology of the Fouquieriaceae. *Aliso* 6: 137–60.

Hickey, L. J. 1971. Evolutionary significance of leaf architectural features in the woody dicots. *Amer. J. Bot.* 58: 469. (Abstr.)

Hickey, L. J. 1973. Classification of the architecture of dicotyledonous leaves. *Amer. J. Bot.* 60: 17–33.

Hickey, L. J., and Doyle, J. A. 1972. Fossil evidence on evolution of angiosperm leaf venation. *Amer. J. Bot.* 59: 661. (Abstr.)

Hickey, L. J., and Wolfe, J. A. 1975. The basis of angiosperm phylogeny: Vegetative morphology. *Ann. Missouri Bot. Gard.*, in press.

Hollick, A. 1906. *The Cretaceous Flora of Southern New York and New England.* U.S. Geological Survey Monograph 50. Washington, D.C.

Horn, H. S. 1971. *The Adaptive Geometry of Trees*. Princeton Univ. Press, Princeton, N. J.

Hughes, N. F. 1961. Fossil evidence and angiosperm ancestry. *Sci. Progr.* 49: 84–102.

Hughes, N. F., and Moody-Stuart, J. C. 1969. A method of stratigraphic correlation using early Cretaceous spores. *Palaeontology* 12: 84–111.

Jardiné, S., and Magloire, L. 1965. Palynologie et stratigraphie du Crétacé des bassins du Sénégal et de Côte d'Ivoire. *Mém. Bur. Rech. Géol. Minières* 32: 187–245.

Jordan, R. R. 1962. Stratigraphy of the sedimentary rocks of Delaware. *Bull. Delaware Geol. Surv.* 9: 1–51.

Jordan, R. R. 1968. Observations on the distribution of sands within the Potomac Formation of northern Delaware. *Southeastern Geol.* 9: 77–85.

Kemp, E. M. 1968. Probable angiosperm pollen from British Barremian to Albian strata. *Palaeontology* 11: 421–34.

Kemp, E. M. 1970. Aptian and Albian miospores from southern England. *Palaeontographica* 131B: 73–143.

Krassilov, V. A. 1967. *Rannemelovaya flora Yuzhnogo Primorya i yeye znacheniye dlya stratigrafii* [Early Cretaceous flora of the southern Primorye and its significance for stratigraphy]. Nauka, Moscow.

Krutzsch, W. 1970a. *Atlas der mittel- und jungtertiären dispersen Sporen-*

und Pollen sowie der Mikroplanktonformen des nördlichen Mitteleuropas. Pt. VII. Gustav Fischer, Jena.

Krutzsch, W. 1970b. Zur Kenntnis fossiler disperser Tetraden pollen. *Paläontol. Abhandl.* 3B: 399–433.

Lesquereux, L. 1883. *Contributions to the Fossil Flora of the Western Territories. Part III. The Cretaceous and Tertiary Floras.* U.S. Geological Survey Territorial Report 8. Washington.

Lesquereux, L. 1892. *The Flora of the Dakota Group.* U.S. Geological Survey Monograph 17. Washington.

Maheshwari, P. 1950. *An Introduction to the Embryology of Angiosperms.* McGraw, New York.

Mamay, S. H. 1969. Cycads: Fossil evidence of Late Paleozoic origin. *Science* 164: 295–96.

Mayr, E. 1963. *Animal Species and Evolution.* Harvard Univ. Press, Cambridge.

McGee, W. J. 1888. Three formations of the Middle Atlantic Slope. *Amer. J. Sci., 3d ser.,* 35: 120–43.

Médus, J., and Pons, A. 1967. Étude palynologique du Crétacé Pyrénéo-Provençal. *Rev. Palaeobot. Palynol.* 2: 111–17.

Mersky, M. L. 1973. Lower Cretaceous (Potomac Group) angiosperm cuticles. *Amer. J. Bot.* 60: 17–18. (Abstr.)

Müller, H. 1966. Palynological investigations of Cretaceous sediments in northeastern Brazil. In *Proceedings of the 2nd West African Micropaleontological Colloquium* (Ibadan), ed. J. E. van Hinte, pp. 123–36. Brill, Leiden.

Muller, J. 1969. A palynological study of the genus *Sonneratia* (Sonneratiaceae). *Pollen Spores* 11: 223–98.

Muller, J. 1970. Palynological evidence on early differentiation of angiosperms. *Biol. Rev. Cambridge Phil. Soc.* 45: 417–50.

Němejc, F. 1956. On the problem of the origin and phylogenetic development of the angiosperms. *Sb. Národ. Musea Praze* 12B: 59–143.

Newberry, J. S. 1895. *The Flora of the Amboy Clays.* U.S. Geological Survey Monograph 26. Washington, D.C.

Norris, G. 1967. Spores and pollen from the Lower Colorado Group (Albian-?Cenomanian) of central Alberta. *Palaeontographica* 120B: 72–115.

Owens, J. P. 1969. Coastal Plain rocks. In *The Geology of Harford County, Maryland,* pp. 77–103. Maryland Geological Survey. Baltimore.

Owens, J. P., and Sohl, N. F. 1969. Shelf and deltaic paleoenvironments in the Cretaceous-Tertiary formations of the New Jersey Coastal Plain. In *Geology of Selected Areas in New Jersey and Eastern Pennsylvania and Guidebook of Excursions,* ed. S. Subitzky, pp. 235–78. Rutgers Univ. Press, New Brunswick, N. J.

Pacltová, B. 1961. Zur Frage der Gattung *Eucalyptus* in der böhmischen Kreideformation. *Preslia* 33: 113–29.

Pacltová, B. 1971. Palynological study of Angiospermae from the Peruc Formation (?Albian–Lower Cenomanian) of Bohemia. *Ústřední Ústav Geol., Sb. geol. Věd, Paleontol., řada P,* 13: 105–41.

Pacltová, B., and Mazancová, M. 1966. Nachweis von Pollen der "Normapolles"-Gruppe in den Peruc-Schichten (Perutzer Schichten) des böhmischen Cenomans. *Věstn. Ústředního Ústavu Geol.* 41: 51–54.

Phillips, P., and Felix, C. J. 1972. A study of Lower and Middle Cretaceous spores and pollen from the southeastern United States. II. Pollen. *Pollen Spores* 13: 447–73.

Pierce, R. L. 1961. Lower Upper Cretaceous plant microfossils from Minnesota. *Bull. Minnesota Geol. Surv.* 42: 1–86.

Pocock, S. A. J. 1962. Microfloral analysis and age determination of strata at the Jurassic-Cretaceous boundary in the western Canada Plains. *Palaeontographica* 111B: 1–95.

Pocock, S. A. J. 1970. Palynology of the Jurassic sediments of western Canada. Part II (continued): Terrestrial species. *Palaeontographica* 130B: 73–136.

Pray, T. R. 1955. Foliar venation of angiosperms. II. Histogenesis of the venation of *Liriodendron*. *Amer. J. Bot.* 42: 18–27.

Pray, T. R. 1960. Ontogeny of the open dichotomous venation in the pinna of the fern *Nephrolepis*. *Amer. J. Bot.* 47: 319–28.

Pray, T. R. 1962. Ontogeny of the closed dichotomous venation of *Regnellidium*. *Amer. J. Bot.* 49: 464–72.

Pray, T. R. 1963. Origin of vein endings in angiosperm leaves. *Phytomorphology* 13: 60–81.

Raven, P. H., and Axelrod, D. I. 1972. Plate tectonics and Australasian paleobiogeography. *Science* 176: 1379–86.

Rushforth, S. R. 1971. A flora from the Dakota Sandstone Formation (Cenomanian) near Westwater, Grand County, Utah. *Brigham Young Univ. Sci. Bull., biol. ser.*, 14: 1–44.

Samylina, V. A. 1960. Pokrytosemennye rasteniya iz nizhnemelovykh otlozheniy Kolymy. *Bot. Zh.* 45: 335–52.

Samylina, V. A. 1968. Early Cretaceous angiosperms of the Soviet Union based on leaf and fruit remains. *J. Linn. Soc., Bot.*, 61: 207–18.

Saporta, G. de. 1894. *Flore fossile du Portugal.* Direction des Travaux géologiques du Portugal, Lisbon.

Schaeffer, B., and Hecht, M. K. 1965. Introduction and historical résumé (Symposium: The origin of higher levels of organization). *Syst. Zool.* 14: 245–48.

Schulz, E. 1967. Sporenpaläontologische Untersuchungen rätoliassischer Schichten im Zentralteil des germanischen Beckens. *Paläontol. Abhandl.* 2B: 541–633.

Schuster, R. M. 1972. Continental movements, "Wallace's Line" and Indomalayan-Australasian dispersal of land plants: Some eclectic concepts. *Bot. Rev.* 38: 3–86.

Scott, R. A., Barghoorn, E. S., and Leopold, E. B. 1960. How old are the angiosperms? *Amer. J. Sci.* 258-A (Bradley vol.): 284–99.

Sen, L. 1971. The geometric structure of an optimal transport network in a limited city-hinterland case. *Geogr. Anal.* 3: 1–14.

Seward, A. C. 1931. *Plant Life through the Ages.* Cambridge Univ. Press, Cambridge.

Slade, B. F. 1957. Leaf development in relation to venation as shown in *Cercis siliquastrum* L., *Prunus serrulata* Lindl., and *Acer pseudoplatanus* L. *New Phytol.* 56: 281–300.
Smart, J., and Hughes, N. F. 1972. The insect and the plant: Progressive palaeoecological integration. In *Insect Plant Relationships,* ed. H. F. van Emden. Symposium of the Royal Entomological Society, London, no. 6, pp. 143–55.
Smith, A. C. 1970. *The Pacific as a Key to Flowering Plant History.* Harold L. Lyon Arboretum Lecture 1, University of Hawaii, Honolulu.
Sporne, K. R. 1972. Some observations on the evolution of pollen types in dicotyledons. *New Phytol.* 71: 181–85.
Stebbins, G. L. 1950. *Variation and Evolution in Plants.* Columbia Univ. Press, New York.
Stebbins, G. L. 1965. The probable growth habit of the earliest flowering plants. *Ann. Missouri Bot. Garden* 52: 457–68.
Stebbins, G. L. 1970a. Biosystematics: An avenue toward understanding evolution. *Taxon* 19: 205–14.
Stebbins, G. L. 1970b. Transference of function as a factor in the evolution of seeds and their accessory structures. *Israel J. Bot.* 19: 59–70.
Strasburger, E. 1900. Einige Bemerkungen zur Frage nach der doppelten Befruchtung bei Angiospermen. *Bot. Zeitung, ser.* II, 58: 293–316.
Takhtajan, A. L. 1969. *Flowering Plants: Origin and Dispersal.* Oliver, Edinburgh.
Teixeira, C. 1948. *Flora mesozóica portuguesa.* Pt. I. Serviços Geol. Portugal, Lisbon.
Tralau, H. 1968. Botanical investigations into the fossil flora of Eriksdal in Fyledalen, Scania. II. The Middle Jurassic microflora. *Sveriges Geol. Undersökn. Årsbok* 62: 1–185.
Vakhrameev, V. A. 1952. Stratigraphy and fossil flora of Cretaceous deposits of western Kazakhstan [in Russian]. *Regional'naya Stratigr. SSSR* 1: 1–340.
Vakhrameev, V. A. 1973. Angiosperms and the boundary of the Lower and Upper Cretaceous [in Russian]. In *Palinologiya mezofita* (*Trudy III Mezhdunarodnoy Palinologicheskoy Konferentsii*) [Palynology of the Mesophytic (Proceedings of the 3rd International Palynological Conference)], ed. A. F. Chlonova, pp. 131–37. Nauka, Moscow.
Van Campo, M. 1971. Précisions nouvelles sur les structures comparées des pollens de Gymnospermes et d'Angiospermes. *Compt. Rend. Acad. Sci. Paris, ser. D,* 272: 2071–74.
Van der Pijl, L. 1972. *Principles of Dispersal in Higher Plants.* Springer, Berlin.
Waage, K. M. 1959. *Stratigraphy of the Inyan Kara Group in the Black Hills.* U.S. Geological Survey Bulletin 1081-B. Washington, D.C.
Wagner, W. H., Jr. 1964. The evolutionary patterns of living ferns. *Mem. Torrey Bot. Club* 21: 86–95.
Walker, J. W. 1971. Pollen morphology, phytogeography, and phylogeny of the Annonaceae. *Contrib. Gray Herbarium* 202: 1–131.

Ward, L. F. 1888. Evidence of the fossil plants as to the age of the Potomac Formation. *Amer. J. Sci., 3d ser.*, 36: 119–31.

Ward, L. F. 1895. The Potomac Formation. *U.S. Geol. Surv. 15th Annu. Rept.*, pp. 307–97.

Ward, L. F. 1899. The Cretaceous formation of the Black Hills as indicated by the fossil plants. *U.S. Geol. Surv. 19th Annu. Rept.*, pt. 2, pp. 523–712.

Ward, L. F. 1905. *Status of the Mesozoic Floras of the United States.* U.S. Geological Survey Monograph 48. Washington, D.C.

Weaver, K. N., Cleaves, E. T., Edwards, J., and Glaser, J. D. 1968. *Geologic Map of Maryland.* Maryland Geological Survey, Baltimore.

Whitehead, D. R. 1969. Wind pollination in the angiosperms: Evolutionary and environmental considerations. *Evolution* 23: 28–35.

Wodehouse, R. P. 1935. *Pollen Grains.* McGraw, New York.

Wolfe, J. A. 1972a. Significance of comparative foliar morphology to paleobotany and neobotany. *Amer. J. Bot.* 59: 664 (Abstr.)

Wolfe, J. A. 1972b. Phyletic significance of Lower Cretaceous dicotyledonous leaves from the Patuxent Formation, Virginia. *Amer. J. Bot.* 59: 664. (Abstr.)

Wolfe, J. A., Doyle, J. A., and Page, V. M. 1975. The bases of angiosperm phylogeny: The fossil record. *Ann. Missouri Bot. Gard.*, in press.

Wolfe, J. A., and Pakiser, H. M. 1971. Stratigraphic interpretations of some Cretaceous microfossil floras of the Middle Atlantic states. *U.S. Geol. Surv. Profess. Paper* 750-B: B35–47.

Young, K. 1966. *Texas Mojsisovicziinae (Ammonoidea) and the Zonation of the Fredericksburg.* Geological Society of America Memoir 100.

Zaklinskaya, Ye. D. 1962a. Importance of angiosperm pollen for the stratigraphy of Upper Cretaceous and Lower Paleogene deposits and botanical-geographical provinces at the boundary between the Cretaceous and Tertiary systems. *Pollen Spores* 4: 389.

Zaklinskaya, Ye. D. 1962b. Significance of angiosperm pollen for stratigraphy of the Upper Cretaceous and Paleogene and botanical-geographic provinces at the boundary of the Cretaceous and Paleogene systems [in Russian]. In *K Pervoy Mezhdunarodnoy Palinologicheskoy Konferentsii (Tucson, U.S.A.): Doklady sovetskikh palinologov* [To the First International Palynological Conference (Tucson, U.S.A.): Reports of Soviet palynologists], pp. 105–13. Nauka, Moscow.

Neoteny and the Origin of Flowering Plants

ARMEN TAKHTAJAN, ***Komarov Botanical Institute the USSR Academy of Sciences, Leningrad***

In its broader sense the evolutionary term "neoteny" means an extension of early developmental phases into maturity, the former adult stages being omitted from the life-cycle. There are a few more or less complete synonyms of neoteny, such as "fetalization," "juvenilization," "juvenilism," and "paedomorphosis." Although some of these terms have slightly different shades of meaning, all of them express the idea of the terminal abbreviation of ontogeny (the loss of late stages) and a premature completion of development of the whole organism (total neoteny) or of parts of it (partial neoteny).

As a result of a relative acceleration of the processes leading to maturity, and a retardation of other developmental processes, early phases of ontogeny are turned into the definitive phases of the neotenous descendants. Therefore, juvenile characters are retained in the adult, and "the adult of the descendant resembles the young of the ancestors" (de Beer, 1959). The shift of juvenile characters toward adult stages (juvenilization) may not involve the whole organism but may confine itself to individual organs only.[1]

The evolutionary importance of neoteny (including Garstang's concept of paedomorphosis) depends on the despecialization of the neotenous organism or of parts of it. Neotenous "rejuvenation" increases evolutionary plasticity and opens new evolutionary avenues. "It is this possibility of escaping from the blind alleys of specialization into a new period of plasticity and adaptive radiation which makes the idea of paedomorphosis so attractive in evolutionary theory," says Huxley (1954, p. 20). Hardy (1954, p. 128) comes to an analogous conclusion: "However specialized the *adults* of a stock may have become in relation to life in some particular environment,

---

[1] Haldane (1932a) explains the genetic mechanism of neoteny by mutations of the genes controlling the speed of developmental processes. It is appropriate to mention here that retardation of the action of even a single gene "would affect the time of action of many genes, with consequential effects on development" (Wardlaw, 1952, p. 134). Neoteny is remarkable for the maximum phenotypic effect being achieved by way of the minimum alteration of the genotype.

it is still open to their young stages to become modified in some quite other way and then by neoteny to produce a new paedomorphic line leading perhaps to a quite new type of animal—perhaps a new Order, Class or even Phylum." The genetic basis of the increase in evolutionary plasticity lies in the fact, indicated by Koltsov (1936), that abrupt neoteny involves at first a great simplification of the phenotype alone, whereas the genotype maintains its complexity. Conservation of this genetic complexity is of considerable significance for the further evolution of the neotenous organisms. "The rich reservoir of genes that are not manifest in the development of the neotenous forms (but which are able to mutate into new active genes) leads to a high degree of variability of the neotenous forms and sometimes enables them to display an exuberant outburst of further progressive evolution" (Kol'tsov, 1936, p. 520). In fact, the evolutionary significance of neoteny is due to this combination of phenotypic simplification and despecialization with the maintenance of the ancestral level of genotypic complexity. Such a combination is likely to lead to evolutionary novelties which may give rise to new major lines of evolution.

The outstanding role of neoteny in the origin and evolution of many genera and families and even classes and phyla of the animal kingdom (including the genus *Homo* and the whole class Mammalia) has been appreciated by many zoologists, as well as some anthropologists and animal geneticists (see Sushkin, 1915; Garstang, 1922, 1928; Hadži, 1923, 1963; Bolk, 1926; de Beer, 1930, 1954, 1958, 1959, 1964; Haldane, 1932a,b; Kol'tsov, 1936; Schindewolf, 1936; Huxley, 1942, 1953, 1954; Keith, 1949; Young, 1950; Hardy, 1954; Montagu, 1955; Remane, 1956; Smith, 1958; Rensch, 1959; Morris, 1967; Vandel, 1968; Codreanu, 1970; Ohno, 1970). In contrast, the evolutionary significance of neoteny has been underestimated in botany. As a rule, botanists have not acknowledged any major role for neoteny in the origin and evolution of the higher systematic groups of the plant kingdom and usually attach to it only a role in the origin of certain species and genera (for literature see Takhtajan, 1954b; Vasil'chenko, 1965), and very rarely of families (e.g., Lemnaceae [Rostovtsev, 1905]). As far as I know, no botanist has applied this idea on such a scale as the zoologists, although some botanists (e.g., Arber, 1937, 1950; Kozo-Polyanskiy, 1937; Zimmermann, 1959; Asama, 1960; Davis and Heywood, 1963) have accepted some role for neoteny in the morphological evolution of plants.

It is 30 years since I attempted to demonstrate that the concept of neoteny in its broader sense has far-reaching importance for evolutionary botany and supplies the most reasonable hypothesis of the origin of some higher systematic units. In a series of publications beginning in 1943 I tried to develop the principle of neoteny on the basis of botanical material and put forward the opinion that the appearance of some large and successful groups of the plant kingdom is correlated with the neotenic mode of evolution. In particular, I tried to show that the concept of neoteny gave the clue to the origin of the Bryophyta from the Rhyniophyta, the Welwitschiaceae [2] from a hypothetical bennettitalean ancestor, and the flowering plants (Magnoliophyta) from the gymnosperms (Pinophyta). I also regarded the monocots (Liliopsida) as descended from dicots (Magnoliopsida) by neotenous retention of many of the juvenile features of the ancestor.

Neoteny in the Origin of Flowering Plants

Neotenous origin of flowering plants is more than a hypothetical possibility. The basic morphological characters of flowering plants find their most plausible explanation in the hypothesis of neotenous origin of the whole group. It is not difficult to show that both vegetative and reproductive organs, especially leaves, flowers, and female and male gametophytes, bear the stamp of neoteny.

VEGETATIVE ORGANS

Traces of neotenous origin are visible in the leaves of some primitive flowering plants: Magnoliales and related orders. The Magnoliales are the modern representatives of the ancestral stock of the Magnoliophyta, doubtless specialized in many ways but still the most primitive flowering plants. Their leaves are always simple, usually entire, and have pinnate venation. The simple, entire leaf is considered by many authors as the most primitive among the modern flowering plants (Parkin, 1953; Takhtajan, 1959, 1969; Eames, 1961; Cronquist, 1968; Hickey, 1971; Doyle and Hickey, 1972). The simple, entire leaves of flowering plants could arise from the frondlike leaves of seed ferns (Lyginopteridopsida), which are considered by many authors to be the most probable ancestors of the magnoliophytes. The entire, pinnately veined leaves of flowering plants can only be imagined as having originated as a result of marked simplification brought about by cessation of development at an early juvenile stage and subsequent modifications of leaf structure (Takhtajan, 1954b). Leaves of seed ferns were the less specialized gymnosperm leaves and therefore provided an easy starting-point for such a transforma-

[2] As early as 1863, I. D. Hooker regarded the adult *Welwitschia* plant as "a seedling arrested in development." As R. Rodin (1953) points out, "it might be described more accurately as a seedling whose apical growth has been arrested."

tion. They also gave rise to other basic types of gymnosperm leaf architecture. Němejc (1956, p. 73) derived the angiosperm leaf from primitive leaves "belonging to some Devonian or Carboniferous archaic plant, of a psilophytoid character, from which both Ferns and Pteridosperms originated." Owing to strictly terrestrial conditions, such leaves "became suddenly arrested in their growth at some initial [neotenic] stage from which the course of the further development of the blade took quite a different direction, with quite a different pattern of a much denser venation, far more appropriate for the distribution of water solutions in a dry terrestrial environment and therefore bearing no resemblance whatever to that of the original hygrophilous plants." Axelrod (1960, p. 133) and Asama (1960) also concluded that the angiosperm leaf arose by way of neoteny. Axelrod believed that the leaves of angiosperms were "derived from naked phyllophores of a primitive fern or seed-fern alliance which were arrested early in growth, presumably as an adaptation to growing in more exposed upland regions." Thus all these authors acknowledge the neotenous origin of the leaves of flowering plants.

The most primitive type of leaf architecture in early flowering plants was probably of the cladodromous type (according to the terminology of Hickey, 1973) and most probably with irregular (disorganized) "first-rank" venation (Hickey, 1971; Hickey and Doyle, 1972). The initially disorganized venation may be related to the neotenous origin.

The primitive structure of the vesselless wood of the Winteraceae and Trochodendrales may be also related to the neotenous origin of the flowering plants (Takhtajan, 1961, p. 19). The mature wood of *Trochodendron*, *Tetracentron*, *Tasmannia*, and *Drimys* corresponds in its anatomical structure to the early wood, with scalariform tracheids, of more primitive gymnosperms (Lyginopteridopsida, Cycadopsida, and Bennettitopsida) rather than to their mature wood. We may therefore conclude that the primitive vesselless wood of the earliest magnoliophytes might have originated from the primitive early wood with scalariform tracheids of the hypothetical gymnospermous ancestors by way of neoteny.

There are many examples of persistent juvenile features in vascular anatomy among living seed plants, both gymnosperms and angiosperms. The tuberous species of *Zamia* are regarded by Chrysler (1937) as persistent juveniles with respect to both their growth habit and their vascular tissue; that is, they remain immature vegetatively although they reproduce freely. Whereas in species of *Zamia* possessing a trunk the secondary xylem consists of tracheids with circular bordered pits, in species with a tuberous stem it consists of scalariform tracheids. "Perhaps the tuberous species represent members of the genus that have met evil days in respect to climate, or an extension of the genus into less favourable climatic regions" (Chrysler, 1937, p. 705). *Encephalartos brachyphyllus* Lehm., "a strictly tuberous species from Zululand, shows nothing more advanced than scalariform markings." *Stangeria* illustrates the same juvenile features.

In his most interesting paper on paedomorphosis, or neoteny, in dicotyledonous xylem, Carlquist (1962) showed that in some cases "juvenile characteristics—those of the primary xylem—have been protracted into the secondary xylem." "If paedomorphosis takes place in a wood," says Carlquist, "appearance of more primitive characters in the secondary xylem may result, because primary xylem tends to be a 'refugium' for certain primitive characters, and transference of these characters to the secondary xylem would result in an admixture of primitive with specialized characters." The same may be true in cases where late secondary wood with specialized tracheids is replaced, by way of neoteny, by more primitive early secondary wood with scalariform tracheids. The protraction of primitive elements of the primary xylem (tracheids with annular and helical thickenings) into the secondary xylem was also described by Melet (1968) for three cushionlike high-mountain plants of eastern Pamir, *Gypsophila capituliflora* Bunge, *Oxytropis immersa* (Baker) Bunge, and *Sibbaldia tetrandra* Bunge. Melet came to the conclusion that paedomorphosis occurs under extreme environmental conditions such as those found in deserts or on rocks.

THE FLOWER

The distinct "infantile" or juvenile characters in the organization of the flower have more than once been mentioned in botanical literature (e.g., Arber, 1937, 1950; Croizat, 1947). As Arber (1950, p. 50) concluded, "the best term of comparison for the flower is, not a mature vegetative shoot, but a vegetative bud; the flower might, indeed, be described as corresponding to a vegetative shoot in a condition of permanent infantilism." In fact the flower may be regarded as a neotenic variant of the strobile of the hypothetical gymnospermous ancestor (Takhtajan, 1943), which in its turn was a neotenic derivative of an initial reproductive long shoot of some seed ferns. Marks of neoteny characterize not only the flower as a whole but also its component parts.

The sepals clearly represent modified juvenile bractlike leaves. All stages in the evolutionary development of the sepals from vegetative leaves by way of progressive juvenilization are present in the genus *Paeonia*, especially in P. *delavayi* Franch. and other species of the primitive section Moutan, as well as in the Calycanthaceae and in some Dilleniaceae.

Juvenilism is even more clearly expressed in sporophylls—stamens and carpels. The stamens of the most primitive angiosperms, especially those of *Degeneria, Himantandra,* and some Magnoliaceae, are infantile structures. Both their external morphology and venation suggest a neotenous origin. These simple laminar structures with palmate venation (three main veins and three leaf traces) resemble scales on underground organs, prophylls, bud scales, and bracts. In *Degeneria, Himantandra,* and certain members of the Magnoliaceae and Nymphaeaceae, and in some other primitive taxa, the microsporangia are embedded in the tissues of the stamen, most probably also as a result of neoteny. In this connection it is interesting that in some staminodia (which are more juvenilized structures than stamens themselves) the vestigial microsporangia are frequently sunken, even in taxa with protuberant sporangia. The sunken microsporangia in primitive taxa are indeed one of the most convincing types of evidence of neotenous origin of the stamens.

The carpels of the most primitive angiosperms, especially those of *Tasmannia* and *Degeneria,* have a clearly expressed appearance of juvenilized, infantile structures. They are conduplicate during the early stages of their development and thus closely resemble young leaves folded adaxially along the midrib (Bailey and Nast, 1943; Bailey and Swamy, 1951). Like primitive stamens and many bracts and cataphylls, they are characterized by palmate venation. It was therefore not difficult to conclude that the evolutionary transformation of the ancestral open megasporophyll into the carpel by folding and gradual closure along its midrib could have occurred in a juvenile stage. It probably occurred the more easily because conduplicate vernation is characteristic of many flowering plants, including the most primitive taxa. It is likely that the transformation of open megasporophylls into closed ones occurred at an early ontogenetic stage at which they were still folded (Takhtajan, 1948, 1954b, 1959, 1964, 1969).

Thus, the flower as a whole and its basic members originated most probably as a result of neoteny and subsequent modifications and specializations.

Male and Female Gametophytes

The sexual generation is even more infantile in its organization than the asexual one. Gametophytes of flowering plants, both male and female, are simplified to the extreme and miniaturized. They are formed as a result of a minimum number of mitotic divisions and from a minimum amount of building material. Even the female gametophyte (which is considerably less simplified than the male) develops by only three mitotic divisions preceded by two meiotic divisions of the megasporocyte, whereas in the gymnosperms the female gametophyte develops as a result of at least nine divisions. Due to a sharp abbreviation of their ontogeny, both male and female gametophytes of angiosperms completely lost their gametangia (antheridia and archegonia). This simplification of gametophytes resulted from neoteny and subsequent specialization (Takhtajan, 1948, 1954a, 1959, 1964, 1969).

The simplified gametophytes of flowering plants could emerge only by strong acceleration of the processes leading to sexual maturity and a corresponding retardation of other developmental processes in the ancestral gametophytes. The premature completion of the ontogeny of gametophytes, having started at relatively late developmental stages, gradually shifted to earlier and earlier stages. Consequently, gametogenesis also shifted to ever-earlier phases of development. This led, finally, in the early angiosperms or their immediate ancestors, to the loss of those phases of development at which, in most gymnosperms, the gametangia are formed. Gametogenesis in flowering plants takes place at such an early phase of development that the gametangia cannot even be formed, and the gametes are formed without them. Moreover, the development of the gametes themselves is also cut short, and they become extremely simplified. Owing to such fundamental ontogenetic alterations, there arose in angiosperms very simplified "gametangialess gametophytes" sharply different from the mature gametophytes of the primitive gymnosperms. It should be noted that the general direction of the evolution toward an abbreviation of gametophyte development, the simplification of the gametophyte and gametes, and acceleration of gametogenesis, started in the gymnosperms, among which certain conifers came close to the angiosperms in development and structure of the male gametophyte, and the genera *Welwitschia* and *Gnetum* came close in some important characteristics of their female gametophytes.

The male gametophyte of the flowering plants, which consists of two cells only, a tube cell ("vegetative" cell) and a generative cell, is the most simplified among the higher plants. It has neither prothalial cells nor antheridia and is even devoid of true spermatogenous cells (which are still present in all living gymnosperms except *Welwitschia* and *Gnetum*). The function of the spermatogenous cells has been transferred to the generative cell, which divides to form two nonmotile male gametes. Thus the male gametes in angiosperms also undergo reduced development and constitute completely new structures which replace the stalk cell and body cell (spermatogenous cell) of the ancestral gametophyte.[3] In a certain sense we can even say that the stalk cell and body cell were transformed into male gametes. There is therefore every reason to suppose that the male gametophyte of angiosperms originated by way of both neoteny (terminal abbreviation and subsequent structural and functional modifications) and basal abbreviation.

The female gametophyte of the flowering plants is somewhat similar to those of the gymnospermous genera *Welwitschia* and *Gnetum*, although more simplified. The resemblance consists not only in the complete absence of any traces of archegonia, but also in the tendency toward a decreased number of nuclei. Both of these types of nonarchegoniate gametophytes resemble the early stages of the female gametophytes of archegoniate gymnosperms, possessing a peripheral layer of free nuclei arranged around a large central vacuole. It is therefore quite possible that in both cases the nonarchegoniate gametophyte originated by way of progressive acceleration of gametogenesis and retardation of all other developmental processes, including the formation of archegonia.

Similar ideas were expressed long ago and can be found, for example, in the works of Strasburger (1900), Coulter (1909, 1914), and even some earlier authors. Coulter (1914, p. 73) says that the "complete elimination of the archegonium, begun in gymnosperms, is a feature of all angiosperms." The female gametophyte "begins to develop archegonia earlier and earlier in its history, until finally eggs are matured before there is any tissue to develop the sterile jacket. In other words, the archegonium is reduced to its essential sexual structure, the egg, which means that the distinguishing feature of an archegonium, the jacket, has disappeared" (p. 73). The only correction we should make is that, strictly speaking, the angiosperm egg is not the former egg of the archegonium but a cell of the initial developmental phase of the gametophyte. But this was clear already to Hoffmeister and Strasburger and has been convincingly shown lately by Gerasimova-Navashina (1958, 1971).[4]

---

[3] The stalk cell and the body cell are the only remnants of an antheridium in gymnospermous plants (except in *Gnetum* and *Welwitschia*, which have lost both of these cells).

[4] According to Gerasimova-Navashina (1958, p. 131), the egg of angiosperms is homologous to only one of the very first cells of the gametophyte of gymnosperms but by no means to their egg.

All these facts and considerations lead to the conclusion that the nonarchegoniate female gametophyte originated by neoteny (Romanov, 1944; Takhtajan, 1948, 1954a,b, 1959, 1964; Zimmerman, 1959), that is to say, through the delay in the development of the ancestral gametophyte at its initial free nuclear stage and subsequent formation of cellular structures of an entirely new type. It was certainly not a mere hereditary fixation of the early stage but a fundamental change of the entire course of development of the gametophyte. The main change here consisted in abrupt terminal abbreviation, as a result of which all the late stages of development were omitted. If neoteny is a sort of "unfinished ontogeny," to use the expression of P. Suchkin (1915), the origin of the female gametophyte is one of its striking examples. The egg is differentiated not later than the third division of the megaspore nucleus, i.e., at such an early stage that the initiation of archegonia is not yet possible. So it is natural that the archegonium is eliminated, leaving no trace.

At first, basal abbreviation did not take part in the origin of the female gametophyte of flowering plants. But having originated, the female gametophyte underwent reduction in its development by this means. The basal abbreviation is realized in quite an original manner. It takes place owing to the inclusion of the products of megasporogenesis in the formation of the gametophyte, which is observed in bisporic and particularly in tetrasporic types. In the latter type the entire development of the gametophyte consists of only one division of the four megaspore nuclei.

The simplification of the female gametophyte and sharply expressed acceleration of its development are the consequence of a rapid maturation of the comparatively small ovules of angiosperms. Again, the early maturation of the latter and their small size are connected with the rapid development of the flower and with the presence of a closed, protecting carpel. Here is a definite chain of interconnected evolutionary changes, at the base of which lies the neotenous transformation of the strobilus of gymnospermous ancestors into the flower. There is no doubt that so-called double fertilization, which is one of the most characteristic features of the flowering plants, also originated as a result of neotenous simplification of the female gametophyte.

Concluding Remarks

All the available evidence shows that genetically neotenous forms arise under some kind of environmental stress. They arise in extreme environmental conditions—on rocks, in both cold and warm deserts, arctic tundra, cold mountains, and aquatic environments. Under such conditions the neotenous populations are able to mature and to produce offspring sooner than the ancestral plants and therefore they are better adapted to the shorter growing season. In trying to explain the origin of neoteny in the ancestors of flowering plants, we inevitably come to the conclusion that flowering plants originated under environmental stress. They probably originated as a result of adaptation to moderate drought on rocky, mountain slopes in an area with a monsoon climate.

References

Arber, A. 1937. The interpretation of the flower: A study of some aspects of morphological thought. *Biol. Rev. Cambridge Philos. Soc.* 12: 157–84.

Arber, A. 1950. *The Natural Philosophy of Plant Form.* Cambridge Univ. Press, Cambridge.

Asama, K. 1960. Evolution of the leaf forms through the ages explained by the successive retardation and neoteny. *Sci. Rep. Tôhoku Univ., ser.* 2, *Special vol.* (4): 252–80.

Axelrod, D. I. 1960. The evolution of flowering plants. In *Evolution after Darwin,* ed. S. Tax, vol. 1, pp. 227–305. Univ. of Chicago Press, Chicago.

Bailey, I. W., and Nast, C. G. 1943. The comparative morphology of the Winteraceae. II. Carpels. *J. Arnold Arbor.* 24: 472–81.

Bailey, I. W., and Swamy, B. G. L. 1951. The conduplicate carpel of dicotyledons and initial trends of specialization. *Amer. J. Bot.* 38: 373–79.

Bolk, L. 1926. *Das Problem der Nenschenwerdung.* Jena.

Carlquist, S. 1962. A theory of paedomorphosis in dicotyledonous wood. *Phytomorphology* 12: 30–45.

Chrysler, M. A. 1937. Persistent juveniles among the cycads. *Bot. Gaz.* 98: 696–710.

Codreanu, R. 1970. Grands problèmes controversés de l'évolution phylogénétique des Métazoaires. *Ann. Biol., Paris* 9: 671–709.

Coulter, J. M. 1909. Evolutionary tendencies among gymnopserms. *Bot. Gaz.* 48: 81–97.

Coulter, J. M. 1914. *The Evolution of Sex in Plants.* Univ. of Chicago Press, Chicago.

Croizat, L. 1947. A study in the Celastraceae, Siphonoideae subfam. nov. *Lilloa* 13: 31–43.

Cronquist, A. 1968. *The Evolution and Classification of Flowering Plants.* Houghton, New York.

Davis, P. H., and Heywood, V. H. 1963. *Principles of Angiosperm Taxonomy.* Van Nostrand, Princeton.

De Beer, G. R. 1930. *Embryology and Evolution.* Clarendon Press, Oxford.

De Beer, G. R. 1954. The evolution of Metazoa. In *Evolution as a Process,* ed. J. Huxley, A. C. Hardy, and E. B. Ford, pp. 24–33. G. Allen, London.

De Beer, G. R. 1958. *Embryos and Ancestors.* 3d ed. Clarendon Press, Oxford.
De Beer, G. R. 1959. Paedomorphosis. *Proc. XV Internat. Congr. Zool.*, London, pp. 927–30.
De Beer, G. R. 1964. *A Handbook on Evolution.* British Museum (Natural History), London.
Doyle, J. A., and Hickey, L. J. 1972. Coordinated evolution in Potomac Group angiosperm pollen and leaves. *Amer. J. Bot.* 59: 660. (Abstr.)
Eames, A. J. 1961. *Morphology of Vascular Plants.* McGraw, New York.
Garstang, W. 1922. The theory of recapitulation. A critical restatement of the biogenetic law. *J. Linn. Soc., Zool.* 35: 81–101.
Garstang, W. 1928. The morphology of the Tunicata, and its bearing on the Phylogeny of the Chordata. *Quart. J. Microscop. Sci.* 72: 51–187.
Gerasimova-Navashina, E. N. 1958. On the gametophyte and on salient features of development and functioning of reproducing elements in angiospermous plants [in Russian]. In *Probl. Bot.* (*Leningrad*) 3: 125–167.
Gerasimova-Navashina, E. N. 1971. Double fertilization in angiosperms and some of its theoretical aspects [in Russian]. In *Problemy embriologii* [Problems in embryology], ed. V. P. Zosimovich, pp. 113–52. Naukova dumka, Kiev.
Hadži, J. 1923. Über den Ursprung, die Verwandschaftverhältnisse und die systematisch Position der Ktenophoren. *Razpr. Slovensk. akad. Mat. Prirod., Zagreb* 17: 53–62.
Hadži, J. 1963. *The Evolution of the Metazoa.* Pergamon Press, Oxford.
Haldane, J. B. S. 1932a. The time of action of genes, and its bearing on some evolutionary problems. *Amer. Naturalist* 66: 5–24.
Haldane, J. B. S. 1932b. *The Causes of Evolution.* Longmans, London.
Hardy, A. C. 1954. Escape from specialization. *In Evolution as a Process,* ed. J. Huxley, A. C. Hardy, and E. B. Ford, pp. 122–42. G. Allen, London.
Hickey, L. J. 1971. Evolutionary significance of leaf architectural features in the woody dicots. *Amer. J. Bot.* 58: 469. (Abstr.)
Hickey, L. J. 1973. Classification of the architecture of dicotyledonous leaves. *Amer. J. Bot.* 60: 17–33.
Hickey, L. J., and Doyle, J. A. 1972. Fossil evidence on evolution of angiosperm leaf venation. *Amer. J. Bot.* 58: 661. (Abstr.)
Hooker, J. D. 1863. On *Welwitschia,* a new genus of Gnetaceae. *Trans. Linn. Soc. London* 24: 1–48.
Huxley, J. S. 1942. *Evolution, the Modern Synthesis.* Harper, New York.
Huxley, J. S. 1953. *Evolution in Action.* Harper, New York.
Huxley, J. S. 1954. The evolutionary process. In *Evolution as a Process,* ed. J. Huxley, A. C. Hardy, and E. B. Ford, pp. 1–23. London.
Keith, A. 1949. *A New Theory of Human Evolution.* Watts, London.
Kol'tsov, N. K. 1936. *Organizatsiya kletki* [The organization of the cell]. Biomedgiz, Moscow-Leningrad.
Kozo-Polyanskiy, B. M. 1937. *Osnovnoy biogeneticheskiy zakon s botanicheskoy tochki zreniya* [The basic biogenetic law from the botanical point of view]. Voronezhskoye oblastnoye knigoizdatel'stvo, Voronezh.

Melet, L. S. 1968. The phenomenon of paedomorphosis in the secondary wood of some cushion-plants of the eastern Pamir [in Russian]. *Uzvestia Div. Biol. Sci. Tadjikistan Acad. Sci.* 2: 19–22.

Montagu, A. 1955. Time, morphology, and neoteny in the evolution of man. *Amer. Anthropol.* 57: 13–27.

Morris, D. 1967. *The Naked Ape.* McGraw, New York.

Němejc, F. 1956. On the problem of the origin and phylogenetic development of the angiosperms. *Sborn. Nár. Mis. v Praze, Řada B, Přir. Vědy* 12: 59–144.

Ohno, S. 1970. *Evolution by Gene Duplication.* Springer-Verlag, Berlin.

Parkin, J. 1953. The strobilus theory of angiospermous descent. *Proc. Linn. Soc. London* 153: 51–64.

Remane, A. 1956. *Die Grundlagen des natürlichen Systems, der vergleichenden Anatomie und der Phylogenetik.* 2d ed. Leipzig.

Rensch, B. 1959. *Evolution above the Species Level.* Columbia Univ. Press, New York.

Rodin, R. 1953. Seedling morphology of *Welwitschia. Amer. J. Bot.* 40: 371–78.

Romanov, I. D. 1944. Evolyutsiya zarodyshevogo meshka tsvetkovykh rasteniy [Evolution of the embryo sac of flowering plants]. Doctoral dissertation, Tashkent University.

Rostovtsev, S. I. 1905 *Biologo-morfologicheskiy ocherk ryasok* [A biologomorphological outline of duckweeds]. Tipographiya V. Rikhter, Moscow.

Schindewolf, O. H. 1936. *Paläontologie, Entwicklungslehre und Genetik: Kritik und Synthese.* Berlin.

Smith, J. M. 1958. *The Theory of Evolution.* Edinburgh.

Strasburger, E. 1900. Einige Bemerkungen zur Frage nach der "Doppelten Befruchting" bei den Angiospermen. *Bot. Zeitung* 58: 293–316.

Sushkin, P. 1915. Is evolution reversible? [in Russian] In *Novye idei v biologii* [New ideas in biology], ed. V. A. Wagner, vol. 8, pp. 1–39. Obrazovaniye, Petrograd.

Takhtajan, A. 1943. Correlations of ontogeny and phylogeny in the higher plants [in Russian with English summary]. *Tr. Erevansk. Gos. Univ.* 22: 71–176.

Takhtajan, A. 1948. *Morfologicheskaya evolyutsiya pokrytosemennykh* [Morphological evolution of the angiosperms]. Izdatel'stvo Moskovskogo obshchestva ispytateley prirody, Moscow.

Takhtajan, A. 1954a. Quelques problèmes de la morphologie évolutive des angiospermes. [in Russian and French]. *Essais Bot., Vopr. Botan.* 2: 763–93.

Takhtajan, A. 1954b. *Voprosy evolyutsionnoy morfologii rasteniy* [Problems in evolutionary morphology of plants]. Leningrad University, Leningrad.

Takhtajan, A. 1959. *Die Evolution der Angiospermen.* G. Fischer, Jena.

Takhtajan, A. 1961. *Proiskhozhdeniye pokrytosemennykh rasteniy* [The origin of angiospermous plants]. 2d. ed. Vysshaya shkola, Moscow.

Takhtajan, A. 1964. *Osnovy evolyutsionnoy morfologii pokrytosemennykh* [Foundations of the evolutionary morphology of the angiosperms]. Nauka, Moscow-Leningrad.
Takhtajan, A. 1969. *Flowering Plants: Origin and Dispersal.* Oliver and Boyd, Edinburgh.
Vandel, A. 1968. *La Genèse du Vivant.* Paris.
Vasil'chenko, I. T. 1965. *Neotenicheskiye izmeneniya u rasteniy* [Neotenous alterations in plants]. Nauka, Moscow-Leningrad.
Wardlaw, C. W. 1952. *Morphogenesis in Plants.* Methuen, London.
Young, J. Z. 1950. *The Life of Vertebrates.* Clarendon Press, Oxford.
Zimmermann, W. 1959. *Die Phylogenie der Pflanzen.* G. Fischer, Stuttgart.

Evolutionary Significance of Chromosomal Differentiation Patterns in Gymnosperms and Primitive Angiosperms

F. EHRENDORFER, *Botanical Institute University of Vienna, Austria*

INFORMATION on chromosome number and structure has for decades been important in clarifying relationships and differentiation processes on the microevolutionary level in higher plants. The fact that chromosome numbers may change rapidly even within genera, the fragmentary nature or even lack of karyological information about many groups (particularly woody tropical groups), and the deplorable rate of erroneous chromosome records have contributed to the widespread opinion that such data are hardly applicable to problems of macrosystematics and macroevolution. The present contribution may lead to reconsideration of that opinion.[1]

Karyological information included here has been taken from reference books by Darlington and Wylie (1955), Bolkhovskikh et al. (1969), the annual "Index to Plant Chromosome Numbers" (Moore, 1972), and from publications cited therein. Such citations are not repeated here. In addition, several more recent contributions (including those from our own research group) are incorporated. Guidelines for the systematic presentation have been mainly obtained from the contributions by Engler (1954–1964), Cronquist (1968), Takhtajan (1969), and Ehrendorfer (1971).

The present cytosystematic considerations are limited to woody seed plants. Growing experience has substantiated well the generalization that evolutionary rates of change of chromosome structure and number are much slower in woody than in herbaceous groups;

[1] After this paper was delivered, Peter Raven kindly brought a manuscript to my attention: "Cytology and the bases of angiosperm phylogeny," scheduled to be published in *Annals of the Missouri Botanical Garden*, 1975. Our two papers seem to complement each other well, each taking a positive attitude toward the usefulness of chromosome number information in problems of angiosperm macrosystematics and major evolution.

macrosystematic conclusions are therefore more feasible in the former. Incorporation of chromosome data from the literature has been critical and selective. Many obviously erroneous records as well as references to occasional aneuploid or polyploid individuals and to B chromosomes have been omitted.

Graphical presentation is through chromosome-number diagrams for orders, subclasses, and so on (figs. 1, 3–6, 8–10): subordinate taxa, usually families (or subfamilies) appear in the left column; suggestions for close affinities among them are indicated by connecting lines at the left margin. Chromosome numbers recorded are arranged into an upper block; possibly extinct numbers (that is, numbers that may not exist among living groups) appear in brackets, gaps are marked by dots. Grouping of numbers into ploidy levels (indicated by $2x$, $4x$, $6x$ . . . on the left side) is not always unquestionable, but gaps or reduced frequencies of numbers at the switch from one ploidy level to the other often help (see figs. 3, 6, 9). All numbers recorded for a particular (sub)family are connected by horizontal underlines (for dysploidy at the same ploidy level) or by vertical sidelines (for polyploidy); broken lines indicate numbers not yet recorded, or missing. The chromosome-number range for the (sub)families is shown in the lower right block. Where representation of numbers is obviously uneven, this is brought out by thicker lines for widespread and thinner for rare numbers. These chromosome diagrams give as objective an impression as possible of our present knowledge of the dispersion and relative frequency of chromosome numbers within families, orders, subclasses, etc. The diagrams convey the idea of "chromosome differentiation patterns" in a graphic form, facilitate comparison of such patterns between groups, and help in further interpretations (e.g., the search for original base numbers).

Gymnosperms

CONIFEROPHYTINA

Within gymnosperms, the extant (sub)classes and orders of Coniferophytina—Ginkgoales, Pinales, and Taxales—follow a relatively uniform pattern of karyological and chromosomal differentiation. Chromosomes are relatively large, DNA content of nuclei is therefore high (2 C nuclei with about 30–80 picograms of DNA [Miksche, 1967; El-Lakany and Dugle, 1972; Price, Sparrow, and Nauman, 1973]). There is a relatively high proportion of repetitious DNA (Miksche and Hotta, 1973). Rather symmetrical karyotypes with predominantly (sub)metacentric chromosomes of not very different

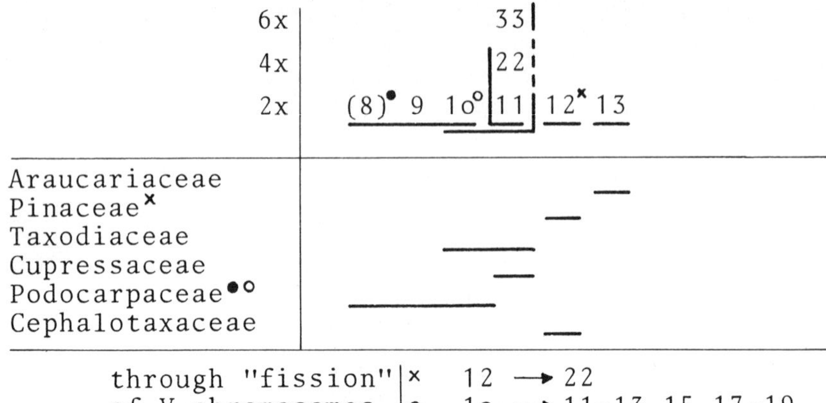

Fig. 1. Chromosome-number diagram for Pinales.

length appear in nearly all families (for recent contributions on Pinaceae see Saylor, 1972; Terasmaa, 1971); according to a widely accepted interpretation, such karyotypes should be regarded as relatively primitive (Khoshoo, 1962).

The chromosome number $n = 12$ is found in *Ginkgo* (Ginkgoales), in the widely divergent Pinaceae and Cephalotaxaceae of Pinales (fig. 1), and in *Taxus* (Taxales). It is usually linked to rather symmetrical karyotypes and can be assumed to be the base number of Coniferophytina. Changes of chromosome number in the group can be related to three main cytogenetic mechanisms: dysploidy through unequal translocations, chromosome fission (or fusion) at the centromere, and polyploidy. These mechanisms are discussed in the paragraphs that follow.

In Pinales, descending dysploidy seems to have been responsible for the origin of $n = 11$ in Cupressaceae and Taxodiaceae, $n = 10$ in the strongly divergent *Sciadopitys*, and $n = 10$, 9 (and 8?) in Podocarpaceae. In Taxaceae, the numbers $n = 11$ in *Torreya* and the somewhat controversial $n = 7$ in *Amentotaxus* of Amentotaxaceae (Chuang and Hu, 1963; a conflicting and possibly erroneous indication of $n = 11$: Sugihara, 1943) also seem to have resulted from this type of numerical change. No truly telocentric chromosomes appear among these dysploid groups.

Change of chromosome numbers by fission at the centromere, leading from one metacentric (V) to two telocentrics (1 + 1) (or by fusion of two telocentrics to one metacentric), i.e., Robertsonian changes, are well documented for several groups of Pinales. In Pina-

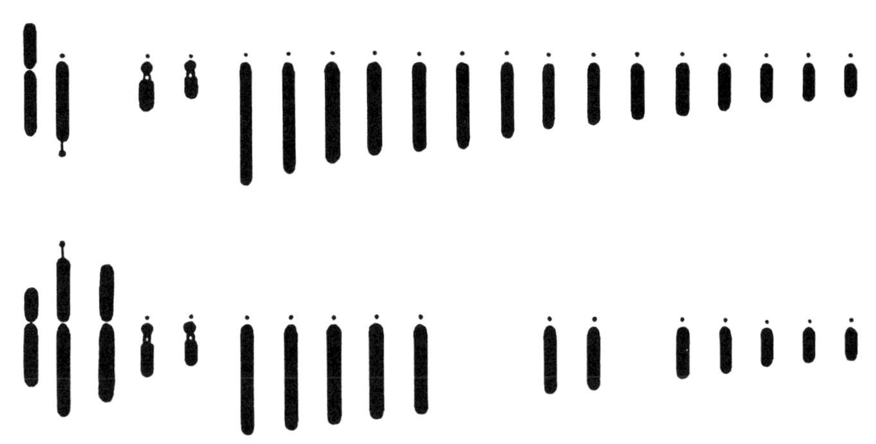

Fig. 2. Karyotypes of *Podocarpus hallii* (*bottom:* 3 V + 14 I) and *P. nivalis* (*top:* 1 V + 18 I). (Modified from Hair and Beuzenberg, 1958.)

ceae, *Pseudolarix* with 2 metacentric and 20 telocentric chromosome pairs obviously has originated from progenitors with 12 metacentric pairs, and the increase to $n = 13$ in *Pseudotsuga* and Araucariaceae seems to have followed similar lines. The ample data on species and hybrids with corresponding numbers of meta- and telocentric chromosomes in Podocarpaceae (fig. 2), presented by Hair and Beuzenberg (1958) and Hair (1963), evidently have to be interpreted in the same way, as increases of chromosome numbers through progressive Robertsonian changes from $n = 10$ to $n = 11$–13, 15, and 17–19, and from 8 (not yet recorded) to 9 and 11–12. (The reverse interpretation of numerical decreases, favored by Hair, is refuted by karyotype sequences in other Pinales and by morphological-geographical progressions in Podocarpaceae; it leaves the origin of primary telocentric karyotypes unexplained.)

Polyploidy has been a very rare event in Coniferophytina. The only evident examples are $6x$ in *Sequoia* (Taxodiaceae) and $4x$ in *Fitzroya* and a few *Juniperus* cytotypes (Cupressaceae). Both families have somewhat smaller chromosomes and lower DNA values than other conifers.

CYCADOPHYTINA

A chromosomal pattern rather similar to the one outlined for Coniferophytina has been revealed for the relic Cycadales, the main stock of the gymnospermous Cycadophytina. (For a careful recent contribution see Marchant [1968].) Chromosomes are also large. As in Coniferophytina, relatively symmetrical karyotypes without (or with few) telocentric and few subterminal chromosomes and a base number of 9 are favored as primitive (for example, in *Bowenia* or *Encephalartos* of Zamiaceae). Again (following Khoshoo, 1969), one can suggest dysploid decreases to 8 chromosome pairs by unequal translocations for *Zamia* and *Ceratozamia* (Zamiaceae) and *Stangeria* (Stangeriaceae), and Robertsonian increase with appearance of corresponding numbers of telocentrics for *Cycas* (with $n = 11$ from $n = 9$) and *Microcycas* (with $n = 13$ from $n = 8$). Dispersed nucleolar organizers may be a basic feature of Cycadales, becoming localized in most Zamiaceae. There is no polyploidy in Cycadales.

As in many other ancient relic groups, heterobathmy (uneven rates of evolutionary change in different characters) seems to have played a role in the differentiation of Cycadales. One therefore must not be surprised that progressions of chromosomal and morphological-anatomical characters do not always coincide, nor should one expect such coincidence.

DISCUSSION

Summing up characteristic aspects of chromosomal differentiation patterns in Coniferophytina and Cycadophytina-Cycadales, one can emphasize: (1) Large chromosomes and high DNA content. (2) Basic changes of chromosome numbers by unequal translocations, i.e., mostly (always?) descending dysploidy. This pattern underlies early phases of evolutionary divergence and today is often characteristic for families or genera. (3) Mostly (always?) ascending Robertsonian changes of chromosome numbers by centromere fission. This pattern underlies early or often later evolutionary divergence and characterizes infrageneric taxa or related genera. (4) Very little polyploidy. These similarities point to common origin of the two gymnosperm groups. Coniferophytina and Cycadophytina-Cycadales are clearly separated by their different ranges of chromosome base numbers, thus corroborating their independent evolution since the early Paleozoic.

The most advanced group of gymnosperms, the Cycadophytina-Gnetatae with the monotypic orders of Ephedrales, Gnetales, and Welwitschiales, do not fit well into the karyological pattern outlined above. All have much smaller chromosomes, and polyploidy appears more prominent. *Ephedra* with $x = 7$ has $2x$ and $4x$ species, *Gnetum* has $n = 12$ (and 11), and *Welwitschia* characteristically has $n = 21$, with predominantly telocentric chromosomes (Khoshoo and Ahuja, 1963). These characteristics substantiate the deep evolutionary hiatus between Gnetatae and other gymnosperms, but tend to approximate the sometimes totally separated subgroups of Gnetatae.

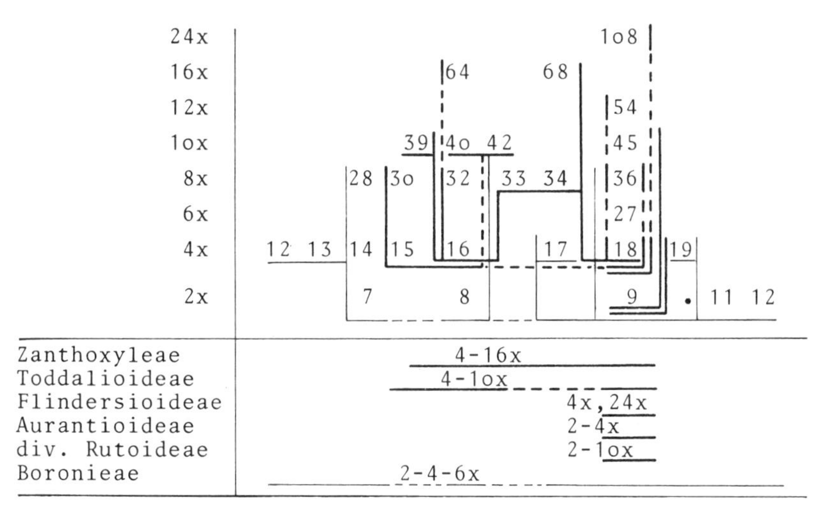

Fig. 3. Chromosome-number diagram for Rutaceae.

Angiosperms

RUTACEAE AND RUTALES

What are the chromosomal differentiation patterns typical for woody angiosperms, and what is their systematic and evolutionary significance? To answer these questions a rather large and diversified family, Rutaceae, has been chosen (fig. 3). A relatively thorough morphologic and systematic treatment (Engler, 1931, 1964) and extensive phytochemical data recently summarized by Fish and Waterman (1973) afford an excellent background for our evolutionary and karyological considerations.

Rutaceae (Rutales, Rosidae) is the only moderately advanced family of dicotyledonous angiosperms clearly linked to the Magnoliidae (generally regarded as the most primitive angiosperm subclass). Genera of the warm temperate and pantropical Rutaceae-Zanthoxyleae and -Toddalioideae share the primitive isoquinoline alkaloid pathway and other characteristic phytochemical features with Ranunculales, particularly Berberidaceae, and evidently have evolved from similar Magnoliidae. Within the Rutaceae, Zanthoxyleae and Toddalioideae have been placed far apart because they have dehiscent and indehiscent carpels, respectively, but such a systematic separation is clearly refuted by common phytochemical, morphological (woody habit, inconspicuous flowers, etc.), and cytological characters. Both groups are represented only by paleopolyploids, and range from a tetraploid level, with n = 15, 16, 17, and 18, in several parallel lines to

$8x$, $10x$, and even $16x$. This pattern points to extinct diploid progenitors with $n = 7$, 8, and 9, and tends to confirm the connection with Magnoliidae (including Ranunculales), for which the base number $x = 7$ is well documented. The close affinity of Zanthoxyleae and Toddalioideae, and their basic position within Rutaceae therefore appear certain.

Another ancient and rather isolated Australasian group of woody Rutaceae is the Flindersioideae, paleo-$4x$ (and in one species $24x$) on $x = 9$. They have lost the primitive isoquinoline pathway and are advanced in their carpel fusion.

The remaining groups of Rutaceae are characterized by tendencies toward more progressive phytochemical and morphological characters (e.g., reduction of woody to herbaceous habit, fruits with pulp, reduction of endosperm). Aurantioideae (= Citroideae) and the majority of Rutoideae (including Ruteae, Diosmeae, and Cusparieae) are predominantly $2x$, and $4x$–$10x$, on $x = 9$. For the Boronieae, Smith-White (1959) has demonstrated explosive divergent evolutionary radiation, mainly in Australia, correlated with descending dysploidy by unequal translocations: $n = 9$–8–7; with polyploidy: $n = 9$–18–36, 8–16–32, 7–14–28; and with further dysploidy: $n = 19$, 17, 13, 12, 11.

Three phases in the evolution of Rutaceae are clearly apparent: (1) Primary dysploid radiation, probably ascending from $n = 7$ to $n = 9$ on the $2x$ level, occurred among (sub)tropical progenitors with primitive phytochemical and morphological characters. Diploids of this primary radiation are extinct, but their chromosome base numbers have been preserved in relic paleopolyploid genera of Zanthoxyleae, Toddalioideae, and the somewhat more advanced Flindersioideae. (2) Evolutionary radiation has continued on $x = 9$, without dysploidy but with chromosome structural and genic changes, predominantly on the $2x$ level, and with polyploidy in some Aurantioideae and many Rutoideae. (3) Secondary explosive dysploid and polyploid radiation has occurred in Boronieae, accompanying expansion into new environments under reduced selective pressures, and the establishment of new barriers to interbreeding.

Obvious similarities can be seen between the chromosomal pattern and evolutionary phase of the Rutaceae and several other woody and pan-(sub)tropical families like Meliaceae (for extensive recent cytosystematic studies see Styles and Vosa [1971], and Mehra, Sareen, and Khosla [1972]), Simaroubaceae, and Anacardiacese (including Podaceae) (fig. 4). Diploids are scarce, represented only by single genera of Meliaceae and Anacardiaceae. The main represen-

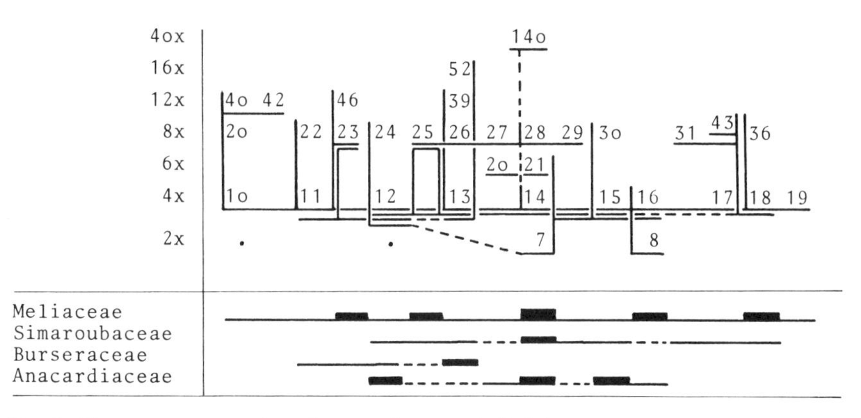

Fig. 4. Chromosome-number diagram for woody Rutales.

tation on paleopolyploid levels is $4x$, $6x$, and $8x$, but some ploidy lines extend to $12x$, $16x$, and even $40x$. Secondary dysploid changes are apparent on these ploidy levels, so that a rather complete and unbroken spectrum of chromosome numbers results. If one takes representation of numbers in diverse groups, and length of ploidy lines as indications of original and ancient base numbers, it again appears that these families of the Rutales had diploid, now nearly extinct progenitors with $n = 7$, and primary dysploid radiation to $n = 8$, 9 (and 10?) as well as to 6 (and 5?). Since Sapindaceae and other woody families of the Sapindales have similar chromosomal patterns, close relationships with Rutales and common affinities to Magnoliidae are suggested.

MAGNOLIIDAE

If we turn to the chromosomal differentiation pattern of woody Magnoliidae, e.g., Magnoliales (*s. str.*), Laurales, Illiciales, and some related but isolated families (figs. 5, 6), quite evident differences as compared with Rutales (figs. 3, 4) become apparent. The range of dysploid diferentiation per family is much narrower, reflecting a decrease of cytogenetic diversification in relation to morphological distinctness. There are obvious gaps in the chromosome-number series. Additionally, paleopolyploids dominate even more, and many families are represented only by $6x$ and higher ploidy levels. All of these aspects can be taken as evidence for great age, a high rate of extinction, and evolutionary stagnation.

From the criteria discussed earlier, and in accordance with many

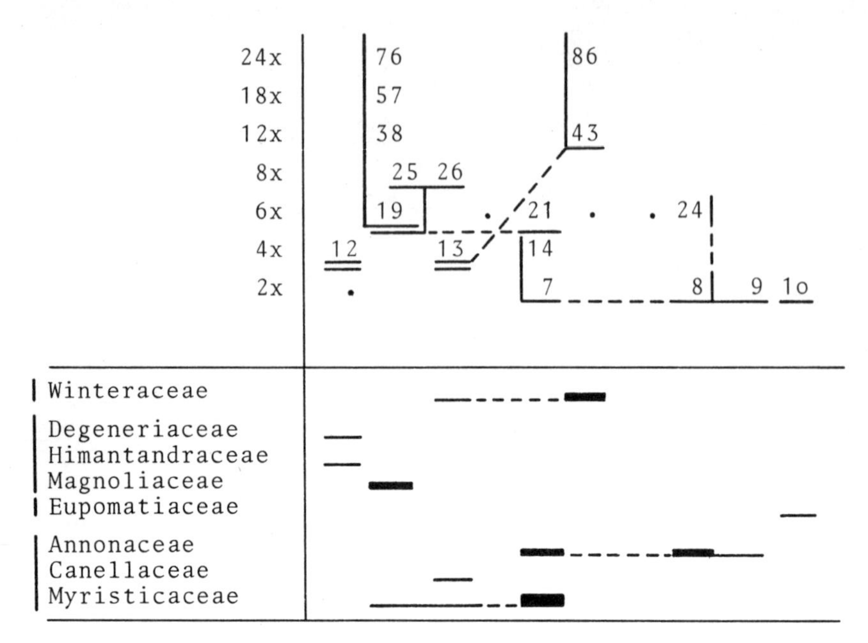

Fig. 5. Chromosome-number diagram for Magnoliales.

other authors, we can assume $n = 7$ to be the original base number for Magnoliidae. Information about the cytogenetic processes which have led to the present chromosomal pattern in woody Magnoliidae can be gained from several sources. Basic (ascending) dysploidy on the $2x$ level and some polyploidy ($4x$, $6x$) is exemplified by Annonaceae (Ehrendorfer et al., 1968; Walker, 1972). Chloranthaceae demonstrate progressive elimination of diploids ($n = 8$), major representation on the $4x$ level ($n = 14$ and 15, the latter from 7 + 8?), and occasional origin of $8x$ ($n = 30$). Within the Magnoliaceae family-group, the monotypic Himantandraceae from New Guinea and northeastern Australia has euchromatic nuclei and 12 pairs of rather symmetrical chromosomes (fig. 7), whose structure seems to reflect an origin from $n = 6 + 6$. Degeneriaceae, also with $n = 12$ and monotypic, is limited to Fiji. These two Western Pacific families evidently are very ancient and relic representatives of a $4x$ group and may have contributed to the origin of the paleo-$6x$ family, Magnoliaceae, which had attained a (now strongly fractionated) Northern Hemisphere distribution by the Cretaceous period (Sauer and Ehrendorfer, 1970); by further polyploidy, *Magnolia* has advanced to $12x$ and $24x$ levels. Finally, the dysploid reduction of the paleo-$4x$

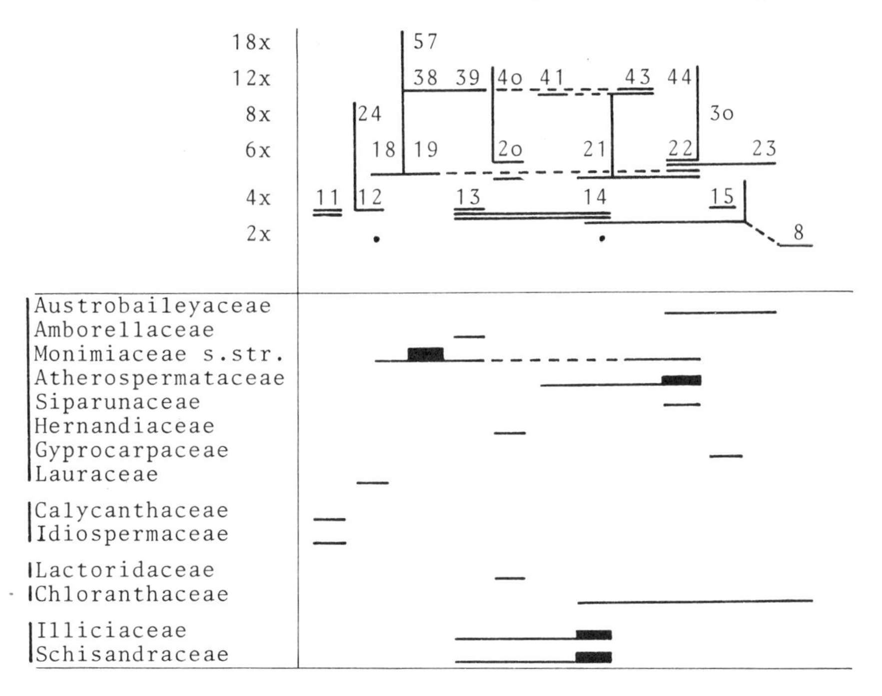

Fig. 6. Chromosome-number diagram for Laurales, Illiciales, and some isolated families.

chromosome number $n = 14$ to $n = 13$ by fusion of telocentrics and unequal translocations has been clearly demonstrated for Illiciaceae (Stone and Freeman, 1968).

Some hints as to possible affinities and systematic grouping of woody Magnoliidae can be gained from available chromosome information. The vesselless Winteraceae, probably the angiosperm family with the highest frequency of primitive attributes, is predominantly represented by paleo-$12x$ members with $n = 43$ and culminates with $24x$ in *Zygogynum. Tasmannia* (formerly *Drimys* sect. Tasmannia) with $n = 13$ deviates in many characters; separation as a distinct family could be considered. Segregation of the family as Winterales (Smith, 1972) is favored by karyological aspects. The same holds for the proposed grouping of Degeneriaceae, Himantandraceae ($x = 12$), and Magnoliaceae ($x = 19$) as Magnoliales *s. str.* (Smith, 1972; see also Baranova, 1972). Only Eupomatiaceae ($x = 10$) may not fit so well here. (Compare $x = 12 \rightarrow 10$ or possible relationships to Idiospermaceae and Calycanthaceae with $x = 11$.) Closer affinities be-

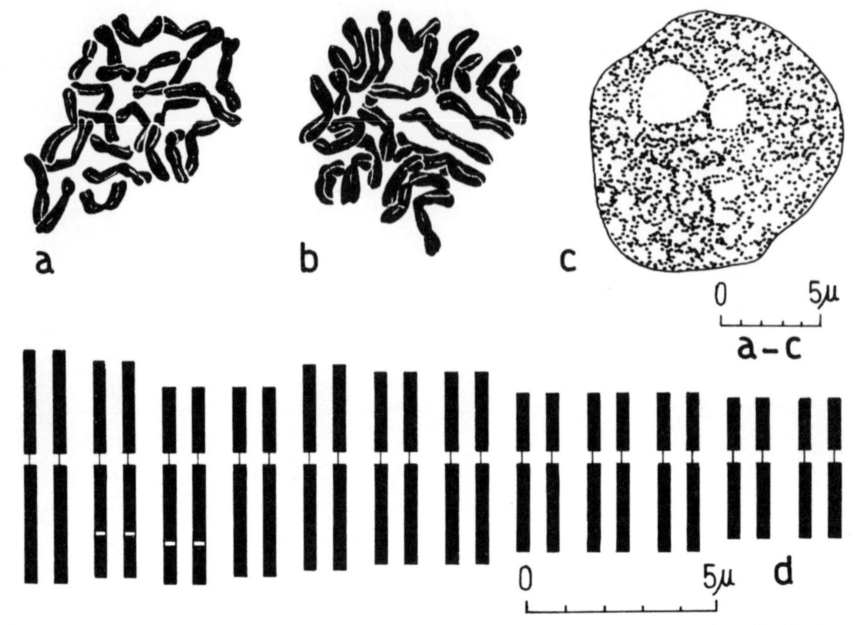

Fig. 7. Mitotic chromosomes, interphase nucleus, and karyotype of *Galbulimima baccata,* Himantandraceae. (From Sauer and Ehrendorfer, 1970; reproduced with permission.)

tween Annonaceae (n = 7, 14; 8, 24; 9), Canellaceae (n = 13, 14),[2] and Myristicaceae (n = 19, 21; 25, 26) are also feasible from a karyological viewpoint. Behnke (1971) has shown that the three families deviate by their P-type sieve-tube plastids, and the name Annonales has been proposed for this assemblage (Smith, 1972).

Laurales are also reasonably uniform in regard to chromosome pattern. Most of the very primitive families, including Austrobaileyaceae, Monimiaceae *s. str.* incl. *Hortonia*,[2] Atherospermataceae, Siparunaceae (the two latter formerly under Monimiaceae), and Hernandiaceae are represented on paleo-6x (or even higher) levels (n = 18–22 or 23; 38–44). A paleo-4x level characterizes the Lauraceae, Amborellaceae, and Gyrocarpaceae (n = 12, 13, 15; Lauraceae has little polyploidy and is very rarely reduced to n = 10). The only diploid recently reported[2] is Trimeniaceae with n = 8. The disputed separation of Hernandiaceae (n = 20) and Gyrocarpaceae (x = 15) is backed by the different chromosome numbers, as there is no clue for a common base with x = 5 in Laurales. Lac-

[2] New references from Goldblatt (1974) and Raven (1975, see fn. 1), not included in figs. 5 and 6.

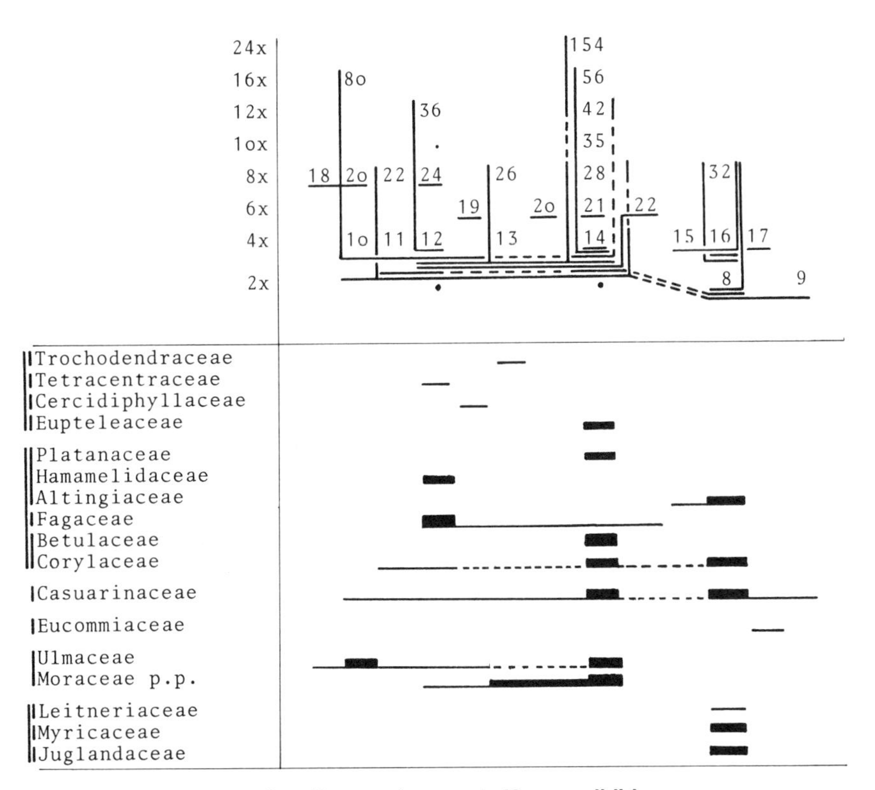

Fig. 8. Chromosome-number diagram for woody Hamamelididae.

toridaceae with $n = 20$ (or 21? [see Raven, Kyhos, and Cave, 1971]) are isolated, but may not be too far from Laurales. On the other hand, Calycanthaceae and the recently separated Idiospermaceae (Blake, 1972), both with $n = 11$, deviate rather grossly (see ultrastructure of P-type sieve-tube plastids [Behnke, 1971]). Assembling Illiciaceae and Schisandraceae into Illiciales (Takhtajan, 1969) is justified because of their similar chromosome pattern.

HAMAMELIDIDAE

The circumscription, grouping, and relationships of the predominantly anemophilous subclass called Hamamelididae are still much disputed (see, e.g., Kubitzki, 1973). Can available data on chromosome patterns in woody Hamamelididae (fig. 8) contribute to a solution of these problems?

Older chromosome counts for the monotypic Trochodendraceae and Tetracentraceae recently were corrected (Ratter and Milne,

1973) as $n = 20$ and $2n = \pm 48$. Like the monotypic Cercidiphyllaceae ($n = 19$) and Eupteleaceae ($n = 14$), these extremely isolated families are best regarded as separate orders. A consideration of the karyology and the morphological and phytochemical attributes of the four genera (*Trochodendron, Tetracentron, Cercidiphyllum* and *Euptelea*) point to some loose affinities between them, and to positions somewhere between Magnoliales/Illiciales and Hamamelidales (see Kubitzki, 1973, for references).

From Hamamelidales, the paleo-6x Platanaceae on $x = 7$, Hamamelidaceae *s. str.* with 4x, 8x, and 12x on $x = 6$, and the Altingiaceae (i.e., subfamilies Liquidambaroideae and Bucklandioideae = Symingtonioideae of Hamamelidaceae) with 4x and 8x on $x = 8$ are included in our chromosome diagram. Fagales (including Betulales) have a similar pattern. For Fagaceae, recent reports (see Soepadmo, 1972) indicate paleo-6x with $n = 22$ for the primitive *Trigonobalanus*, and paleo-4x, $n = 14$(?) apart from the widespread n = 12 for *Quercus; Nothofagus* has $n = 13$. Betulaceae have a basic paleo-4x, $n = 14$, while base numbers in Corylaceae obviously are $n = 14$ and $n = 8$. These similarities in chromosome pattern between Hamamelidales and Fagales (including Betulales) justify the closer approximation of the two orders originally proposed on the basis of other arguments (Endress, 1967).

Casuarinales are quite isolated. In parallel with considerable radiation in Australasia, chromosome numbers have strongly diversified and numerical plasticity may be linked to hybridization between 2x and 4x, with original bases $x = 8$ or 7. Woody Urticales, e.g., Ulmaceae and some Moraceae, exhibit comparable chromosome patterns, characterized by descending dysploidy from paleo-4x, $n = 14$ to 12 and 10. Finally, Leitneriaceae, Myricaceae, and Juglandaceae are probably best placed in separate orders but fit together in many characteristics, including their common and uniform base number, $x = 8$; affinities to Rutales are possible.

In spite of great diversity, signs of long independent evolution, and loss of many connecting groups, the taxa grouped under Hamamelididae seem to be held together by more than just parallel adaptations to anemophily. Chromosome patterns tend to substantiate such an interpretation. A comparison with woody Magnoliidae (figs. 5–6) reveals many similarities. Diploid base numbers have disappeared nearly completely. Only $n = 8$ and $n = 9$ are found in Corylaceae, Casuarinaceae, and Myricaceae; otherwise paleopolyploids clearly dominate. The same original base number as in Magnoliidae, $x = 7$, is apparent for Hamamelididae because of the wide representation of paleopolyploid derivatives of $x = 7$ in very diverse subgroups, and the number and length of polyploid lines originating from them. The range of dysploid differentiation within most families is quite limited, exceptions being Ulmaceae, Moraceae, *Corylus* (Corylaceae) and *Casuarina* (Casuarinaceae). In view of the obviously different chromosome patterns in other orders of woody angiosperms, such similarities clearly indicate a common cytogenetic disposition and some evolutionary affinity between Magnoliidae and Hamamelididae, and must not be lightly dismissed (Meeuse, 1970; Kubitzki, 1973). An independent origin of the two subclasses, and thereby a truly polyphyletic origin of angiosperms, appears quite unlikely under such circumstances.

ROSIDAE: SAXIFRAGALES AND MYRTALES

As a final example of how chromosome patterns and diagrams can illuminate questions of macroevolution and macrosystematics in primitive woody angiosperms, we can briefly discuss relevant information on Saxifragales (fig. 9) and Myrtales (fig. 10).

Woody Saxifragales are often regarded as an ancient basic group from which other, more advanced orders of Rosidae and Dilleniidae have originated (e.g., Kubitzki, 1969; Takhtajan, 1969). Such an assumption is well reinforced by the fact that woody Saxifragales have largely lost their diploid progenitors and are now mainly developed on $4x$ (and rarely $8x$) paleopolyploid levels. But, compared with woody Magnoliidae, there is more dysploid dispersion in most families, and ploidy lines are usually shorter; furthermore, a shift from the main base $x = 7$ to $x = 8$ is obvious.

While chromosome patterns of most families of Saxifragales fit together well, Pittosporaceae are somewhat anomalous, being represented only by $x = 12$. This lends support to other arguments which favor removal of this family from Saxifragales (e.g., Hegnauer, 1969). Another family that usually has been inserted in Saxifragales is the shrubby South African Bruniaceae. The fact that it has (contrary to earlier reports) $x = 11$ and $x = 20$ (with $2x$, $4x$, and $6x$ cytotypes) may also militate against its inclusion in Saxifragales (P. Goldblatt, personal communication, 1974).

On the basis of morphological and anatomical considerations, most authors concur that Myrtales probably have developed from sources near woody Saxifragales. Chromosome patterns actually demonstrate that woody Myrtales are in a more "juvenile" evolutionary phase than woody Saxifragales. This suggestion is substantiated by a shift of base numbers in Myrtales from $x = 8$ to 9 and 10 in relatively primitive families (e.g., Rhizophoraceae), and to 11, 12, and 13 in derived families (e.g., Myrtaceae, Combretaceae). Most

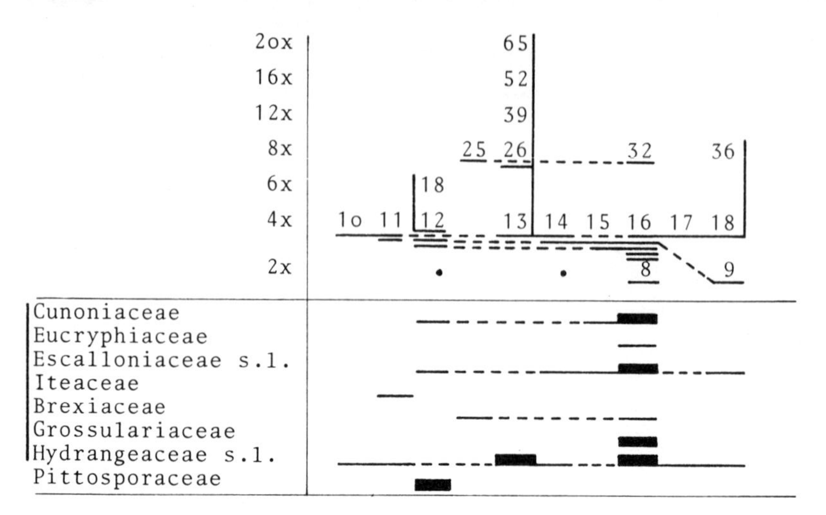

Fig. 9. Chromosome-number diagram for woody Saxifragales.

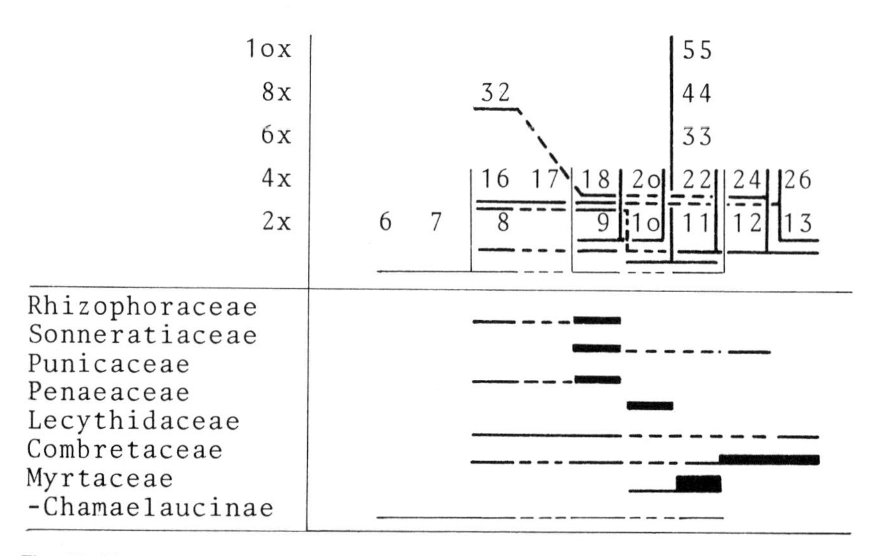

Fig. 10. Chromosome-number diagram for woody Myrtales.

families are represented by a fairly broad dysploid spectrum of diploid chromosome numbers, but ploidy lines are mostly short (except in *Eugenia* and *Pisidium* of Myrtaceae). Radiation into a variety of new habitats on the Australian continent has been paralleled in Myrtaceae-Chamaelaucinae by explosive, descending dysploidy and further polyploidy (Smith-White, 1959), comparable to the situation in

Rutaceae-Boronieae. Another karyologically versatile family of Myrtales is the tropical Melastomataceae, changing from a woody to an herbaceous habit.

DISCUSSION

When we review chromosomal differentiation patterns in woody angiosperms, a number of differences, in comparison with gymnosperms, become apparent: (1) Chromosomes are small to medium sized, with lower DNA content (2C nuclei usually with a range of about 15–35 picograms of DNA [El-Lakany and Dugle, 1972; Rees and Jones, 1972]). (2) Basic changes of chromosome numbers are common and have developed predominantly by descending or ascending dysploidy, involving unequal translocations, or by hybridization between different ploidy levels and subsequent stabilization; in contrast, Robertsonian changes by centromere fission (or fusion), and involving telocentric chromosomes, are evidently rare. (3) Polyploidy is widespread and common, with divergent evolution often continuing on higher ploidy levels. (4) Both dysploidy and polyploidy are generally less frequent in woody than in herbaceous groups; depending on the phase of evolutionary divergence, such changes may parallel the origin of taxa of infraspecific, specific, generic, or family rank.

The chromosome diagrams illustrate such different phases of evolutionary divergence as the following: (1) Rapid basic dysploid differentiation coupled with the incorporation of reproductive barriers seems to correspond to a primary phase of evolutionary diversification, and the occupation of unsaturated ecological niches under reduced selective pressure (anagenesis and early cladogenesis). Examples are Rutaceae-Boronieae or Myrtaceae-Chamaelaucinae. (2) Retarded dysploid differentiation and predominant chromosome structural and genic differentiation, coupled with hybridization and gradual transference to higher ploidy levels, appear to be characteristic for a secondary evolutionary phase with adaptive improvement, but are accompanied by a slowing down of evolutionary divergence in saturated ecosystems under increased selective pressure (late cladogenesis). Examples are Rutaceae-Aurantioideae, various Rutoideae, Annonaceae, Casuarinaceae, Meliaceae, and most woody Myrtales and woody Saxifragales. (3) Extinction and loss of diploids and later of polyploids, with progressive immobilization of relic survivors in shrinking niches and under unfavorable environmental conditions, characterize a final phase of stagnant and regressive

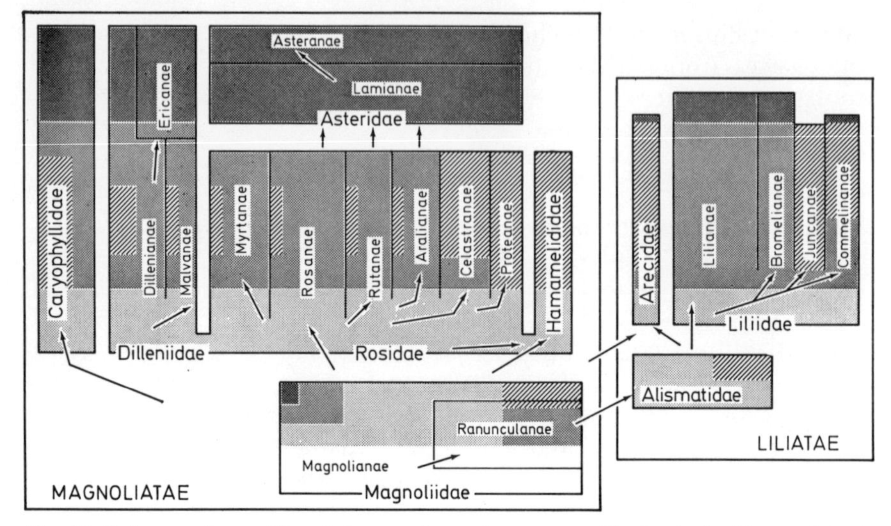

Fig. 11. Scheme of possible phylogeny in angiosperms with dicots (= Magnoliatae) and monocots (= Liliatae), and their subclasses and superorders. Different shades refer to increasing specialization in flower structure (from acyclic perianth, primary polyandry, and choricarpy indicated by white, to sympetalous perianth, reduced androecium, and coenocarpy, indicated by dark gray). (From Ehrendorfer, 1971; reproduced with permission.)

evolution (stasigenesis). Examples are Rutaceae-Zanthoxyleae and -Toddalioideae, many Magnoliales, Laurales, and several Hamamelididae. It is remarkable that recycling and thereby reversal of these phases seems to be possible. Examples are genera or families with active evolution on polyploid levels, like *Tasmannia*, Betulaceae, Ulmaceae, and Meliaceae.

In many respects, differences in chromosome patterns between the various primitive angiosperms discussed favor modern ideas about the possible phylogeny of angiosperms (see fig. 11). Nevertheless, the many general karyological similarities among all woody angiosperms, and particularly the remarkable parallelisms between those most-debated Magnoliidae and Hamamelididae, are strong arguments against a truly polyphyletic origin of angiosperms (as advocated, e.g., by Meeuse, 1970).

Conclusions

Our survey of chromosome numbers in various groups of seed plants, and its illustration by diagrams, has shown quite varied patterns of chromosome differentiation. Important criteria for the evaluation of such differences are: differentiation on diploid and polyploid levels,

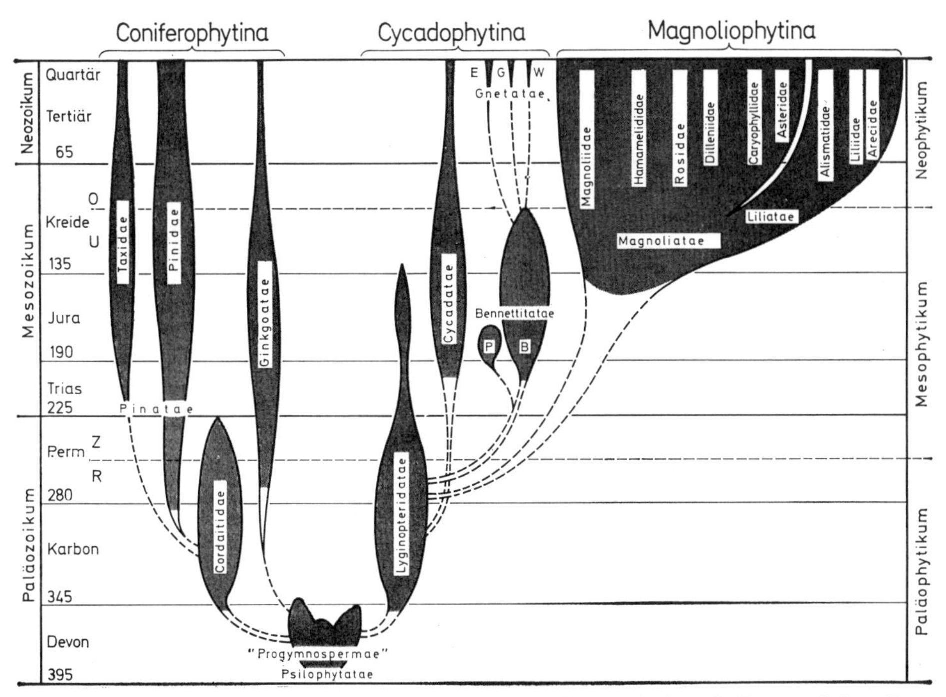

Fig. 12. Possible phylogeny of seed plants. *B*, Bennettitales, *P*, Pentoxylales, *E*, Ephedrales, *G*, Gnetales, *W*, Welwitschiales. (From Ehrendorfer, 1971; reproduced with permission.)

breadth of dysploid differentiation on various ploidy levels, length of polyploid lines, the continuous or interrupted nature of dysploid or polyploid series, and the relationship of these karyological features to the morphological and systematic differentiation of the particular group involved. Evidently, differences in respect to these criteria are due partly to different cytogenetic constitutions and partly to the different evolutionary phases of the groups.

The prominent differences in chromosome patterns between various gymnosperm and angiosperm groups are clear expressions of different cytogenetic potentials and different phases of the evolutionary development of these groups. Chromosome patterns are therefore an important aspect of macrosystematics and macroevolution.

A comparison between the general aspects of chromosome patterns, cytogenetic mechanisms for numerical change, and the differences in average chromosome size, DNA quantity, and redundancy in gymnosperms and angiosperms generally favors ideas about the phylogeny of these groups that were developed on the basis of other arguments (fig. 12). In addition, it appears that gymnosperms and angiosperms have followed different evolutionary strategies. Elaborating on suggestions of Miksche and Hotta (1973), one may speculate that large chromosomes and the greater quantity and redundancy of DNA have prevented (allo)polyploidy and more active divergent evolution in gymnosperms. In angiosperms, on the other hand, small chromosomes and smaller quantities of less repetitive DNA seem to have brought about more cytogenetic and evolutionary versatility, and better avenues for cyclic diversification on diploid and polyploid levels of organization.

Karyological patterns favor a common origin of gymnosperms (Coniferophytina and Cycadophytina), as well as a common origin of all angiosperms. Gnetatae may not be far from the roots of angiosperms.

Remarkably similar chromosome patterns in Magnoliidae and Hamamelididae support affinities between these subclasses and militate against a polyphyletic origin of these ancient angiosperms. Furthermore, karyological aspects confirm the connection between Magnoliidae and Rutales through the primitive Rutaceae-Zanthoxyleae and -Toddalioideae, progressions from woody Saxifragales to Myrtales, and so on.

References

Baranova, M. 1972. Systematic anatomy of the leaf epidermis in the Magnoliaceae and some related families. *Taxon* 21: 447–69.

Behnke, H. D. 1971. Sieve-tube plastids of Magnoliidae and Ranunculidae in relation to systematics. *Taxon* 20: 223–30.

Blake, S. T. 1972. *Idiospermum* (Idospermaceae), a new genus and family for *Calycanthus australiensis*. *Contrib. Queensland Herb.* 12: 1–38.

Bolkhovskikh, Z., Grif, V., Matrejeva, T., and Zakharyeva, O. 1969. *Chromosome Numbers of Flowering Plants.* V. L. Komarov Botanical Institute, Academy of Sciences of the USSR, Leningrad.

Chuang, T. I., and Hu, W. W. L. 1963. Study of *Amentotaxus argotaenia* (Hance) Pilger. *Bot. Bull. Acad. Sinica*, n.s. 4: 10–14.

Cronquist, A. 1968. *The Evolution and Classification of Flowering Plants.* Houghton, Boston.

Darlington, C. D., and Wylie, A. P. 1955. *Chromosome Atlas of Flowering Plants.* G. Allen, London.

Ehrendorfer, F. 1971. Systematik und Evolution: Spermatophyta, Samenpflanzen. In *Lehrbuch der Botanik für Hochschulen* (*"Strasburger"*). G. Fischer, Stuttgart.

Ehrendorfer F., Krendl, F., Habeler, E., and Sauer, W. 1968. Chromosome numbers and evolution in primitive angiosperms. *Taxon* 17: 337–53.

El-Lakany, M. H., and Dugle, J. R. 1972. DNA content in relation to phylogeny of selected boreal forest plants. *Evolution* 26: 427–34.

Endress, P. K. 1967. Systematische Studien über die verwandtschaftlichen Beziehungen zwischen den Hamamelidaceen und Betulaceen. *Bot. Jahrb. Syst.* 87: 431–525.

Engler, A. 1931. Rutaceae. In *Die natürlichen Pflanzenfamilien,* 2d ed., vol. 19a, pp. 187–359. W. Engelmann, Leipzig.

Engler, A. 1954–1964. *Syllabus der Pflanzenfamilien.* 12th ed. Gebrüder Borntraeger, Berlin.

Fish, F., and Watermann, P. G. 1973. Chemosystematics in the Rutaceae II. The chemosystematics of the *Zanthoxylum/Fagara* complex. *Taxon* 22: 177–203.

Goldblatt, P. 1974. A contribution to the knowledge of cytology in Magnoliales. *J. Arnold Arbor.* 55: 453–57.

Hair, J. B. 1963. Cytogeographical relationships of the southern Podocarps. In *Pacific Basin Geography,* ed. J. L. Gressitt, pp. 401–14. Bishop Mus. Press, Honolulu.

Hair, J. B., and Beuzenberg, E. J. 1958. Chromosomal evolution in the Podocarpaceae. *Nature* 181: 1584–86.

Hegnauer, R. 1969. *Chemotaxonomie der Pflanzen,* vol. 5. Birkhauser Verlag, Basel and Stuttgart.

Khoshoo, T. N. 1962. Cytogenetical evolution in the gymnosperms: Karyotype. In *Proceedings of the Summer School of Botany,* pp. 119–35. Ministry of Scientific Research, New Delhi.

Khoshoo, T. N. 1969. Chromosome evolution in cycads. In *Chromosomes Today,* vol. 2, ed. C. D. Darlington and K. R. Lewis, pp. 236–40. Plenum Press, New York.

Khoshoo, T. N., and Ahuja, M. R. 1963. The chromosomes and relationships of *Welwitschia mirabilis. Chromosoma* 14: 522–33.

Kubitzki, K. 1969. Chemosystematische Betrachtungen zur Grossgliederung der Dicotylen. *Taxon* 18: 360–68.

Kubitzki, K. 1973. Probleme der Gross-Systematik der Blütenpflanzen. *Ber. Deut. Bot. Ges.* 85: 259–77.

Marchant, C. J. 1968. Chromosome patterns and nuclear phenomena in the cycad families Stangeriaceae and Zamiaceae. *Chromosoma* 24: 100–134.

Meeuse, A. D. J. 1970. The descent of the flowering plants in the light of new evidence from phytochemistry and from other sources. *Acta Bot. Neerl.* 19: 61–72, 133–40.

Mehra, P. N., Sareen, T. S., and Khosla, P. K. 1972. Cytological studies on Himalayan Meliaceae. *J. Arnold Arbor.* 53: 558–68.

Miksche, J. P. 1967. Variation in DNA content of several gymnosperms. *Can. J. Genet. Cytol.* 9: 717–22.

Miksche, J. P., and Hotta, Y. 1973. DNA base composition and repetitious DNA in several conifers. *Chromosoma* 41: 29–36.

Moore, R. J., ed. 1972. Index to plant chromosome numbers for 1970. *Regnum Veg.* 84: 1–138.

Price, H. J., Sparrow, A. H., and Nauman, A. F. 1973. Evolutionary and developmental considerations of the variability of nuclear parameters in higher plants. I. Genome volume, interphase chromosome volume, and estimated DNA content of 236 gymnosperms. *Brookhaven Symp. Biol.* 25: 390–421.

Ratter, J. A., and Milne, C. 1973. Chromosome numbers of some primitive angiosperms. *Notes Roy. Bot. Gard. Edinburgh* 32: 423–28.

Raven, P. H., Kyhos, D. W., and Cave, M. S. 1971. Chromosome numbers and relationships in Annoniflorae. *Taxon* 20: 479–83.

Rees, H., and Jones, R. N. 1972. The origin of the wide species variation in nuclear DNA content. *Int. Rev. Cytol.* 32: 53–92.

Sauer, W., and Ehrendorfer, F. 1970. Chromosomen, Verwandtschaft und Evolution tropischer Holzpflanzen, II. Himantandraceae. *Österr. Bot. Z.* 118: 38–54.

Saylor, L. C. 1972. Karyotype analysis of the genus *Pinus* subgenus *Pinus*. *Silvae Genet.* 21: 155–63.

Smith, A. C. 1972. An appraisal of the orders and families of primitive extant angiosperms. *J. Indian Bot. Soc.* 50A: 215–26.

Smith-White, S. 1959. Cytological evolution in the Australian flora. *Cold Spring Harbor Symp. Quant. Biol.* 24: 273–89.

Soepadmo, E. 1972. Fagaceae. *Flora Malesiana* 1(7): 265–403.

Stone, D. E., and Freeman, J. L. 1968. Cytotaxonomy of *Illicium floridanum* and *I. parviflorum* (Illiciaceae). *J. Arnold Arbor.* 49: 41–51.

Styles, B. T., and Vosa, C. G. 1971. Chromosome numbers in the Meliaceae. *Taxon* 20: 485–99.

Sugihara, Y. 1943. Notes on *Amentotaxus*. *Bot. Mag. (Tokyo)* 57: 404–5.

Takhtajan, A. 1969. *Flowering Plants: Origin and Dispersal.* Smithsonian Institution Press, Washington, D.C.

Terasmaa, T. 1971. Karyotype analysis of Norway Spruce *Picea abies* (L.) Karst. *Silvae Genet.* 20: 179–82.

Walker, J. W. 1972. Chromosome numbers, phylogeny, phytogeography of the Annonaceae and their bearing on the (original) basic chromosome numbers of angiosperms. *Taxon* 21: 57–65.

Comparative Pollen Morphology and Phylogeny of the Ranalean Complex

JAMES W. WALKER, ***Department of Botany University of Massachusetts, Amherst***

CERTAIN FAMILIES OF dicotyledonous angiosperms that have been grouped in part into formal taxa under names such as Ranales, Polycarpicae, Apocarpicae, Monochlamydeae, Magnoliales, Magnoliidae, and Annoniflorae are now generally recognized as representing the most primitive group of extant flowering plants (Cronquist, 1968; Thorne, 1968; Takhtajan, 1969). Many of these families possess characters which are considered primitive by most phylogenists, including vesselless wood, monosulcate pollen, unsealed carpels, laminar stamens, and free, spirally arranged floral parts indefinite in number. However, since each of the taxa mentioned above has been circumscribed to include or exclude different families of putatively primitive angiosperms, I have chosen to use the loose, informal term "ranalean complex" to refer to those families of primitive (dicotyledonous) angiosperms (see table 1) whose pollen morphology is the subject of this paper. All other families of dicotyledonous angiosperms are referred to as "higher dicotyledons."

This paper represents a preliminary summary of data gathered through light-microscope study of acetolyzed pollen of over 1000 species and 230 genera of primitive, ranalean angiosperms. Pollen representing more than 100 of these genera was also examined with the scanning electron microscope, and thin sections of pollen grains of a number of ranalean species were observed with the light microscope or the transmission electron microscope. A set of the pollen slides upon which the light-microscope part of this study is based is on deposit in the paleobotanical collections of the Botanical Museum of Harvard University. All material studied and all slides made are vouchered. Palynological accession numbers were given to all pollen samples studied except those from material that I personally collected; the latter are designated by my (herbarium) collection number. Accession numbers (preceded by a "P") or my collection numbers are listed in the figure descriptions after each species whose pollen is illustrated. Unless otherwise indicated, all photographs in this paper represent scanning electron micrographs. Pollen was investigated from a total of 35 different angiosperm families which, as a group, constitute all families generally considered to be primitive dicotyledons.

The purpose of this paper is to examine what light the comparative pollen morphology and phylogeny of the ranalean complex sheds on the problem of the origin and early evolution of angiosperms. With this goal in mind, the paper has been divided into four parts. Phylogenetically useful pollen characters are briefly outlined, followed by a family-by-family description of the pollen of the ranalean complex. In the third part of the paper, early evolutionary trends observed in the pollen of extant primitive angiosperms are examined. The paper concludes with a discussion of the nature of ancestral angiosperm pollen and its bearing on the question of the origin of flowering plants.

Phylogenetically Useful Pollen Characters

For purposes of this paper, phylogenetically useful pollen characters have been organized into five categories. Types of apertures and pollen-grain polarity, symmetry, and shape, which are all largely correlated with aperture type, are discussed initially. Then follows a brief review of the characters inherent in the pollen wall itself. Finally, types of pollen-units and pollen-grain size are examined.

APERTURE TYPES

Aperture type (along with the pollen wall) is one of the most important phylogenetic pollen characters. Apertures represent specially delimited, generally thin-walled areas in the outer pollen wall or exine through which the pollen tube usually emerges at the time of germination. In a phylogenetic-evolutionary context, the most significant feature of pollen apertures is that they are not located randomly on the surface of the pollen grain but usually have very definite placement with reference to the grain's poles and equator as defined by its position in the pollen tetrad. The fact that all pollen results from meiosis of pollen mother cells allows certain dimensions of even pollen grains which are shed at maturity as solitary grains or monads to be defined in terms of the spatial relationships of the four grains of the pollen tetrad of which they were previously members (fig. 1). The line passing through the center of each grain from the

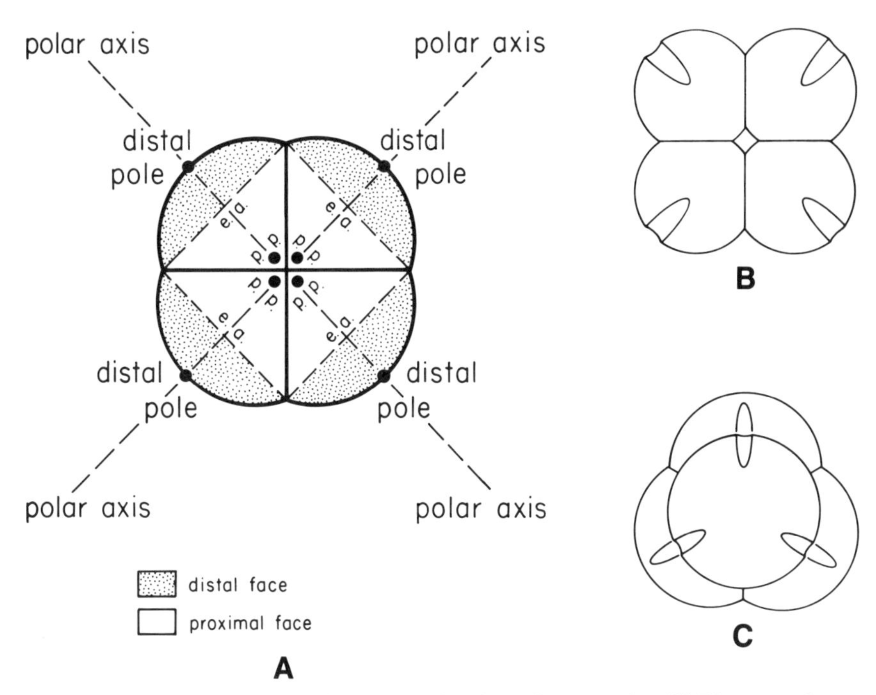

Fig. 1. Spatial relationships of pollen grains in pollen tetrads. (A) Diagram of a tetragonal pollen tetrad (which has all pollen grains in the same plane), showing the polar axis, one equatorial axis of the equatorial plane, the distal face, the proximal face, the distal pole, and the proximal pole of each pollen grain in the tetrad; *e.a.*, equatorial axis; *p.p.*, proximal pole. (B) Diagram of a tetragonal pollen tetrad composed of four anasulcate pollen grains. (C) Diagram of a tetrahedral pollen tetrad (which has pollen grains in more than one plane and arranged at the corners of a tetrahedron) composed of four tricolpate pollen grains. The top pollen grain is shown in distal-polar view, with its distal face up and its polar axis perpendicular to the plane of the figure.

outside to the center of the pollen tetrad (or to the center of the meiotic tetrad at the time of its formation, in the case of solitary grains) is defined as the polar axis of the pollen grain, while the plane which perpendicularly bisects the polar axis is termed the equatorial plane. The equatorial plane contains two or more equatorial axes and forms the boundary between the distal and proximal faces of the pollen grain. The distal face (as well as the distal pole) is directed away from the center of the pollen tetrad, while the proximal face and proximal pole are directed inward, facing the center of the tetrad.

Pollen apertures that occur within the ranalean complex include the following types:

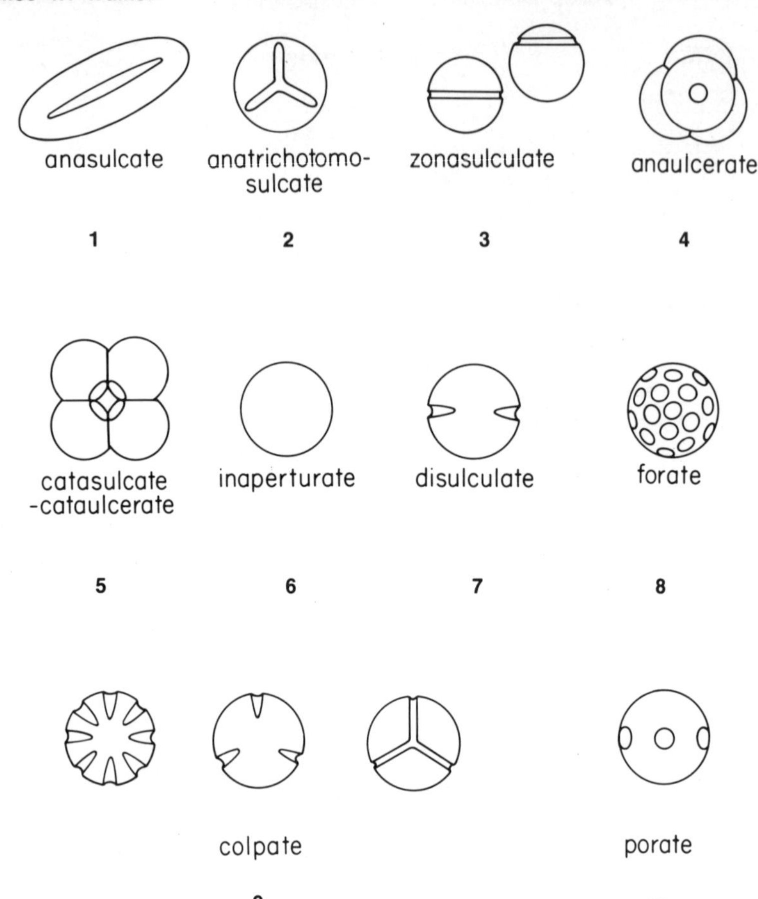

Fig. 2. Pollen aperture types found in the ranalean complex. In (3), *at left,* zonizonasulculate pollen grain; *at right,* ana- or catazonasulculate pollen grain, depending on whether the top of the grain shown represents the distal or proximal pole. In (9) *left to right,* pollen grains that are polycolpate, tricolpate, and tricolpate with colpi fused at the poles.

1. *Anasulcate,* with a single, elongate aperture at the distal pole (fig. 2: 1).
2. *Anatrichotomosulcate,* with a single, three-armed aperture at the distal pole (fig. 2: 2).
3. *Zonasulculate,* with an encircling, ring- or band-like aperture, which may be either around the equator (zonizonasulculate) or around the distal pole (anazonasulculate) or proximal pole (catazonasulculate) (fig. 2: 3).

4. *Anaulcerate,* with a single, rounded aperture at the distal pole (fig. 2: 4).
5. *Catasulcate-cataulcerate,* with a single, elongate (catasulcate) or rounded (cataulcerate) aperture at the proximal pole (fig. 2: 5). Except for some catasulcate-cataulcerate pollen in the Annonaceae (Walker, 1971b, 1972b), all other sulcate ranalean pollen appears to be anasulcate (Bailey and Swamy, 1949; Swamy, 1949; Canright, 1953; Wilson, 1964).
6. *Inaperturate,* without any aperture(s) (fig. 2: 6).
7. *Disulculate,* with two separate, elongate apertures located on and parallel to the equator, i.e., with their long axis perpendicular to the polar axis (fig. 2: 7). Should such apertures become rounded, the pollen may be described as diulculate.
8. *Forate,* with several to numerous rounded apertures distributed over the entire pollen-grain surface (fig. 2: 8).
9. *Colpate,* with elongate apertures that have their long axis perpendicular to the equator, and that are normally bisected by the equatorial plane (fig. 2: 9).
10. *Porate,* with round apertures located on the equator (fig. 2: 10). Unlike ulculate pollen, which generally appears to have evolved from pollen with elongate, sulculate apertures, porate pollen in the ranalean complex has evolved *de novo* from inaperturate pollen grains.

Weakly defined apertures may be indicated by introduction of the syllable *-oid-* into the terms describing the corresponding well-developed apertures, e.g., colpoidate pollen would have poorly developed colpi or colpoids. The prefixes *ana-* and *cata-* are used to refer to distal- and proximal-polar apertures, respectively, and the prefix *zoni-* is used to mean "at the equator."

POLLEN-GRAIN POLARITY, SYMMETRY, AND SHAPE

Largely correlated with aperture type are the interrelated characters of pollen-grain polarity, symmetry, and shape. Pollen may be apolar (fig. 2: 6, 8), without distinct poles once the grains have separated from the meiotic pollen tetrad, or polar, with recognizable poles even after the grains have separated from the tetrad. Polar pollen grains may be divided into two major subcategories: isopolar (fig. 2: 7, 9, 10), with the equatorial plane dividing the grain into similar halves, and heteropolar (fig. 2: 1, 2, 4, 5), with the polar faces markedly dissimilar.

Pollen symmetry is based largely on the number of planes of symmetry that exist in a particular grain as observed from polar view. Rarely, pollen may be asymmetric (fig. 2: 8) because irregular grain shape or the unsymmetrical distribution of apertures renders it incapable of division into similar halves. Most pollen, however, is symmetrical. Symmetric pollen may be further classified as radiosymmetric, bilateral, isobilateral, or biradial. Radiosymmetric pollen (fig. 2: 2–4, 6, 9, 10) is divisible into equal, symmetrical portions by any of three or more equally long planes of symmetry passing through the polar axis. Bilateral pollen grains (fig. 2: 1) are doubly symmetrical, i.e., divisible into two similar halves by either of two (never three or more) mutually perpendicular planes of symmetry passing through the polar axis. The two lines of symmetry are of different length; the equatorial axes are not equally long and the equatorial outline of the grain is elliptical or oval but not circular. Isobilateral and biradial pollen grains are also doubly symmetrical, but unlike bilateral pollen they have two mutually perpendicular planes of symmetry of the same length passing through the polar axis, i.e., the equatorial axes are equally long and the equatorial outline of the grain is usually more or less circular. Isobilateral pollen grains (fig. 5: 4, 5) are heteropolar and monosulcate, their type of symmetry being a function largely of the elongate nature of their polar aperture. They are essentially bilateral pollen grains with a round rather than elongate equatorial outline. Biradial pollen (fig. 2: 7) is isopolar and equatorially diaperturate (i.e., either disulculate-diulculate or dicolpate-diporate), its type of symmetry being chiefly a function of its aperture number (two). Biradial pollen grains are essentially radial pollen with two rather than three or more apertures.

Normal angiosperm pollen (fixiform pollen with a definite shape) may be divided into two basic shape classes: boat-shaped and globose. Boat-shaped pollen (fig. 2: 1) has a short polar axis and one equatorial axis that is longer than the other, resulting in a generally elliptical to oblong equatorial outline. Boat-shaped pollen may be further divided into a number of subtypes depending on the ratio of the longer equatorial axis to the shorter. If this ratio is greater than 1.00 but less than 1.50, the pollen is boat-shaped–elliptic (fig. 4: 3), whereas pollen with a ratio between 1.50 and 1.99 is boat-shaped–oblong (fig. 4: 2). Finally, any pollen grains with a ratio equal to or greater than 2.00 are described as boat-shaped–elongate (fig. 14: 3). Globose pollen (fig. 2: 2–10), on the other hand, is basically spherical; its equatorial axes are all the same length, resulting in a nearly circular equatorial outline or some modification thereof, such as triangular or lobed. Globose pollen may also be further subdivided, depending on the ratio of polar to equatorial axes. If both polar and

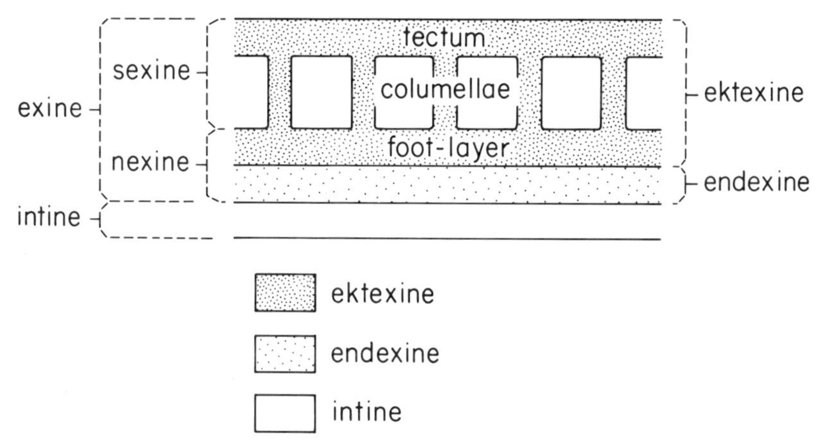

Fig. 3. Pollen-wall stratification observed in a typical angiosperm pollen grain that has an endexine and a foot-layer.

equatorial axes are the same length, the pollen is spherical (fig. 2: 6). If the polar axis is shorter than the equatorial axes, i.e., the grains are compressed along the polar axis, the pollen is flattened or oblate (fig. 4: 5, 6). Finally, if the polar axis is longer than the equatorial axes, i.e., the grains are compressed around the equator rather than along the polar axis, the pollen is prolate.[1] At times, diaperturate, globose pollen such as disulculate pollen may be equatorially elongate instead of spherical, with one equatorial axis longer than the other. Although such pollen may be more or less boat-shaped, the latter term is reserved for monosulcate pollen, and globose pollen of this type (which is also secondarily bilateral) is described as globose-elliptic.

POLLEN-WALL ARCHITECTURE

The architecture of the pollen wall provides a number of phylogenetically useful characters, among which exine stratification, structure, and sculpturing appear to be basic. With reference to stratification, the pollen wall of most angiosperms (fig. 3) appears to consist of two fundamentally different layers: an inner, primarily cellulosic layer (the intine), which is usually destroyed upon acetolysis (Faegri and Iversen, 1964), and an outer, acetolysis-resistant layer (the exine), which is composed of oxidative polymers of carotenoids or carotenoid esters and known as sporopollenin (Shaw, 1971). Since most modern pollen is prepared for study by acetolysis and since the in-

[1] The terms oblate and prolate are used here in a broader sense than that of Erdtman (1966). I suggest that Erdtman's oblate and prolate be replaced by the terms *euoblate* and *euprolate*, so that the former pair may be used as broader grouping terms.

tine is lacking in fossil pollen, for all practical purposes the study of pollen morphology consists of the study of exine. In most angiosperm pollen the exine is three-layered, consisting of a basal layer or nexine ("nonsculptured exine"), internal, upright, rodlike elements known as columellae (bacula), and a rooflike layer, or tectum. Pollen of this type is described as tectate, and the columellae and tectum in such pollen constitute the sexine ("sculptured exine"). In many angiosperms the exine is also chemically differentiated into two distinct layers (the outer known as the ektexine and the inner the endexine), which may or may not be equivalent to the morphologically defined sexine and nexine. Particularly when examined with the transmission electron microscope, an outer layer of the nexine in some angiosperms appears to be similar chemically to the sexine (columellae and tectum, if present). This sexinous, outer zone which occurs in the nexine of some angiosperms has been designated the foot-layer.

In the normal, three-layered or tectate angiosperm pollen grain a distinction may be made between the internal, infratectal columellae located *below the tectum* (between it and the nexine), which constitute "structure," and any external supratectal elements located *upon the tectum,* such as spines or granules, which constitute "sculpturing." With reference to structure, angiosperm pollen may be divided into a number of exine structure types (see fig. 13). Tectate pollen itself may be further categorized as either tectate-imperforate (without any holes, or perforations, in the tectum) or tectate-perforate (with small holes). Should the tectal perforations in a tectate-perforate exine enlarge so that their diameter becomes greater than the breadth of the pollen wall between them, the result may be an open network or reticulum composed of remnants of the once-complete tectum and either held up by otherwise free columellae or lacking columellae; such pollen is described as semitectate. Further, should the tectum be lost entirely and only free, exposed columellae (or their modified but homologous forms) remain, the pollen would be described as intectate. Surface elements present in semitectate and intectate pollen (the reticulum, the rodlike columellae, etc.) are here defined as comprising simultaneously both structure and sculpturing since sculpturing in semitectate and intectate pollen (at least within the ranalean complex) appears to be homologous with previously existing structure in tectate pollen grains. Finally, some ranalean families have pollen grains with a morphologically homogeneous exine that appears to be primitively devoid of columellae (Walker and Skvarla, 1975). Such pollen has been designated "atectate."

In its broadest definition, exine sculpturing consists of any exposed surface details of the pollen wall. The main types of sculptured pollen found in the ranalean complex include pollen grains which are psilate (smooth), foveolate (pitted), fossulate (grooved), scabrate (with very fine projections), verrucate (warty), baculate (with rodlike sculpturing elements), pilate (with rodlike sculpturing elements with swollen heads), echinate (spiny), rugulate (with elongate sculpturing elements that are irregularly distributed), reticulate (with sculpturing elements forming an open network or reticulum), and striate (with elongate, more or less parallel sculpturing elements).

POLLEN-UNIT

The pollen-unit (Walker, 1971b) is the grouping in which pollen is found at maturity within the anther locules of the stamen. Like the majority of angiosperms, most families of the ranalean complex have solitary pollen grains or monads at maturity. However, some ranalean taxa have their pollen grains in permanent tetrads or polyads. Tetrads represent a retention of the four products resulting from meiosis of the pollen mother cell, while polyads are pollen-units generally of a multiple of four; for example, octads or sixteens.

POLLEN-GRAIN SIZE

Order of magnitude, based on defined size classes, is probably the single most useful measurement of whole pollen grains. I have adopted the following size classes (in part after Erdtman, 1945), based on the length of the longest grain axis, exclusive of sculpturing elements in baculate, pilate, and echinate grains: minute grains (less than 10 μ), small grains (10–24 μ), medium-sized grains (25–49 μ), large grains (50–99 μ), very large grains (100–199 μ), and gigantic grains (200 μ or larger).

Comparative Palynology of the Ranalean Complex

The pollen morphology of each ranalean family included in this study is summarized in the form of a generalized palynological description. For each family the following information is given: number of genera and species; geographical distribution; pollen morphology; remarks on the taxonomic and phylogenetic significance of the pollen morphology; list of major palynological references; and list of pollen illustrations (figures) in this paper. The nature of eight phylogenetically important pollen characters (aperture types, pollen polarity, pollen symmetry, pollen shape, exine structure, exine sculpturing, pollen-units, and pollen size) is recorded for each family. Other workers who have also studied the pollen of a number of ranalean families include Agababian (1966–1972), Mitroiu (1963, 1966, 1970), and Nair (1965, 1967, 1968, 1970).

Table 1 represents my current thinking on the classification of the ranalean complex. Although this classification reflects the use of palynological characters, I should emphasize that it is based on the correlation of as many characters as possible, with particular emphasis on characters of floral morphology, wood anatomy, cytotaxonomy, and phytogeography. I must also stress that this is a preliminary classification, which probably will be changed in some details as more phylogenetic information becomes available about the members of the ranalean complex.

MAGNOLIALES

The order Magnoliales consists of seven families that are grouped into the two suborders Magnoliineae and Annonineae. It contains taxa which as a whole retain more primitive characters than any other members of the ranalean complex. Anasulcate pollen is the most common aperture type in the Magnoliales, occurring in some or all members of six of the seven families. Occasionally, anatrichotomosulcate grains are found together with anasulcate pollen in some members of the Annonaceae and Canellaceae. Anasulcate pollen represents the only known aperture type (exclusive of occasional grains which may be anatrichotomosulcate) that is found in four families in the order (Magnoliaceae, Degeneriaceae, Himantandraceae, and Canellaceae). The family Eupomatiaceae is unique in the order in having zonasulculate pollen. Inaperturate pollen is found in some members of Annonaceae and Myristicaceae. Annonaceae appears to be unique within the ranalean complex in having catasulcate-cataulcerate pollen (Walker, 1971a). A few genera in this family have disulculate pollen, an aperture type otherwise unknown in the order.

The Magnoliales have more taxa with boat-shaped pollen than any other ranalean order, and they are the only order with pollen grains that are boat-shaped–elongate. Typically boat-shaped pollen is found in the families Magnoliaceae, Degeneriaceae, and Annonaceae. Pollen is isobilateral and markedly globose-oblate in many Canellaceae and Myristicaceae, whereas it is globose-spherical in the Himantandraceae, many Annonaceae, and some Canellaceae and Myristicaceae. Most pollen in the order is tectate, although some pollen may be semitectate in the family Myristicaceae and, more rarely, in the Annonaceae.

Within the ranalean complex, intectate pollen is restricted to three relatively advanced genera in the magnolialean suborder Annonineae—*Ophrypetalum* and *Trigynaea* in the Annonaceae and

Table 1. Classification of the ranalean complex [a]

- MAGNOLIALES
  - Magnoliineae
    - Magnoliaceae
    - Degeneriaceae
    - Himantandraceae
    - Eupomatiaceae
  - Annonineae
    - Annonaceae
    - Canellaceae
    - Myristicaceae
- LAURALES
  - Austrobaileyineae
    - Austrobaileyaceae
  - Calycanthineae
    - Calycanthaceae
    - Idiospermaceae
  - Trimeniineae
    - Trimeniaceae
  - Monimiineae
    - Amborellaceae
    - Monimiaceae (including Hortoniaceae and Siparunaceae)
    - Atherospermataceae
    - Hernandiaceae
  - Laurineae
    - Gomortegaceae
    - Lauraceae (including Cassythaceae)
    - Gyrocarpaceae
- WINTERALES
  - Winterineae
    - Winteraceae
  - Illiciineae
    - Illiciaceae
    - Schisandraceae
- ARISTOLOCHIALES
  - Aristolochiaceae
- PIPERALES
  - Chloranthineae
    - Chloranthaceae
  - Lactoridineae
    - Lactoridaceae
  - Piperineae
    - Saururaceae
    - Piperaceae (including Peperomiaceae)
- TROCHODENDRALES
  - Trochodendraceae
  - Tetracentraceae
- CERCIDIPHYLLALES
  - Cercidiphyllaceae
- EUPTELEALES
  - Eupteleaceae
- NYMPHAEALES
  - Nymphaeineae
    - Nymphaeaceae (including Euryalaceae and Nupharaceae)
    - Nelumbonaceae
    - Cabombaceae
    - Barclayaceae
  - Ceratophyllineae
    - Ceratophyllaceae

[a] With reference to subclasses of the Takhtajan-Cronquist system of angiosperm classification (Cronquist, 1968; Takhtajan, 1969), the orders Magnoliales, Laurales, Winterales, Aristolochiales, and Piperales would constitute my subclass Magnoliidae, the Trochodendrales, Cercidiphyllales, and Eupteleales would be primitive orders of the subclass Hamamelididae, and I would elevate the Nymphaeales to a subclass Nymphaeidae.

some species of the myristicaceous genus *Horsfieldia*. Pollen which appears to be primitively columellaless or atectate occurs in the families Magnoliaceae, Degeneriaceae, Eupomatiaceae, and Annonaceae (Walker and Skvarla, 1975). Sculpturing in the suborder Magnoliineae is mainly psilate, foveolate, fossulate, scabrate, or more rarely verrucate, whereas in the Annonineae these five sculpturing types occur along with reticulate pollen (in the Myristicaceae and, more rarely, in the Annonaceae), baculate and pilate pollen (rarely in the Annonaceae), and echinate pollen (in several genera of the Annonaceae and in one genus of Myristicaceae). All families in the order shed their pollen in the form of solitary grains, except the Annonaceae, in which many genera have their pollen grains united in tetrads or less frequently in polyads. Pollen size in the order ranges from small to gigantic. The Magnoliaceae, Degeneriaceae, and primitive genera of the Annonaceae all have pollen chiefly in the large size class, whereas pollen in the Canellaceae and Myristicaceae is generally medium sized to small. Again, tremendous diversity exists in the Annonaceae, where the pollen may range from small to gigantic within the taxonomic limits of a single family.

Magnoliaceae

12 genera; ca. 220 species

Distribution. The family is centered in Indo-Malaysia. Nine genera are strictly Indo-Malaysian (*Michelia, Manglietia, Elmerrillia, Aromadendron, Pachylarnax, Alcimandra, Kmeria, Paramichelia,* and *Tsoongiodendron*), and three genera also occur in the New World (*Magnolia, Talauma,* and *Liriodendron*).

Pollen morphology. Pollen grains anasulcate. Heteropolar. Bilateral. Boat-shaped–elongate, –oblong, or –elliptic. Atectate or tectate. Psilate, foveolate, fossulate, scabrate, or verrucate. Monads. Large to very large or medium sized.

Remarks. The Magnoliaceae is the only relatively large ranalean family to have anasulcate pollen exclusively, and as such it represents one of the palynologically most primitive of extant angiosperm families. The pollen of *Liriodendron* stands somewhat apart from that of the rest of the Magnoliaceae in being conspicuously verrucate.

Palynological references. Wodehouse, 1935 (pp. 322–33); Erdtman, 1943 (p. 108), 1966 (pp. 254–56); Canright, 1953; Ikuse, 1956; Hayashi, 1960; Veloso and Barth, 1962; Ueno, 1962, 1963; Mitroiu, 1966, 1970 (p. 9); Huang, 1967; Agababian, 1968b, 1972b; Roland, 1968; Rao and Lee, 1970; Praglowski and Punt, 1973; Praglowski, 1974; Walker and Skvarla, 1975.

Figure. 4: 1, 2.

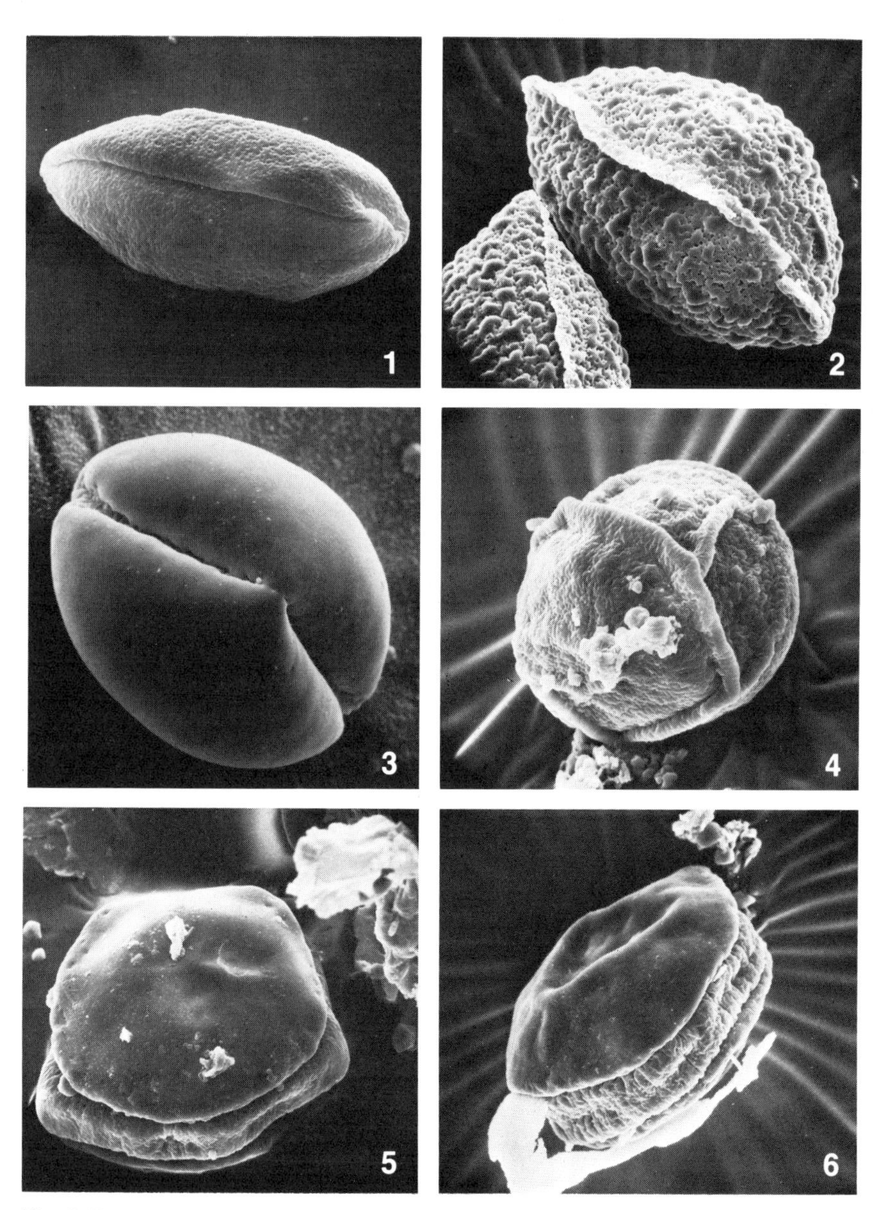

Fig. 4. Pollen of the order Magnoliales, suborder Magnoliineae, families Magnoliaceae (1, 2), Degeneriaceae (3), Himantandraceae (4), and Eupomatiaceae (5, 6). (1) *Magnolia fraseri* Walt. (P-1048) (× ca. 780). (2) *Liriodendron tulipifera* L. (P-30) (× ca. 1200). (3) *Degeneria vitiensis* I. W. Bailey and A. C. Smith (P-12) (× ca. 2160). (4) *Galbulimima* (=*Himantandra*) belgraveana (F. Muell.) F. Muell., aperture not visible (P-13) (× ca. 2400). (5, 6) *Eupomatia laurina* R. Br. (P-7) (both, × ca. 1680). (5) More or less polar view. (6) More or less equatorial view.

Degeneriaceae

1 genus; 1 species

Distribution. The single species, *Degeneria vitiensis* I. W. Bailey and A. C. Smith, is endemic to Fiji.

Pollen morphology. Pollen grains anasulcate. Heteropolar. Bilateral. Boat-shaped–oblong or –elliptic. Atectate. Psilate, with a few foveolae (pits). Monads. Large.

Remarks. The anasulcus of *Degeneria* is distinctive in that it runs more than halfway around the grain.

Palynological references. Bailey and Smith, 1942; Swamy, 1949; Dahl and Rowley, 1965; Erdtman, 1966 (p. 144); Walker and Skvarla, 1975.

Figures. 4: 3; 14: 7.

Himantandraceae

1 genus; 2 species

Distribution. The single genus, *Galbulimima* (=*Himantandra*), has one species in New Guinea (*G. belgraveana* [F. Muell.] F. Muell.), and a second species in Queensland, Australia (*G. baccata* [F. M. Bailey] Diels).

Pollen morphology. Pollen grains anasulcate. Heteropolar. Isobilateral. Globose-spherical. Atectate (?). More or less psilate. Monads. Medium sized.

Remarks. The exine of *Galbulimima* is quite reduced and is frequently destroyed during acetolysis.

Palynological references. Bailey, Nast, and Smith, 1943; Erdtman, 1966 (p. 204); Mitroiu, 1970 (p. 9).

Figure. 4: 4.

Eupomatiaceae

1 genus; 2 species

Distribution. The single genus, *Eupomatia,* has one species, *E. laurina* R. Br., in both New Guinea and eastern Australia (Queensland to New South Wales and Victoria), and a second species, *E. bennettii* F. Muell., in eastern Australia only (Queensland to New South Wales).

Pollen morphology. Pollen grains zonizonasulculate. Isopolar. Radiosymmetric. Globose-oblate. Atectate. Psilate. Monads. Medium sized.

Remarks. Pollen of *Eupomatia* has a very distinctive type of aperture, a zonasulculus, which is found elsewhere in the ranalean complex chiefly in the family Nymphaeaceae.

Palynological references. Hotchkiss, 1958; Canright, 1963; Erdtman, 1966 (pp. 175–76); Mitroiu, 1970 (p. 15); Walker and Skvarla, 1975.

Figure. 4: 5, 6.

Annonaceae

132 genera; ca. 2300 species

Distribution. The family has three centers of distribution—the Americas (mainly tropical South and Central America) with 36 endemic genera, Africa (including Madagascar) with some 40 endemic genera, and Asia (mainly Indo-Malaysia) with approximately 50 endemic genera. One genus (*Xylopia*) is pantropical, 3 occur in both Asia and Africa (*Uvaria, Polyalthia, Artabotrys*), 1 is found in both Asia and America (*Anaxagorea*), and 1 is in both Africa and America (*Annona*).

Pollen morphology. Pollen grains anasulcate, anatrichotomosulcate, catasulcate-cataulcerate, anaulcerate, inaperturate, or disulculate. Heteropolar, apolar, or isopolar. Bilateral, isobilateral, radiosymmetric, or biradial. Boat-shaped–elongate, –oblong, or –elliptic, globose-spherical, or globose-oblate, rarely globose-elliptic. Atectate, tectate, semitectate, or intectate. Psilate, foveolate, fossulate, scabrate, verrucate, baculate, pilate, echinate, or reticulate. Monads, tetrads, or polyads. Large to small or gigantic.

Remarks. The family has been divided informally into three subfamilies (the *Fusaea, Annona,* and *Malmea* subfamilies), largely on the basis of pollen morphology (Walker, 1971b, 1971c, 1972a, 1972b). The Annonaceae, which is the largest family of primitive angiosperms, exhibits more pollen diversity than any other ranalean family. The pollen grains of the annonaceous genus *Cymbopetalum,* which may be 350 μ in diameter, are the largest known among angiosperms (Walker, 1971a).

Palynological references. Veloso and Barth, 1962; Canright, 1963; Erdtman, 1966 (pp. 49–50); Agababian, 1967a,b, 1971a; Mitroiu, 1970 (pp. 11–15); Walker, 1971a–d, 1972b; Le Thomas, 1972; Le Thomas and Lugardon, 1972, 1974; Guinet and Le Thomas, 1973; Van Campo and Lugardon, 1973; Lugardon and Le Thomas, 1974; Walker and Skvarla, 1975.

Figures. 5: 1–3; 14: 3, 4.

Canellaceae

6 genera; ca. 20 species

Distribution. Four genera (*Canella, Capsicodendron, Cinnamodendron,* and *Pleodendron*) are American, occurring from southern Brazil and Venezuela to the West Indies and Florida. The

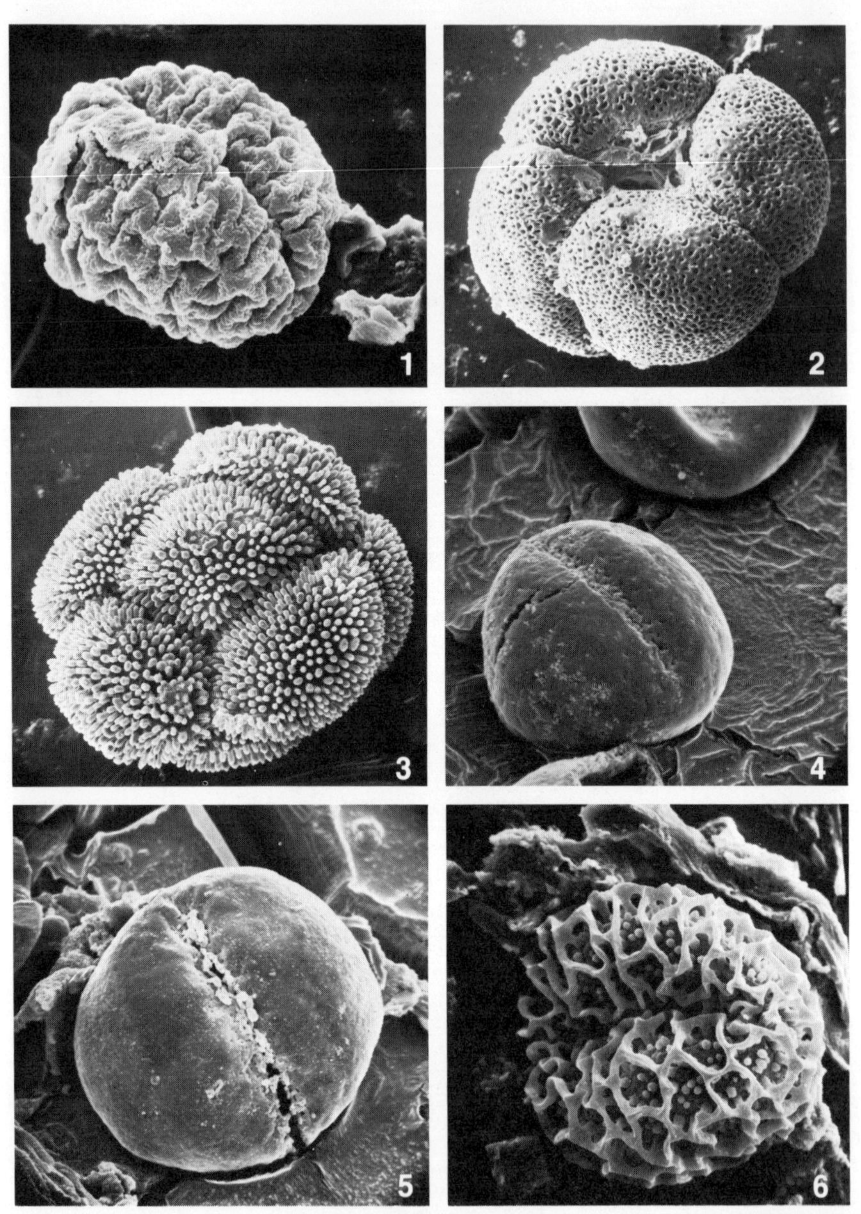

Fig. 5. Pollen of the order Magnoliales, suborder Annonineae, families Annonaceae (1–3), Canellaceae (4), and Myristicaceae (5, 6). (1) *Onychopetalum lucidum* R. E. Fries (P-157) (× ca. 1320). (2) *Annona muricata* L., pollen tetrad (P-4) (× ca. 300). (3) *Trigynaea caudata* (R. E. Fries) R. E. Fries, octad (P-583) (× ca. 600). (4) *Warburgia ugandensis* Sprague (P-981) (× ca. 1200). (5) *Dialyanthera otoba* (Humb. and Bonpl.) Warb. (P-897) (× ca. 1200). (6) *Myristica globosa* Warb. (P-927) (× ca. 1680).

remaining two genera are African, one being restricted to the East African mainland (*Warburgia*) and the other to Madagascar (*Cinnamosma*).

Pollen morphology. Pollen grains anasulcate or anatrichotomosulcate. Heteropolar. More or less isobilateral or radiosymmetric. Globose-oblate or globose-spherical, sometimes boat-shaped–elliptic. Tectate. Psilate, foveolate, scabrate, or weakly verrucate. Monads. Medium sized, sometimes large.

Remarks. Canellaceous pollen shows a marked resemblance to the pollen of certain genera in the Myristicaceae; see pollen of *Warburgia* (fig. 5: 4) and *Dialyanthera* (fig. 5: 5).

Palynological references. Erdtman, 1946, 1966 (pp. 94–95); Wilson, 1964; Agababian, 1966; Mitroiu, 1970 (p. 17).

Figure. 5: 4.

Myristicaceae

17 genera; ca. 300 species

Distribution. Although the family is pantropical, the genera are more restricted in distribution. Five genera of Myristicaceae are endemic to the New World (*Dialyanthera, Compsoneura, Iryanthera, Virola,* and *Osteophloeum*). Eight genera are endemic to Africa (*Staudtia, Coelocaryon, Cepalosphaera, Scyphocepalium, Pycnanthus, Brochoneura, Mauloutchia,* and *Haematodendron*), the last three occurring only on Madagascar. Four genera are endemic to the Asian tropics, centering in Indo-Malaysia (*Gymnacranthera, Knema, Horsfieldia,* and *Myristica*).

Pollen morphology. Pollen grains anasulcate or inaperturate. Heteropolar or apolar. Bilateral, isobilateral, or radiosymmetric. Boat-shaped–elliptic, globose-oblate, or globose-spherical, rarely boat-shaped–oblong. Tectate, semitectate, or intectate. Psilate, foveolate, fossulate, scabrate, verrucate, echinate, or reticulate. Monads. Medium sized to small.

Remarks. The family has diverse pollen types, and most of the genera appear to be separable palynologically. *Pycnanthus* is distinctive in having echinate pollen grains.

Palynological references. Wodehouse, 1938; Joshi, 1946; Veloso and Barth, 1962; Canright, 1963; Erdtman, 1966 (pp. 278–79); Huang, 1967; Agababian, 1970a; Mitroiu, 1970 (p. 16); Rao and Lee, 1970.

Figure. 5: 5, 6.

LAURALES

The order Laurales consists of 11 families, which are grouped into 5 suborders (Austrobaileyineae, Calycanthineae, Trimeniineae, Monimiineae, and Laurineae). The monotypic Austrobaileyineae is the most primitive suborder in the Laurales and has the only family in the order that retains pollen with a well-developed sulcus. Pollen of the Calycanthineae is consistently disulculate, while pollen of the Trimeniineae may be inaperturate, forate, or diulculate. Pollen of the remaining two suborders, the Monimiineae and Laurineae, is largely inaperturate except for sulcoidate pollen in the Amborellaceae, disulculate pollen in the Atherospermataceae, and zonasulculate pollen in a few members of the Monimiaceae. The order is characterized by globose pollen, with no families retaining pollen that is boat-shaped. The exine is reduced and is frequently damaged or destroyed by acetolysis in the more advanced members of the order, e.g., some Monimiaceae, most Lauraceae. Lauralean pollen appears to be consistently tectate. Sculpturing in the order is largely scabrate to verrucate, never baculate, pilate, or reticulate. Prominently echinate pollen occurs in some Monimiaceae and the Hernandiaceae, and pollen grains with relatively small spines are found in some other genera of Monimiaceae, the Gomortegaceae, the Lauraceae, and the Gyrocarpaceae. Monads characterize the order except for the monimiaceous genus *Hedycarya*, which has permanent tetrads. Pollen size ranges from very large (Hernandiaceae) to small (Monimiaceae).

Austrobaileyaceae

1 genus; 2 species

Distribution. The genus *Austrobaileya*, with two species (*A. scandens* C. T. White and *A. maculata* C. T. White) is restricted to Queensland, Australia.

Pollen morphology. Pollen grains anasulcate. Heteropolar. Isobilateral. Globose-spherical. More or less tectate. Verrucate. Monads. Large.

Remarks. The Austrobaileyaceae is the only family in the order Laurales to retain pollen with a well-developed sulcus.

Palynological references. Bailey and Swamy, 1949; Money, Bailey, and Swamy, 1950; Erdtman, 1966 (p. 254).

Figure. 6: 1.

Calycanthaceae

3 genera; 6 species

Distribution. Two genera occur in China (*Chimonanthus*, with three species, and the monotypic *Sinocalycanthus*); the third genus, *Calycanthus*, has one species (*C. floridus* L.) in the southeastern United States and a second species in northern California (*C. occidentalis* Hook. and Arnott).

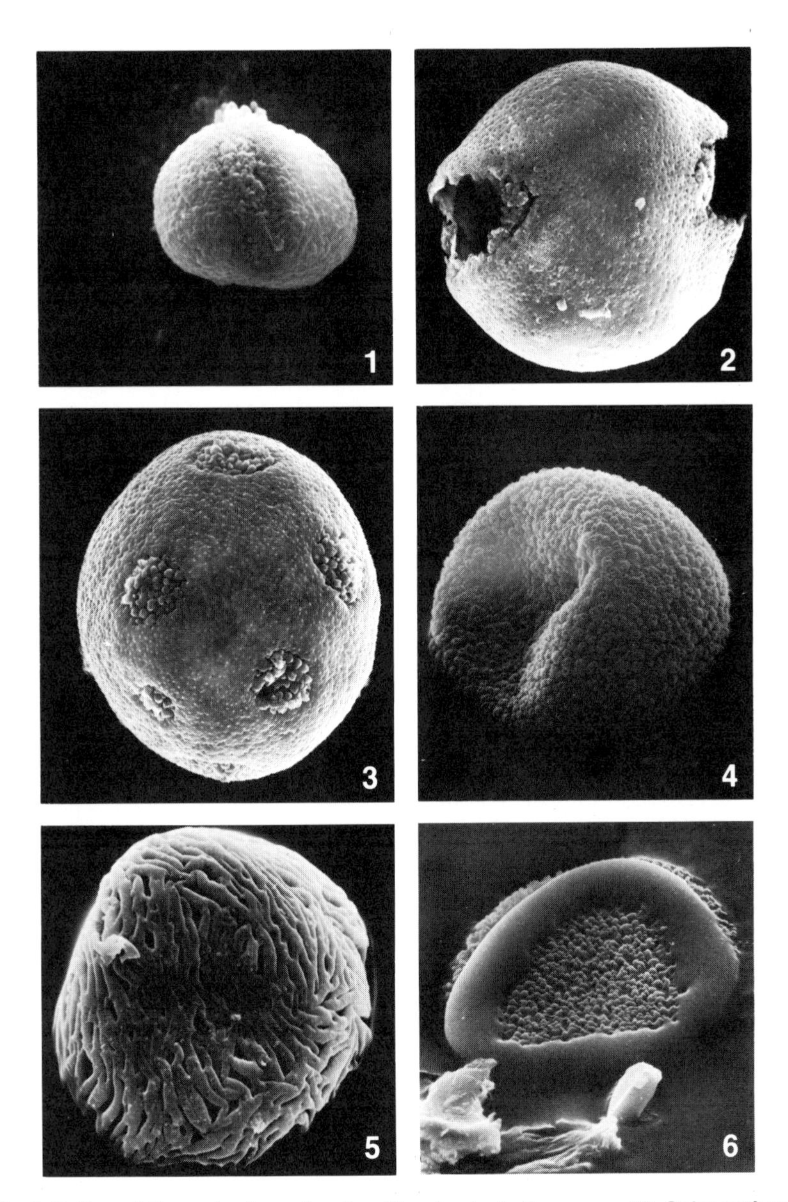

Fig. 6. Pollen of the order Lauraies, families Austrobaileyaceae (1), Calycanthaceae (2), Trimeniaceae (3), Amborellaceae (4), and Monimiaceae (5, 6). (1) *Austrobaileya scandens* C. T. White (P-1453) (× ca. 550). (2) *Calycanthus floridus* L. (P-38) (× ca. 1000). (3) *Trimenia papuana* Ridl. (P-934) (× ca. 2500). (4) *Amborella trichopoda* Baill., some pollen grains sulcoidate but sulcus not visible in this particular grain (P-1034) (× ca. 1750). (5) *Ephippiandra capuronii* Cavaco (P-1456) (× ca. 3750). (6) *Mollinedia elliptica* (Gardn.) A. DC., presumed proximal-polar view (P-1087) (× ca. 2000).

Pollen morphology. Pollen grains disulculate. Isopolar. Biradial to bilateral. Globose-elliptic to globose-oblate, sometimes more or less globose-spherical. Tectate. More or less psilate, sometimes fossulate. Monads. Large to medium sized.

Remarks. Pollen of *Chimonanthus* is tectate-imperforate; that of *Calycanthus* has tectal perforations.

Palynological references. Ikuse, 1956; Canright, 1963; Agababian, 1966; Erdtman, 1966 (p. 89); Mitroiu, 1966, 1970 (p. 21); Walker and Skvarla, 1975.

Figures. 6: 2; 14: 6.

Idiospermaceae

1 genus; 1 species

Distribution. The single species of this monotypic family of primitive angiosperms, *Idiospermum australiense* (Diels) S. T. Blake, is found in Queensland, Australia.

Pollen morphology. Pollen grains disulculate. Isopolar. Biradial. Globose-oblate. Tectate. Psilate. Monads. Small.

Remarks. This new family of primitive angiosperms, whose pollen is quite similar to that of the Calycanthaceae, was recently described by Blake (1972), after rediscovery of plants previously described by Diels as *Calycanthus australiensis.*

Palynological references. Blake, 1972.

Figure. None.

Trimeniaceae

2 genera; 5 species

Distribution. There are two genera in the family, assuming that the African genus *Xymalos* belongs in the Monimiaceae (q.v.). *Trimenia* has three species, whose range includes Celebes, the Moluccas, New Guinea, Bougainville, New Caledonia, Fiji, Samoa, and the Marquesas. The second genus, *Piptocalyx,* has two species, one in New Guinea and the other in Australia (New South Wales).

Pollen morphology. Pollen grains inaperturate, forate, or diulculate. Apolar or isopolar. Radiosymmetric, asymmetric (in forate forms), or biradial to bilateral. Globose-spherical or globose-oblate to globose-elliptic. Tectate. Psilate or scabrate. Monads. Medium sized to small.

Remarks. The two genera differ palynologically with regard to both aperture types and exine sculpturing.

Palynological references. Money, Bailey, and Swamy, 1950; Erdtman, 1966 (pp. 271–73); Agababian, 1968c; Mitroiu, 1970 (pp. 17–18).

Figure. 6: 3.

Amborellaceae

1 genus; 1 species

Distribution. The single species of this monotypic family, *Amborella trichopoda* Baill., is known only from New Caledonia.

Pollen morphology. Pollen grains anasulcoidate (more or less "ulcerate") or inaperturate. Heteropolar or apolar. Radiosymmetric. Globose-spherical. Tectate. Scabrate. Monads. Medium sized to small.

Remarks. Pollen of *Amborella* possesses a distinctive type of sculpturing, which is reminiscent of the pollen of *Piptocalyx* in the Trimeniaceae and *Xymalos* in the Monimiaceae.

Palynological references. Bailey and Swamy, 1948; Money, Bailey, and Swamy, 1950; Erdtman, 1966 (pp. 271–73); Agababian, 1968c.

Figure. 6: 4.

Monimiaceae

28 genera; ca. 450 species

Distribution. The Monimiaceae *s. str.* was divided into four subfamilies by Money, Bailey, and Swamy (1950). All four have been elevated to familial status at one time or another by different authors. In this paper, I have treated the Atherospermatoideae as a distinct family, while retaining the other three subfamilies recognized by Money and her associates (Hortonioideae, Monimioideae, and Siparunoideae) as members of the family Monimiaceae. The ditypic genus *Hortonia*, which is the sole member of the subfamily Hortonioideae, is endemic to Ceylon. This subfamily has been raised to a family (as the Hortoniaceae) by Smith (1971). The subfamily Monimioideae, the largest of the three subfamilies, contains 24 genera. Ten genera (*Wilkiea*, *Tetrasynandra*, *Carnegieodoxa*, *Monimiopsis*, *Anthobembix*, *Lauterbachia*, *Steganthera*, *Levieria*, *Palmeria*, and *Hedycarya*) are restricted to the Australasian region, being widely distributed or quite localized in the area covering Australia, New Zealand, Fiji, New Caledonia, the Solomons, New Guinea, the Moluccas, and Celebes. Two genera (*Matthaea* and *Kibara*) extend from Australasia to the Malay Peninsula and the Nicobar Islands. Seven genera occur on Madagascar and the Mascarenes (*Decarydendron*, *Phanerogonocarpus*, *Hedycaryopsis*, *Ephippiandra*, *Monimia*, *Tambourissa*, and *Schrameckia*). One genus, *Xymalos*, which was retained in the Monimioideae by Money and co-workers but placed in the Trimeniaceae by Hutchinson (1964), occurs on the African mainland. Four genera (*Mollinedia*, *Macropeplus*, *Macrotorus*, and *Peumus*) occur in the New World, the first ranging from South America to southern Mexico. *Peumus,* which is endemic to Chile, has been elevated to subfamilial status by Schodde (1970). The subfamily Siparunoideae, which was raised to a family by Schodde (1970), has 3 genera. The largest, *Siparuna,* with approximately 150 species, is found from tropical South America to Mexico; the monotypic *Bracteanthus* occurs in Brazil. The third genus, *Glossocalyx,* with 4 species, is found in tropical West Africa.

Pollen morphology. Pollen grains inaperturate or catazonasulculate. Apolar or heteropolar. Radiosymmetric. Globose-spherical. Tectate. Fossulate, scabrate, verrucate, or echinate. Monads or tetrads. Medium sized to small.

Remarks. Most genera in the family have inaperturate pollen grains, but some species of *Mollinedia* have catazonasulculate pollen. Pollen of *Hortonia* and most of the genera on Madagascar and the Mascarenes is distinctive in possessing spiral bands on the surface. *Hedycarya* is the only genus in the family with pollen in permanent tetrads. The echinate pollen of *Peumus* is distinctive.

Palynological references. Money, Bailey, and Swamy, 1950; Barth, 1962; Erdtman, 1966 (pp. 271–73); Agababian, 1968c; Mitroiu, 1970 (pp. 18–21); Heusser, 1971.

Figures. 6: 5, 6; 7: 1, 2.

Atherospermataceae

6 genera; ca. 12 species

Distribution. Three genera (*Daphnandra, Doryphora,* and *Atherosperma*) are found in eastern Australia, with *Daphnandra* reaching New Guinea and *Atherosperma* on Tasmania. The monotypic *Dryadodaphne* is endemic to New Guinea, while *Nemuaron* is found only on New Caledonia. *Laurelia* has one species in New Zealand and two in Chile.

Pollen morphology. Pollen grains disulculate, sometimes with a narrower, ringlike apertural zone connecting the two sulculi. Isopolar. Biradial to bilateral. Globose-oblate to globose-elliptic. Tectate. Psilate, fossulate, scabrate, or verrucate. Monads. Large to medium sized.

Remarks. The family is at once set apart from the Monimiaceae *s. str.* by its disulculate pollen.

Palynological references. Money, Bailey, and Swamy, 1950; Erdtman, 1966 (pp. 271–73); Agababian, 1968c; Mitroiu, 1970 (p. 18); Heusser, 1971.

Figure. 7: 3.

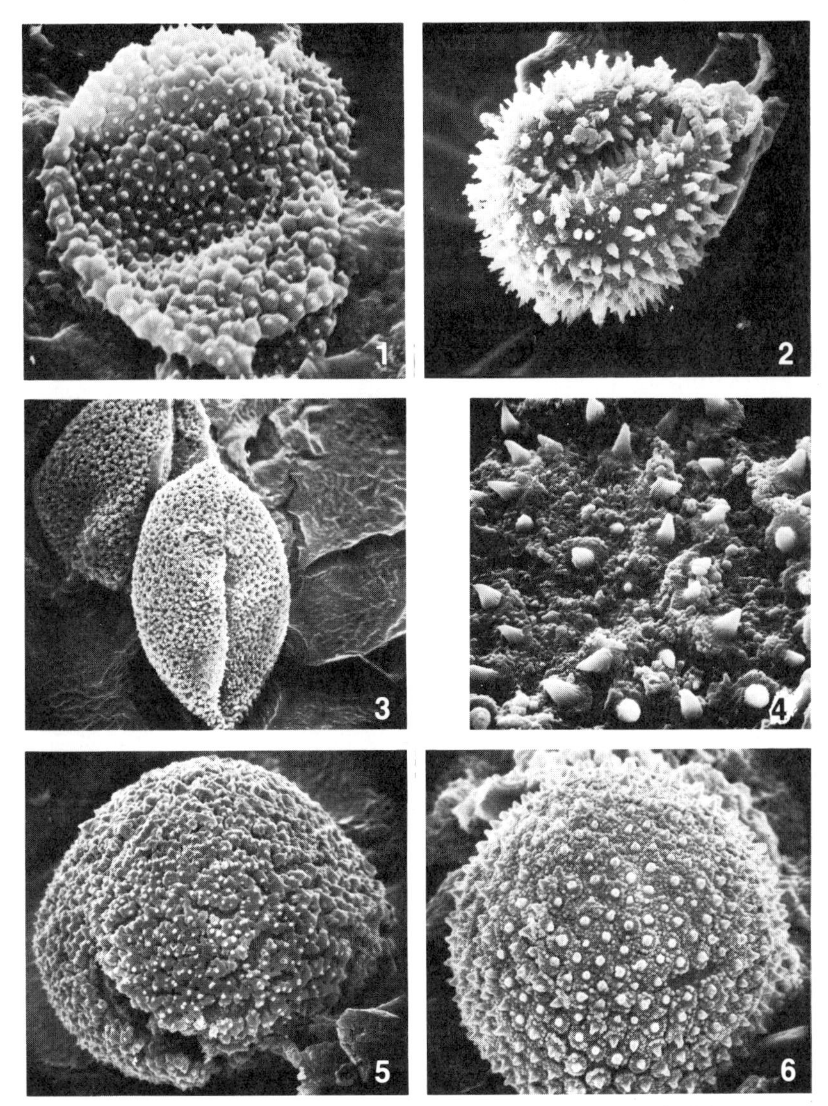

Fig. 7. Pollen of the order Laurales, families Monimiaceae (1, 2), Atherospermataceae (3), Hernandiaceae (4), Gomortegaceae (5), and Lauraceae (6). (1) *Siparuna glabrescens* (Pressl.) A. DC. (P-1395) (× ca. 5000). (2) *Peumus boldus* Mol., pollen inaperturate, grain with fold apparent (P-40) (× ca. 1800). (3) *Atherosperma moschatum* Labill., equatorial view (P-995) (× ca. 1000). (4) *Illigera appendiculata* Bl. (P-1053) (× ca. 925). (5) *Gomortega keule* (Molina) I. M. Johnst. (P-43) (× ca. 1700). (6) *Lindera benzoin* Bl. (P-1000) (× ca. 2500).

Hernandiaceae

2 genera; 42 species

Distribution. One genus, *Hernandia,* is pantropical; the second, *Illigera,* is found in southeastern Asia from Assam to the Philippines, in central Africa, and on Madagascar.

Pollen morphology. Pollen grains inaperturate. Apolar. Radiosymmetric. Globose-spherical. Tectate. Echinate. Monads. Very large.

Remarks. Spines that occur on pollen in this family closely resemble those on the pollen of the monimiaceous genus *Peumus* (see fig. 7: 2, 4).

Palynological references. Erdtman, 1966 (pp. 202–3); Agababian, 1969a; Kubitzki, 1969; Mitroiu, 1970 (p. 29).

Figure. 7: 4.

Gomortegaceae

1 genus; 1 species

Distribution. The single species, *Gomortega keule* (Molina) I. M. Johnston, is known only from the environs of Concepcion, Chile.

Pollen morphology. Pollen grains inaperturate. Apolar. Radiosymmetric. Globose-spherical. Tectate. Echinate. Monads. Medium sized.

Remarks. The spines of the pollen of *Gomortega* are reminiscent of those found in the Lauraceae.

Palynological references. Erdtman, 1966 (p. 191); Heusser, 1971.

Figure. 7: 5.

Lauraceae

30 (?) genera; ca. 2000 species

Distribution. The family is mainly tropical and subtropical, with major centers of distribution in southeastern Asia and Brazil.

Pollen morphology. Pollen grains inaperturate. Apolar. Radiosymmetric. Globose-spherical. Tectate. Echinate. Monads. Medium sized to large.

Remarks. A poorly understood family taxonomically. Lauraceous pollen has a reduced exine that is destroyed during acetolysis.

Palynological references. Selling, 1947; Ikuse, 1956; Veloso and Barth, 1962; Erdtman, 1966 (pp. 221–22); Agababian, 1969a; Wang, 1969; Mitroiu, 1970 (pp. 21–29); Rao and Lee, 1970; Heusser, 1971.

Figure. 7: 6.

Gyrocarpaceae

2 genera; 16 species

Distribution. One genus, *Gyrocarpus,* is pantropical; the second,

Sparattanthelium, is restricted to the New World, occurring from South America to southern Mexico.

Pollen morphology. Pollen grains inaperturate. Apolar. Radiosymmetric. Globose-spherical. Tectate. Echinate. Monads. Medium sized.

Remarks. The pollen of this family resembles that of the Lauraceae.

Palynological references. Erdtman, 1966 (pp. 202–3); Agababian, 1969a; Kubitzki, 1969; Mitroiu, 1970 (p. 29).

Figure. None.

WINTERALES

The order Winterales consists of three families and two suborders (the Winterineae and Illiciineae). All members of the order possess semitectate-reticulate pollen. The monotypic Winterineae has anaulcerate pollen grains in permanent tetrads, whereas pollen of the suborder Illiciineae, which contains the families Illiciaceae and Schisandraceae, is solitary and tri- or hexacolpate. All pollen in the order is globose (never boat-shaped) and medium sized.

Winteraceae

7 genera; ca. 90 species

Distribution. The seven genera are distributed as follows: *Drimys*—southern Mexico to Cape Horn and the Juan Fernandez Islands. *Bubbia*—New Guinea, New Caledonia, Lord Howe Island, Queensland, and Madagascar. *Belliolum*—New Caledonia and the Solomon Islands. *Pseudowintera*—New Zealand. *Tasmannia*—Australia, Tasmania, New Guinea, and one wide-ranging species in the Philippines, Borneo, and Celebes. *Exospermum* and *Zygogynum* are both endemic to New Caledonia.

Pollen morphology. Pollen grains anaulcerate (or rarely anatrichotomosulcate). Heteropolar. Radiosymmetric. Globose-spherical. Semitectate (rarely tectate-perforate). Reticulate. Tetrads (rarely monads). Medium sized.

Remarks. Winteraceous pollen can be told at once from all other ranalean pollen because it is always anaulcerate, semitectate, and reticulate, and consistently occurs in permanent tetrads. Some genera of Winteraceae can be recognized palynologically (Bailey and Nast, 1943), but pollen characters do not appear to be very useful at the species level (Fiser and Walker, 1967). Recently Sampson (1974) found that the pollen was in monads in a few species.

Palynological references. Wodehouse, 1935 (pp. 322–31, 333–35); Bailey and Nast, 1943; Erdtman, 1943 (p. 107), 1964, 1966 (pp. 254,

256, 258); Hotchkiss, 1955; Straka, 1963; Agababian, 1966; Mitroiu, 1966, 1970 (p. 11); Fiser and Walker, 1967; Roland, 1969, 1971a; Heusser, 1971; Sampson, 1974.

Figure. 8: 1.

Illiciaceae

1 genus; 42 species

Distribution. The single genus, *Illicium,* has five species in the southeastern United States, eastern Mexico, Cuba, and Haiti; the remainder of its species are endemic to southeastern Asia (northeastern India to China and Japan and south to the Malay Peninsula, Sumatra, Borneo, and the northern Philippines).

Pollen morphology. Pollen grains tricolpate, usually with the colpi fused at both poles. Isopolar. Radiosymmetric. Globose-oblate. Semitectate. Reticulate. Monads. Medium sized.

Remarks. Pollen of the Illiciaceae greatly resembles the pollen of the Schisandraceae, and to a lesser extent the pollen of the Winteraceae.

Palynological references. Wodehouse, 1935 (pp. 322–31, 335–37); Ikuse, 1956; Hayashi, 1960; Ueno, 1962, 1963; Agababian, 1966; Erdtman, 1966 (pp. 254–56); Mitroiu, 1966, 1970 (p. 17); Tschudy, 1970.

Figure. 8: 2.

Schisandraceae

2 genera; 47 species

Distribution. One genus, *Schisandra,* has a single species in the southeastern United States and the remainder in eastern Asia (Himalayan India to China, Korea, and Japan, and southward with two species in Sumatra and Java). The other genus, *Kadsura,* is endemic to southeastern Asia, ranging from Japan and southern Korea to China, Sikkim, and peninsular India–Ceylon and then south and eastward to Java and Amboina.

Pollen morphology. Pollen grains tricolpate with the three colpi fused only at one pole or hexacolpate with three colpi fused only at one pole alternating with three colpi which are not fused at either pole; colpi frequently with a peculiar linear median thickening. Heteropolar. Radiosymmetric. Globose-oblate. Semitectate. Reticulate. Monads. Medium sized.

Remarks. The pollen of the Schisandraceae resembles pollen found in the families Illiciaceae and Menispermaceae.

Palynological references. Wodehouse, 1935 (pp. 322–31, 337–40); Ikuse, 1956; Hayashi, 1960; Jalan and Kapil, 1964; Agababian, 1966;

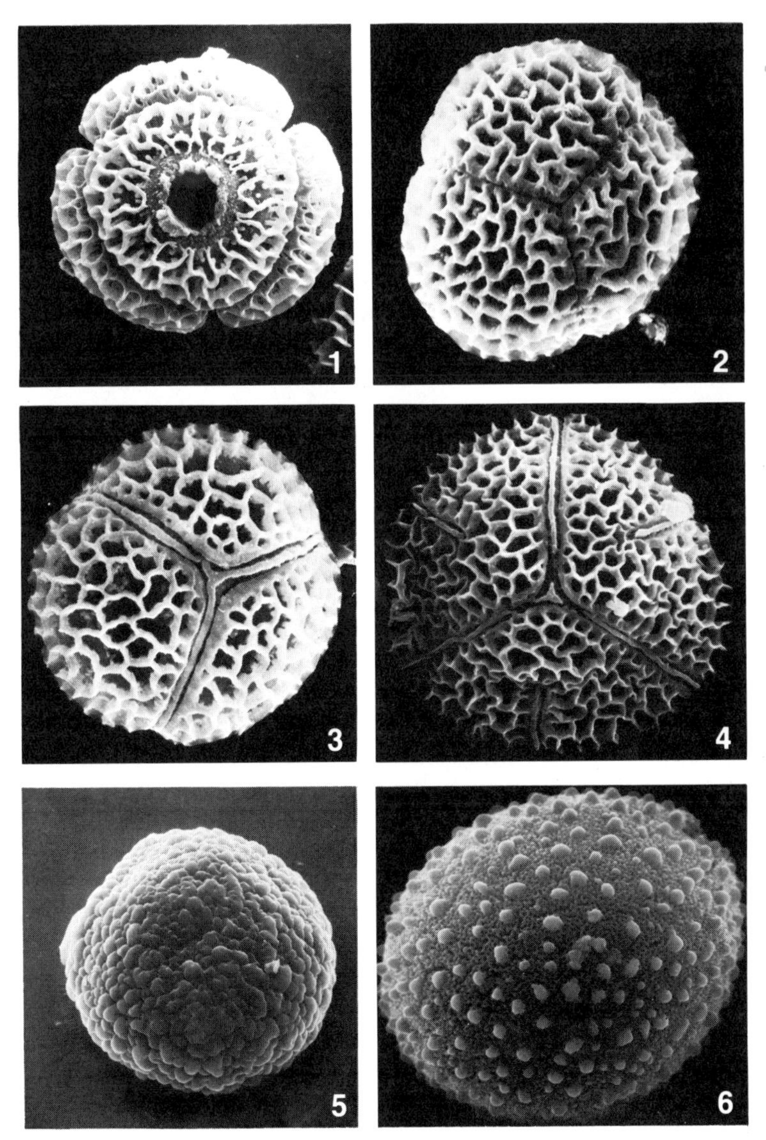

Fig. 8. Pollen of the orders Winterales-Aristolochiales, families Winteraceae (1), Illiciaceae (2), Schisandraceae (3, 4), and Aristolochiaceae (5, 6). (1) *Drimys confertifolia* Phil., pollen tetrad (P-10) (× ca. 1000). (2) *Illicium dunnianum* Tutcher, pollen syncolpate, grain in polar view (P-958) (× ca. 1500). (3) *Schisandra grandiflora* (Wall.) Hk. f. and Th., pollen syncolpate, grain in polar view (P-961) (× ca. 1750). (4) *Kadsura longepedunculata* Finet and Gagnep., grain in polar view, with three fused colpi and three normal colpi (P-967) (× ca. 2000). (5) *Aristolochia grandiflora* Sw. (*Walker 309)* (× ca. 1000). (6) *Asarum caudatum* Lindl. (P-31) (× ca. 900).

Erdtman, 1966 (pp. 254–58); Mitroiu, 1966, 1970 (p. 17); Ueno, 1966; Roland, 1971b.

Figure. 8: 3, 4.

ARISTOLOCHIALES

Aristolochiales is a monotypic order containing the family Aristolochiaceae. Aristolochiaceous pollen is largely inaperturate, although the genus *Saruma* has sulcate pollen and some species of *Asarum* have polyporate or, more rarely, polycolpate pollen grains. Most pollen in the family is tectate, but *Saruma* is unusual in possessing pollen which is semitectate. Pollen in the order is always globose (never boat-shaped).

Aristolochiaceae

7 genera; ca. 400 species

Distribution. Three genera, *Saruma, Thottea,* and *Apama,* are Indo-Malaysian–Chinese, and one, *Asarum,* is Eurasian–North American. The largest genus, *Aristolochia,* occurs in Eurasia, Africa, and the Americas; the monotypic genera *Holostylis* and *Euglypha* are confined to South America. Some authors recognize some of the African species of *Aristolochia* as the genus *Pararistolochia*, and a number of genera have been segregated from *Asarum*, including *Hexastylis*, *Heterotropa*, and *Japonasarum*.

Pollen morphology. Pollen grains anasulcate, inaperturate, polycolpate, polyporoidate, or polyporate. Heteropolar, apolar, or isopolar. Isobilateral or radiosymmetric. Globose-spherical or globose-oblate. Tectate or semitectate. Psilate, fossulate, scabrate, verrucate, or reticulate. Monads. Large to medium sized.

Remarks. Saruma is remarkable in retaining a reduced sulcus and in having the only semitectate-reticulate pollen in the family. The *de novo* development of polycolpate/polyporate pollen from inaperturate pollen grains may be observed in various species of the genus *Asarum.*

Palynological references. Ikuse, 1956; Erdtman, 1966 (pp. 61–62); Agababian, 1969b; Mitroiu, 1970 (pp. 60–61); Heusser, 1971.

Figure. 8: 5, 6.

PIPERALES

The order Piperales contains four families and three suborders (the Chloranthineae, Lactoridineae, and Piperineae). Sulcate pollen is found in at least some of the members of each family in the order. Pollen of the Chloranthaceae is sulcate, inaperturate, or polycolpate, and may be semitectate. The family Lactoridaceae has anasulcate pollen in permanent tetrads. Pollen grains of the Piperaceae-Saururaceae are small to minute, and may be sulcate or inaperturate. Boat-shaped pollen occurs in the Chloranthaceae and Saururaceae, but most members of the order have globose pollen.

Chloranthaceae

5 genera; ca. 65 species

Distribution. One genus, *Ascarina,* with seven species, is widespread among the Pacific islands, occurring on New Zealand, the Marquesas, Society Islands, Cook Islands, Samoan Islands, Fiji, New Caledonia, New Hebrides, the Solomons, New Britain, and New Guinea. One species of *Ascarina,* which is also found on New Guinea, reaches the Philippines and Borneo. The monotypic genus *Ascarinopsis* is endemic to Madagascar. *Hedyosmum* is the only New World genus, occurring in South and Central America to Mexico and the West Indies. One species, *H. orientale* Merr. and Chun, is found in southeastern Asia. The primitively vesselless genus *Sarcandra* has three species. One ranges from southern India and Ceylon to southern China, Japan, Formosa, and the Philippine Islands; the second is endemic to southern India and the third to Hainan. The fifth genus in the family, *Chloranthus,* occurs from Ceylon and eastern India to southern China, southeastern Asia, and the Philippines.

Pollen morphology. Pollen grains anasulcate, inaperturate, with "colpoid complexes" or "colpoid streaks," polycolpoidate, or polycolpate. Heteropolar, apolar, or isopolar. Bilateral, isobilateral, or radiosymmetric. Boat-shaped–elliptic, globose-oblate, or globose-spherical. Tectate or semitectate. More or less psilate, fossulate, scabrate, rugulate, or reticulate. Monads. Medium sized to small.

Remarks. Some of the oldest fossil angiosperm pollen, e.g., *Clavatipollinites,* appears to be similar to pollen found in extant members of the Chloranthaceae (see Kuprianova, 1967). *De novo* development of polycolpate pollen from inaperturate pollen can be observed in the pollen of this family.

Palynological references. Swamy and Bailey, 1950; Swamy, 1953; Ikuse, 1956; Erdtman, 1966 (pp. 111–12); Huang, 1966; Straka, 1966; Kuprianova, 1967; Agababian, 1968c.

Figure. 9: 1–3.

Lactoridaceae

1 genus; 1 species

Distribution. The single species, *Lactoris fernandeziana* Phil., is confined to Masatierra of the Juan Fernandez Islands.

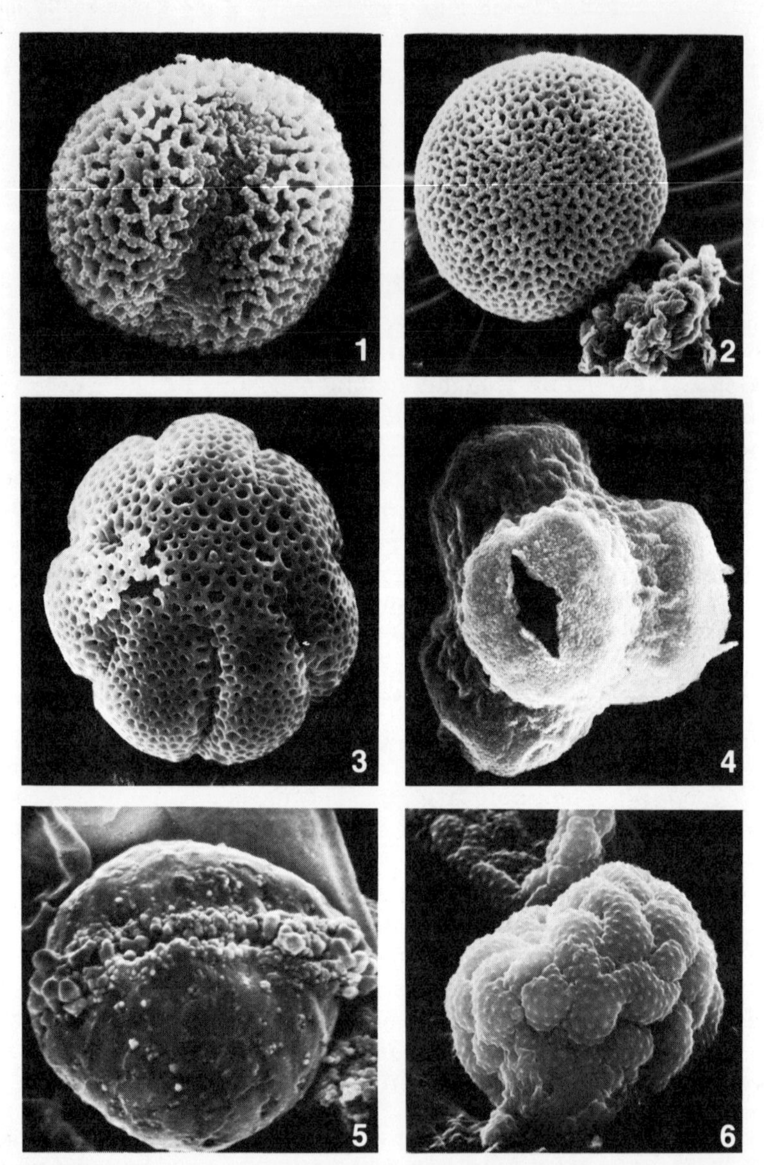

Fig. 9. Pollen of the order Piperales, families Chloranthaceae (1–3), Lactoridaceae (4), Saururaceae (5), and Piperaceae (6). (1) *Ascarina rubricaulis* Solms (P-945) (× ca. 1750). (2) *Hedyosmum domingense* Urb. (P-949) (× ca. 1100). (3) *Chloranthus fortunei* (A. Gray) Solms (P-937) (× ca. 2000). (4) *Lactoris fernandeziana* Phil., pollen tetrad (P-44) (× ca. 1250). (5) *Saururus cernuus* L. (P-45) (× ca. 5000). (6) *Peperomia* sp. (P-52) (× ca. 4000).

Pollen morphology. Pollen grains anasulcate. Heteropolar. Isobilateral. Globose-spherical. Tectate. Scabrate and with loose folds due to localized uplifting of the exine. Tetrads. Medium sized.

Remarks. Pollen of *Lactoris* is as distinctive as its floral morphology, and like the latter appears to offer few clues to the phylogenetic relationships of the family.

Palynological references. Carlquist, 1964; Erdtman, 1964, 1966 (p. 220); Agababian, 1966; Heusser, 1971.

Figure. 9: 4.

Saururaceae

5 genera; 7 species

Distribution. One genus, *Saururus,* has a species in Asia, which ranges from China, Korea, Japan, Taiwan, Hainan, and Tonkin to the Philippines, and a second species in eastern North America. Three genera (*Houttuynia, Gymnotheca,* and *Circaeocarpus*) are restricted to Asia, ranging from the Himalayas to China and Japan. The fifth genus, *Anemopsis,* is found in the southwestern United States and adjacent Mexico.

Pollen morphology. Pollen grains anasulcate, reportedly also anatrichotomosulcate. Heteropolar. Bilateral to isobilateral. Boat-shaped–elliptic to globose-spherical. Tectate. Psilate. Monads. Small to minute.

Remarks. The pollen of the Saururaceae closely resembles that found in the Piperaceae.

Palynological references. Ikuse, 1956; Dufau, 1961; Erdtman, 1966 (pp. 398–99); Huang, 1966; Agababian, 1969b; Mitroiu, 1970 (p. 58).

Figure. 9: 5.

Piperaceae

8–13 genera; ca. 2000 species

Distribution. The family has two distinct subfamilies (Piperoideae and Peperomioideae), each with one large pantropical genus (*Piper* and *Peperomia* respectively), from which a number of generic segregates have been recognized by different authors. Generic segregates of *Piper* include the pantropical genus *Lepianthes* (=*Pothomorphe, Heckeria*), five genera of the New World tropics (*Ottonia, Sarcorhachis, Trianaeopiper, Pleiostachyopiper,* and *Lindeniopiper*), one Malaysian genus (*Zippelia*), and one Australasian genus (*Macropiper*). Three main segregate genera have been recognized from *Peperomia,* all of which occur in the West Indies (*Verhuellia, Manekia,* and *Piperanthera*).

Pollen morphology. Pollen grains anasulcate, reportedly also anatrichotomosulcate, or inaperturate. Heteropolar or apolar. Isobilateral or radiosymmetric. Globose-spherical. Tectate. Psilate, scabrate, verrucate, or echinate. Monads. Small to minute.

Remarks. A taxonomically difficult family (see Burger, 1972; Howard, 1973), whose pollen closely resembles that of the Saururaceae.

Palynological references. Selling, 1947; Ikuse, 1956; Erdtman, 1966 (pp. 321–22); Huang, 1966; Straka, 1966; Agababian, 1969b; Mitroiu, 1970 (pp. 58–60); Heusser, 1971.

Figure. 9: 6.

TROCHODENDRALES, CERCIDIPHYLLALES, AND EUPTELEALES

Trochodendrales, Cercidiphyllales, and Eupteleales are all small orders with tricolpate or tricolpate-derived pollen grains. The Cercidiphyllales and Eupteleales are both monotypic, while the order Trochodendrales contains the families Trochodendraceae and Tetracentraceae. Pollen of these three orders is always globose and ranges from medium sized to small.

Trochodendraceae

1 genus; 1 species

Distribution. The single species, *Trochodendron aralioides* Sieb. and Zucc., is restricted to Japan and Formosa.

Pollen morphology. Pollen grains tricolpate. Isopolar. Radiosymmetric. Globose-oblate to slightly prolate. Tectate. Rugulate. Monads. Medium sized to small.

Remarks. The pollen of *Trochodendron* shows some resemblance to the pollen of *Tetracentron.*

Palynological references. Nast and Bailey, 1945; Ikuse, 1956; Mitroiu, 1963, 1970 (p. 30); Erdtman, 1966 (pp. 439–40); Huang, 1967; Agababian, 1968b; Rao and Lee, 1970.

Figure. 10: 1.

Tetracentraceae

1 genus; 1 species

Distribution. The single species, *Tetracentron sinense* Oliv. in Hook., is found in China and Burma.

Pollen morphology. Pollen grains tricolpate. Isopolar. Radiosymmetric. Globose-spherical to slightly prolate. Tectate. Striato-rugulate. Monads. Small.

Remarks. Pollen of *Tetracentron* possesses a distinctive type of sculpturing.

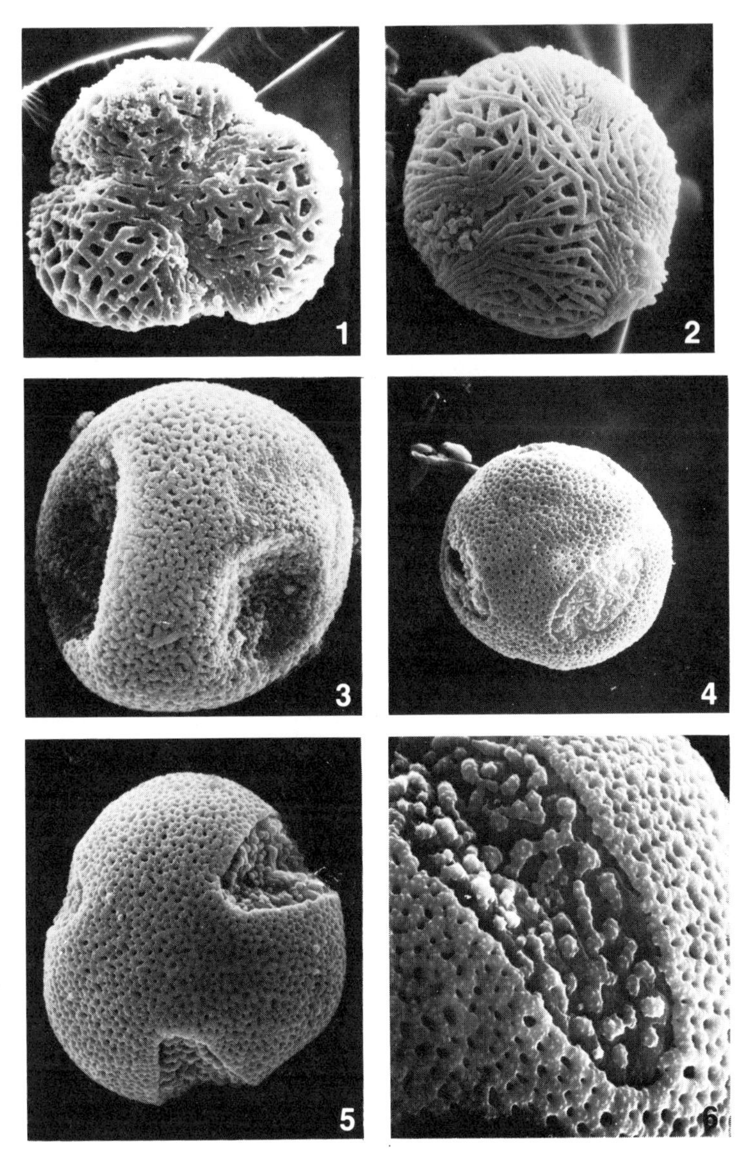

Fig. 10. Pollen of the orders Trochodendrales-Cercidiphyllales-Eupteleales, families Trochodendraceae (1), Tetracentraceae (2), Cercidiphyllaceae (3), and Eupteleaceae (4–6). (1) *Trochodendron aralioides* Sieb. and Zucc. (P-953) (× ca. 2500). (2) *Tetracentron sinense* Oliv. in Hook. (P-954) (× ca. 4250). (3) *Cercidiphyllum japonicum* Sieb. and Zucc., equatorial view (P-971) (× ca. 1500). (4) *Euptelea polyandra* Sieb. and Zucc. (P-955) (× ca. 1600). (5, 6) *E. pleiosperma* Hk. f. and Th. (P-956) (5, × ca. 1500; 6, × ca. 4000). (6) A close-up of the aperture membrane.

Palynological references. Nast and Bailey, 1945; Erdtman, 1966 (pp. 426–27); Agababian, 1968b.
Figure. 10: 2.

Cercidiphyllaceae

1 genus; 1 species
Distribution. The single species, *Cercidiphyllum japonicum* Sieb. and Zucc., is endemic to Japan.
Pollen morphology. Pollen grains tricolpoidate. Isopolar. Radiosymmetric. Globose-spherical to slightly prolate. Tectate. Scabrate. Monads. Medium sized.
Remarks. Pollen of *Cercidiphyllum* is quite similar to the pollen of certain species of the chloranthaceous genus *Hedyosmum.*
Palynological references. Swamy and Bailey, 1949; Ikuse, 1956; Erdtman, 1966 (pp. 106–7); Agababian, 1968b; Mitroiu, 1970 (p. 30).
Figure. 10: 3.

Eupteleaceae

1 genus; 2 species
Distribution. One species, *Euptelea pleiosperma* Hk.f. and Th., is found in China and northeastern India (Assam); the second species, *E. polyandra* Sieb. and Zucc., is endemic to Japan.
Pollen morphology. Pollen grains tricolpate or with furrowlike apertures distributed over the entire surface (rugate). Isopolar or apolar. Radiosymmetric or asymmetric (in the rugate forms). Globose-oblate or globose-spherical to slightly prolate. Tectate. Scabrate. Monads. Medium sized.
Remarks. The two species may be distinct palynomorphologically, but Nast and Bailey (1946) report that the type of aperture (tricolpate versus rugate) does not always correlate with one species or the other.
Palynological references. Nast and Bailey, 1946; Ikuse, 1956; Erdtman, 1966 (pp. 176, 439); Agababian, 1968b.
Figure. 10: 4–6.

NYMPHAEALES

The order Nymphaeales contains five families in two suborders (the Nymphaeineae and Ceratophyllineae). Monosulcate pollen is found in the families Nymphaeaceae and Cabombaceae, whereas pollen of the Barclayaceae and Ceratophyllaceae is inaperturate. Most members of the Nymphaeaceae have zonasulculate pollen. The pollen of the Nelumbonaceae is tricolpate. Boat-shaped pollen occurs in the families Nymphaeaceae and Cabombaceae, but most pollen in the order is globose. Exine sculpturing in the Nymphaeales is diverse and may be baculate, echinate, or even striate. One genus in the family Nymphaeaceae has pollen in permanent tetrads. Pollen in the order ranges from large to medium sized, and, rarely, may be very large.

Nymphaeaceae

5 genera; ca. 70 species

Distribution. The largest genus, *Nymphaea,* with approximately 40 species, is cosmopolitan. The recently segregated monotypic genus *Ondinea* is endemic to Australia (den Hartog, 1970). *Victoria,* with 2 species, is found in South America. *Euryale,* with 1 species, is found in China and southeastern Asia. The fifth genus, *Nuphar,* with approximately 25 species, occurs in North America, Europe, and eastern Asia.

Pollen morphology. Pollen grains anasulcate, anazonasulculate, or zonizonasulculate. Heteropolar or isopolar. Bilateral or radiosymmetric. Boat-shaped–elliptic, globose-oblate, or globose-spherical. Atectate(?) or tectate. Psilate, fossulate, scabrate, verrucate, baculate, or echinate. Monads or tetrads. Large to medium sized.

Remarks. The zonasulculate pollen of *Nymphaea, Ondinea, Victoria,* and *Euryale* contrasts strongly with the monosulcate pollen of the genus *Nuphar.* The pollen of *Nuphar* is also distinctive with its extremely well developed spines. Pollen of *Victoria* stands apart from that of the remainder of the family because it is in permanent tetrads, of an unusual type (Roland, 1965). *Euryale* pollen possesses small spines which are quite different from those of *Nuphar.* The pollen of the genus *Nymphaea* is extremely diverse, and many of the species in this genus can be distinguished palynologically.

Palynological references. Wodehouse, 1935 (pp. 340–46); Erdtman, 1943 (pp. 110–12), 1966 (pp. 287–89); Snegirevskaja, 1955; Ikuse, 1956; Katz and Katz, 1961; Ueno and Kitaguchi, 1961; Ueno, 1962, 1963; Meyer, 1964, 1966a,b; Roland, 1965, 1968, 1969; Rowley, 1967; Rowley and Flynn, 1968; Muller, 1970b.

Figure. 11: 1–4.

Nelumbonaceae

1 genus; 2 species

Distribution. One species, *Nelumbo lutea* (Willd.) Pers., is found in eastern North America, Central America, and the West Indies. The second species, *N. nucifera* Gaertn., occurs in India, Iran, and from China to Australia.

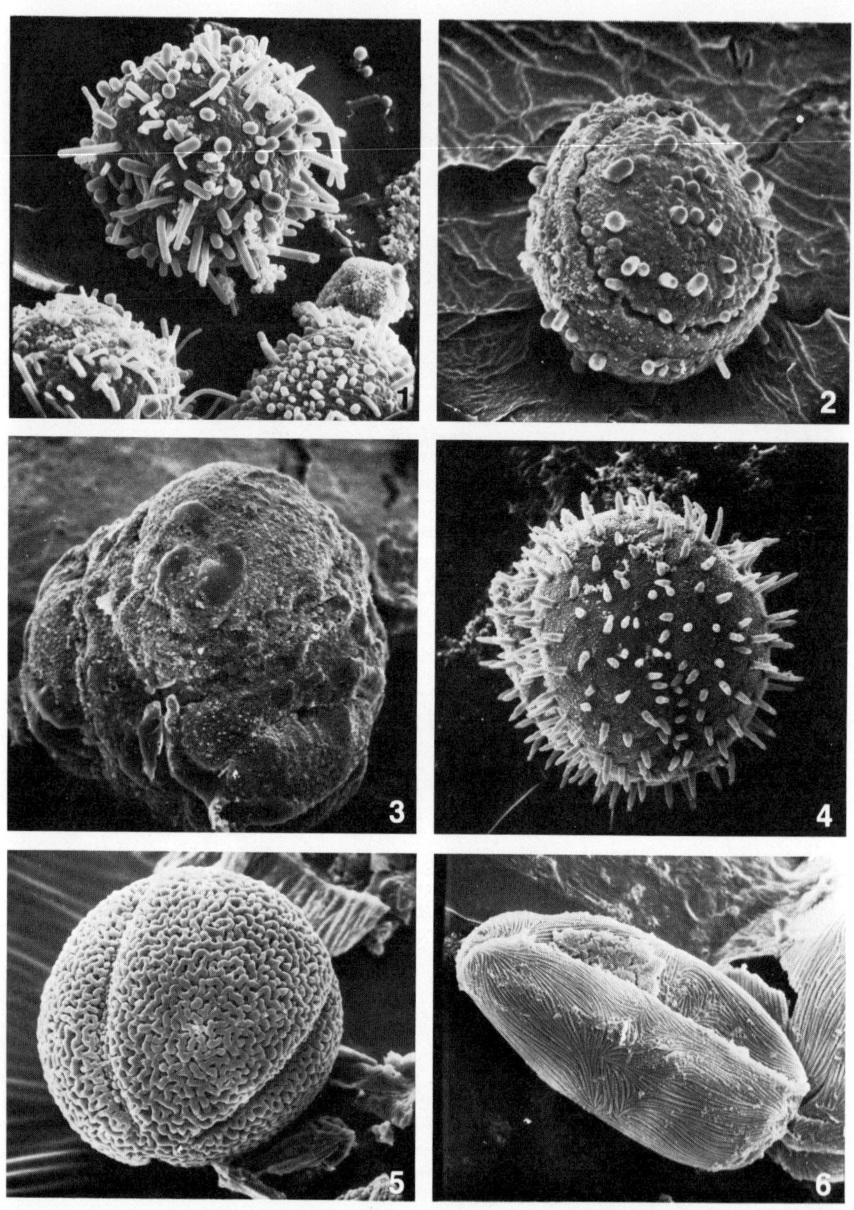

Fig. 11. Pollen of the order Nymphaeales, families Nymphaeaceae (1–4), Nelumbonaceae (5), and Cabombaceae (6). (1) *Nymphaea odorata* Ait., aperture not visible (P-975) (× ca. 1200). (2) *N. amazonum* Mart. and Zucc. (P-986) (× ca. 1200). (3) *Victoria amazonica* (Poepp.) Sow., pollen tetrad (P-983) (× ca. 600). (4) *Nuphar variegatum* Engelm. ex Clinton (P-974) (× ca. 1080). (5) *Nelumbo lutea* (Willd.) Pers. (P-973) (× ca. 1140). (6) *Cabomba caroliniana* Gray (P-972) (× ca. 1080).

Pollen morphology. Pollen grains tricolpate. Isopolar. Radiosymmetric. Globose-spherical. Tectate. Rugulate. Monads. Large.

Remarks. Nelumbonaceae is the only family in the order Nymphaeales which has tricolpate pollen.

Palynological references. Erdtman, 1943 (p. 110), 1966 (pp. 287–89); Ikuse, 1956; Ueno and Kitaguchi, 1961; Ueno, 1963; Meyer, 1964, 1966a,b.

Figure. 11: 5.

Cabombaceae

2 genera; 8 species

Distribution. One genus, *Cabomba,* with seven species, occurs in the New World, ranging from the United States to Argentina; the second genus, *Brasenia,* with a single species, *B. schreberi* Gmel., is nearly cosmopolitan, occurring in North America, Mexico, Central America, the West Indies, eastern Asia, Australia, and Africa.

Pollen morphology. Pollen grains anasulcate or anatrichotomosulcate. Heteropolar. Bilateral or radiosymmetric. Boat-shaped–oblong to boat-shaped–elliptic. Tectate. Striate, or more or less psilate. Monads. Large to very large.

Remarks. Pollen of *Brasenia* has a reduced exine that is scarcely acetolysis-resistant. The highly developed striate sculpturing of *Cabomba* pollen is unique within the ranalean complex. Sections show that this sculpturing is supratectal.

Palynological references. Erdtman, 1943 (p. 110), 1966 (pp. 287–89); Ikuse, 1956; Ueno and Kitaguchi, 1961; Ueno, 1963; Meyer, 1964, 1966a,b; Roland, 1968.

Figure. 11: 6.

Barclayaceae

1 genus; 4 species

Distribution. The single genus, *Barclaya,* occurs in Indo-Malaysia.

Pollen morphology. Pollen grains inaperturate. Apolar. Radiosymmetric. Globose-spherical. Tectate. More or less psilate. Monads. Medium sized.

Remarks. Pollen of *Barclaya* has a reduced exine which is not very acetolysis-resistant.

Palynological references. Erdtman, 1943 (p. 110), 1966 (pp. 287–89); Meyer, 1964, 1966a,b.

Figure. None.

Ceratophyllaceae

1 genus; ca. 6 species

Distribution. The single genus, *Ceratophyllum,* is cosmopolitan.

Pollen morphology. Pollen grains inaperturate. Apolar. Radiosymmetric. Globose-spherical. Tectate. Psilate. Monads. Medium sized.

Remarks. The exine of *Ceratophyllum* pollen is very reduced, and the pollen grains do not survive acetolysis.

Palynological references. Erdtman, 1943 (p. 80), 1966 (p. 106); Mitroiu, 1963, 1970 (pp. 57–58); Heusser, 1971.

Figure. None.

Early Evolution of Angiosperm Pollen

Comparative study of ranalean pollen has revealed a number of evolutionary trends which probably also characterized the pollen of early angiosperms (table 2). Two different kinds of observations were made in order to determine the phylogenetic relationships of pollen characters in extant primitive angiosperms. First, correlations of pollen characters with nonpalynological characters such as flower types, wood structure, and chromosome numbers were noted. Second, in many instances the transitional stages which led from one pollen character to another were observed directly in ranalean pollen. In the pages that follow, each major evolutionary trend in the pollen of the ranalean complex is discussed.

APERTURE TYPES

The evolution of pollen apertures in primitive angiosperms probably followed the course outlined in figure 12. Anasulcate pollen appears to have produced a number of aperture types directly, including anaulcerate pollen in the Winteraceae, zonasulculate pollen in the Nymphaeaceae and Eupomatiaceae, and catasulcate-cataulcerate pollen in the Annonaceae. Trichotomosulcate pollen grains have evolved occasionally in a few species in some ranalean families (such as the Canellaceae), but they always occur together with more numerous sulcate pollen grains. The most important evolutionary trend in early angiosperm aperture types seems to have been the development of inaperturate pollen grains in many families of primitive angiosperms. From such pollen there appears to have been a second major radiation of aperture types, much as took place previously in the initial radiation of anasulcate pollen. Disulculate pollen, which occurs in families such as the Calycanthaceae and Atherospermataceae, evolved *de novo* from such inaperturate pollen, whereas in the Trimeniaceae inaperturate pollen gave rise to forate pollen grains. Finally, colpate pollen arose in relatively specialized ranalean families such as the Chloranthaceae, Aristolochiaceae, Cercidiphyllaceae, Schisandraceae, and Nelumbonaceae, apparently also *de novo* from inaperturate pollen.

Table 2. Major evolutionary trends in pollen of the ranalean complex

APERTURE TYPES

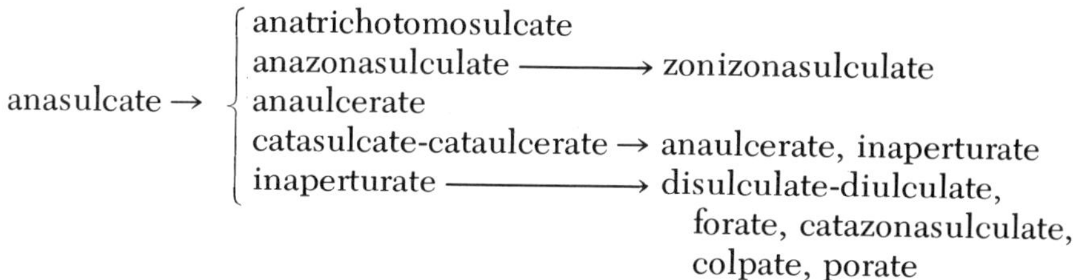

POLARITY

heteropolar → apolar → isopolar

SYMMETRY

bilateral → isobilateral → radiosymmetric → biradial
↓
secondarily bilateral

SHAPE

boat-shaped–elongate → boat-shaped–oblong →
boat-shaped–elliptic → globose-oblate →
globose-spherical → globose-oblate, globose-prolate, globose-elliptic

EXINE STRUCTURE

atectate → tectate-imperforate → tectate-perforate →
semitectate → intectate

EXINE SCULPTURING

psilate → { foveolate, fossulate; scabrate, verrucate; baculate, pilate, echinate; rugulate, reticulate, striate }

POLLEN-UNITS

monads → tetrads → polyads

POLLEN SIZE

large → { medium-sized → small → minute; very large → gigantic }

With regard to aperture types, it must be emphasized that all angiosperm pollen can be divided into two fundamentally different types—that which is monosulcate or monosulcate-derived versus that which is tricolpate or tricolpate-derived, depending on whether the

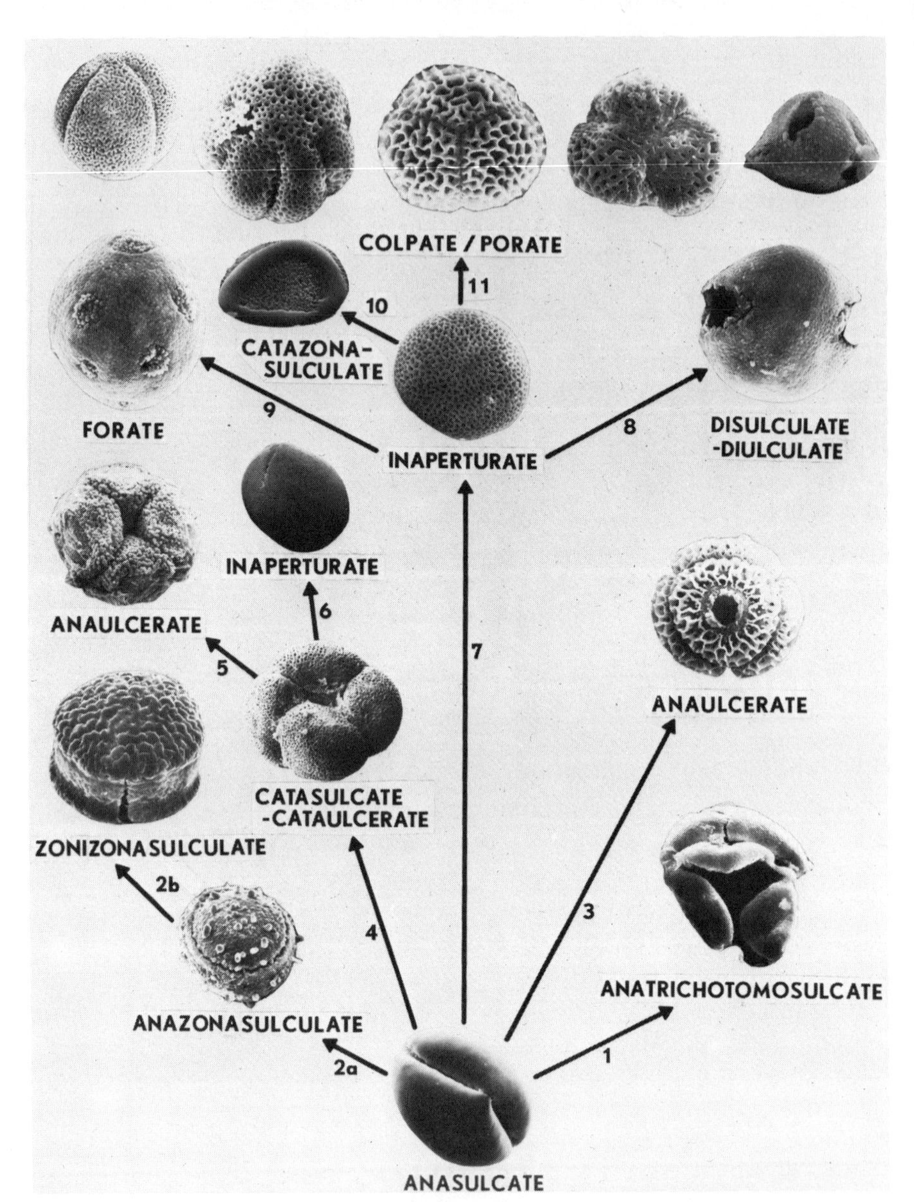

Fig. 12. Evolution of pollen aperture types in the ranalean complex. The anasulcate pollen grain is from *Degeneria* (Degeneriaceae). (1) *Pseudoxandra* (Annonaceae). (2a,b) *Nymphaea* (Nymphaeaceae). (3) *Drimys* (Winteraceae). (4) *Annona* (Annonaceae). (5) *Ophrypetalum* (Annonaceae). (6) *Rollinia* (Annonaceae). (7) *Hedyosmum* (Chloranthaceae). (8) *Calycanthus* (Calycanthaceae). (9) *Trimenia* (Trimeniaceae). (10) *Mollinedia* (Monimiaceae). (11, *left to right*) *Nelumbo* (Nelumbonaceae), *Chloranthus* (Chloranthaceae), *Illicium* (Illiciaceae), *Trochodendron* (Trochodendraceae), *Asarum* (Aristolochiaceae).

ancestral forms are believed to have been monosulcate or tricolpate. Although in a sense all angiosperm pollen other than monosulcate pollen itself is "monosulcate-derived," tricolpate and tricolpate-derived pollen is recognized as distinct because tricolpate pollen (with its derivative forms) is the main type found in most dicotyledons, is essentially restricted to dicotyledonous angiosperms, and has served as the basis for a radiation of "tricolpate-derived" aperture types. In this connection the polycolpate/polyporate pollen found in the Chloranthaceae and Aristolochiaceae must be considered unique in its clear monosulcate derivation; all other colpate pollen in dicotyledonous angiosperms appears to be tricolpate or tricolpate-derived. By contrast, monosulcate or monosulcate-derived pollen is unknown among higher dicotyledons and occurs only in the ranalean dicots, the monocotyledons, and gymnosperms. Evolutionary relationships of ranalean aperture types are discussed in greater detail in another paper (Walker, 1974b).

POLARITY, SYMMETRY, AND SHAPE

Since polarity is largely determined by aperture condition, the most common type of polarity in ranalean pollen—heteropolarity—is directly related to possession of a sulcate aperture. From such heteropolar pollen the main evolutionary trend has been toward apolar inaperturate grains, and from pollen of this type to the basic isopolar colpate pollen of the higher dicotyledons.

Pollen grain symmetry, on the other hand, is largely determined by aperture type and pollen shape. Most sulcate pollen has bilateral symmetry. From this type of symmetry has evolved the common radiosymmetry of dicotyledonous colpate pollen, apparently through an isobilateral intermediate stage which retains a sulcus combined with a globose shape. In some families, biradial pollen has evolved from radially symmetrical grains because of the presence of only two apertures. Such biradial pollen may change its shape from spherical to equatorially elongate, thus producing pollen which is secondarily bilateral. Although pollen shape is somewhat correlated with aperture type, it represents a more independent character than either polarity or symmetry.

The major phylogenetic trend in the form of boat-shaped, monosulcate ranalean pollen appears to go from elongate to oblong to elliptic. Typical boat-shaped pollen characterizes the Magnoliaceae, Degeneriaceae, some Annonaceae, and some Cabombaceae (*Cabomba*). Globose-spherical, inaperturate pollen, which represents one of the most common types in the ranalean complex, has developed from boat-shaped–elliptic pollen, generally by a more or less globose-oblate intermediate stage in which the pollen is still sulcate. Pollen of the families Canellaceae and Myristicaceae is frequently at this level of evolution. Globose-spherical pollen in turn may develop into pollen which is either oblate again or, more rarely, prolate.

EXINE STRUCTURE

Studies in progress on ranalean pollen suggest that a great amount of phylogenetic information exists in the stratification of the exine itself. Preliminary investigation (Walker and Kemp, 1972) has shown that the ratio of nexine to sexine is phylogenetically important. Current studies indicate that presence or absence of endexine and footlayers as well as their thickness relative to other layers of the pollen wall are also phylogenetically useful characters. However, since these studies are only in their initial stage, evolutionary trends in exine stratification will not be discussed here.

With regard to exine structure of ranalean pollen, it is apparent that a recurrent and major evolutionary trend goes from tectate-imperforate pollen to tectate-perforate to semitectate pollen, and thence more rarely to pollen which is intectate (fig. 13). A few ranalean families appear to be at an even more primitive, columellaless or "atectate" stage of exine structure (Walker and Skvarla, 1975). The overwhelming majority of primitive angiosperms are characterized by tectate pollen. Semitectate pollen is characteristic of only three ranalean families (Winteraceae, Illiciaceae, and Schisandraceae), and it is rare to frequent in four other families (Annonaceae, Myristicaceae, Aristolochiaceae, and Chloranthaceae). The rarest exine structure type in the ranalean complex is represented by intectate pollen. Such pollen has been observed in a total of three ranalean genera in two families (Annonaceae, Myristicaceae). The evolution of exine structure in primitive angiosperms is discussed in more detail in a separate paper (Walker, 1974a).

EXINE SCULPTURING

A majority of the most primitive families in the ranalean complex have pollen grains which are remarkably psilate (Dahl and Rowley, 1965). This group includes the Magnoliaceae, Degeneriaceae, Eupomatiaceae, Annonaceae, Canellaceae, Myristicaceae, Aristolochiaceae, Saururaceae, Piperaceae, and Nymphaeaceae. A trend from psilate to foveolate pollen is apparent within certain members of the family Annonaceae. Scabrate or verrucate pollen has developed in many ranalean taxa, whereas baculate, pilate, and echinate pollen are less frequently encountered.

Ranalean pollen which is baculate to pilate is best observed in a

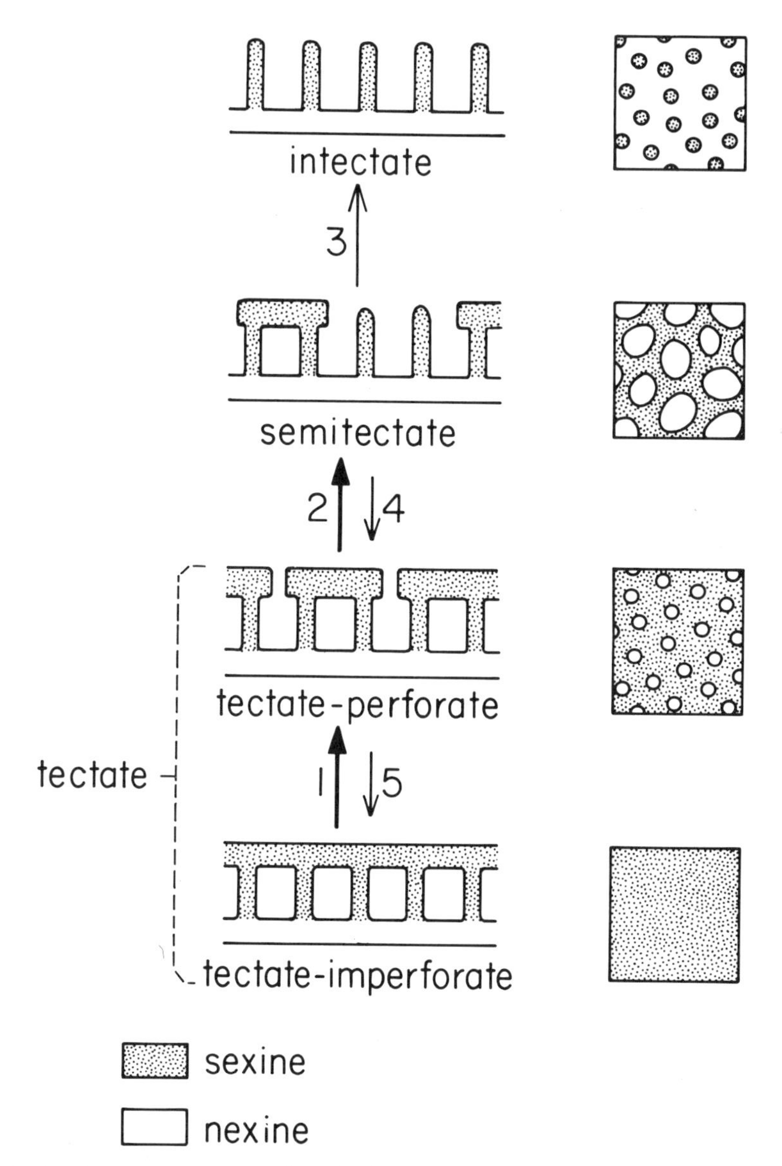

Fig. 13. Evolution of exine structure in the pollen of the ranalean complex. Cross-sectional views to the left, surface views to the right. Strong arrows 1 and 2 indicate more common stages, weaker arrow 3 indicates a less frequently observed stage. Arrows 4 and 5 indicate possible reversibility of the trend, particularly in its earlier stages. The most primitive atectate type of exine structure is not shown.

few Annonaceae (e.g., *Trigynaea*) and in some Nymphaeaceae (*Nymphaea* spp.). Echinate pollen occurs sporadically in a number of ranalean families, including the Annonaceae (e.g., *Desmos, Dasymaschalon, Friesodielsia, Schefferomitra*), Myristicaceae (*Pycnanthus*), Monimiaceae (*Peumus*), Hernandiaceae, and Nymphaeaceae (*Nuphar, Euryale*). Reticulate pollen grains, which are almost synonymous with semitectate pollen, are best developed in the ranalean families Annonaceae (*Deeringothamnus*), Myristicaceae (e.g., *Myristica, Osteophloeum, Horsfieldia* spp.), Winteraceae, Illiciaceae, Schisandraceae, Aristolochiaceae (*Saruma*), and Chloranthaceae (*Sarcandra, Chloranthus* spp.). The markedly striate sculpturing of pollen of *Cabomba* (Cabombaceae) is unique within the ranalean complex. Exine sculpturing types conform less to a straight-line evolutionary pattern than do most pollen characters (see table 2).

POLLEN-UNITS

The evolutionary trend in types of pollen-units is clearly from monads (solitary grains) to tetrads to polyads. In the Annonaceae (Walker, 1971b), monads have secondarily evolved from tetrads within the genus *Annona*. Pollen tetrads are known in five ranalean families. They characterize all members of the Winteraceae [2] and Lactoridaceae, and one genus each in two other ranalean families—*Hedycarya* in the Monimiaceae and *Victoria* in the Nymphaeaceae. Finally, they are present in nearly one-third of the genera in the Annonaceae, where they constitute an important taxonomic character (Walker, 1971b, 1972b). Tetrahedral tetrads are characteristic of four genera of the Annonaceae (*Pseudoxandra, Mitrephora, Pseuduvaria,* and *Petalolophus*), all genera of the Winteraceae, *Lactoris* (Lactoridaceae), and *Victoria* (Nymphaeaceae). In contrast, most tetrad-producing genera in the Annonaceae and *Hedycarya* in the Monimiaceae have tetragonal tetrads. Polyads, most commonly in the form of octads, have been found in eight annonaceous genera, including *Xylopia, Cymbopetalum, Cardiopetalum, Froesiodendron, Porcelia, Trigynaea, Hornschuchia,* and *Disepalum* (Walker, 1971b, 1972b). Polyads are unknown elsewhere within the ranalean complex.

POLLEN SIZE

Most ranalean pollen falls into the medium-sized to large grain categories. Saururaceae and Piperaceae have the smallest pollen grains

[2] A few advanced species within the family Winteraceae appear to have developed monads secondarily (see Sampson, 1974).

in the ranalean complex (often around 10μ), and Monimiaceae also has genera with rather small pollen. Exceedingly large pollen occurs in the Annonaceae, which has the largest fixiform pollen grain known among angiosperms (Walker, 1971a,b). A trend is evident within the order Magnoliales from large pollen in some primitive Magnoliaceae, the Degeneriaceae, and certain primitive Annonaceae to medium-sized or small grains in the families Canellaceae and Myristicaceae. There also appears to be a trend toward reduction in pollen size within many members of the order Laurales. In addition, there is a decided decrease in pollen size in the Winterales and Piperales as compared with the more primitive families in the order Magnoliales (the Magnoliaceae, Degeneriaceae, and Annonaceae). Thus, a general trend from large pollen to medium-sized pollen is discernible within the ranalean complex. In some ranalean taxa, continuation of this trend has resulted in the evolution of small or even minute pollen grains (in the Saururaceae-Piperaceae), while in a few taxa, large pollen has evolved into very large pollen (Annonaceae, Hernandiaceae, Cabombaceae) or even pollen which is gigantic (in certain Annonaceae).

Ancestral Angiosperm Pollen and the Origin of Flowering Plants

Primitive or ancestral angiosperm characters may be determined by several methods. First, if an angiosperm character also occurs in gymnosperms or even pteridophytes, its state may be observed in these lower, nonangiospermous groups. Caution must be exercised, however, lest advanced characters peculiar to gymnosperms or pteridophytes be interpreted as basic for all spermatophytes or vascular plants. Second, correlation of characters may be studied within the angiosperms themselves. The results may then be used in conjunction with "spiral reasoning" (see Walker, 1974b) or "reciprocal illumination," as it has been called by Hennig (1966). In view of the wholly inadequate fossil record of angiosperms, a combination of character correlation and spiral reasoning frequently represents the only available way of determining the primitive character-state of characters peculiar to angiosperms.

The study of character correlation may be random or directional. The former consists of the study of the random correlation of characters at one particular taxonomic level. A study of this type has been carried out in angiosperms by Sporne (1949, 1972) at the taxonomic level of the family. Directional character correlation consists of independently observing the *directionality* of character correlation within a number of different lower taxa included in the taxon for which primitive character-states are to be determined. Correlations are sought between demonstrable transitions from one character-state to another and occurrence in taxa which are thought to be primitive and advanced (as determined from as many other characters as possible). If in numerous taxa there is a high independent correlation of a large number of characters, some of which are thought to be primitive on other grounds (such as occurrence in lower, nonangiospermous groups or in the fossil record), there is strong validity in suggesting that each character represents an ancestral (primitive) character. This is the essence of spiral reasoning, which is inherently noncircular (see Hull, 1967).

Although a primitive character is usually also ancestral, a primitive taxon is not. Due to mosaic evolution, primitive taxa are rarely ancestral. A primitive taxon is simply one that retains a large number of primitive characters *relative to some other taxon.* Such retention of primitive characters is frequently used to infer a comparatively early evolutionary origin, but this may not always be the case.

The most definitive (but, unfortunately, often least available) means of determining primitive characters is a study of character-states sequentially present in the geologic (fossil) record, i.e., use of paleobotanically time-sequenced characters. The fossil record unquestionably provides the most decisive evidence on primitive characters. However, one must never forget that the fossil record can be misinterpreted because of gaps, and that in the case of fossil pollen grains it tells nothing about the nonpalynological characters of the plant that produced a given type of pollen. This situation makes it much more difficult (if not sometimes impossible) to tell whether a given sequence of pollen types observed in the fossil record represents a clade or true phylogenetic line, or merely represents a collection of grade taxa that are at the same evolutionary level. It is much easier to distinguish grades from clades in the pollen of living plants because all other characters of the plant that produced the pollen are known and can be used to determine phylogenetic relationships. Because of these inherent weaknesses in paleopalynology, comparative study of the pollen of living angiosperms combined with the judicious use of character correlation and spiral reasoning may yield results unobtainable from paleobotanical studies alone.

With the methods outlined above, it can be shown that the anasulcus unquestionably represents the primitive (ancestral) aperture type in the ranalean complex and indeed among angiosperms as a whole (see Walker, 1974b). First, anasulcate pollen is found in, and is widespread among, extant and fossil gymnosperms. Second, within angiosperms the anasulcate pollen is restricted to monocots and the otherwise primitive ranalean dicots. Virtually every ranalean order has some taxa with anasulcate pollen, and such pollen is found almost invariably in the more primitive members of each order (as independently determined by other characters). For example, in the Magnoliales, which is the most primitive order in the ranalean complex, six out of seven families have at least some genera with anasulcate pollen, while within the more advanced order Laurales such pollen occurs only in the families Austrobaileyaceae (with a highly primitive floral morphology) and Amborellaceae (with vesselless wood). In the Aristolochiaceae, sulcate pollen is found only in *Saruma*, the only genus in the family with well-developed petals and a semisuperior, apocarpous gynoecium. Finally, monosulcate pollen has been shown to precede other angiospermous aperture types in the fossil record (Doyle, 1969; Muller, 1970a). Thus the primitive status of anasulcate pollen appears to be well established.

If one examines the shape of such anasulcate pollen, in both ranalean and gymnospermous spermatophytes, one is immediately struck by the obvious trend from pollen which is extremely boat-shaped or elongate (fig. 14: 1, 3) to pollen which is oblong (fig. 4: 2) or elliptic (fig. 4: 3). Again, most taxa with boat-shaped pollen are found in the order Magnoliales, particularly in the families Magnoliaceae, Degeneriaceae, and Annonaceae. The more primitive members of two of these families (the Magnoliaceae and Annonaceae) possess boat-shaped pollen which has the longest equatorial axis found in the ranalean complex. In both of these families the direction of character correlation is clearly from primitive taxa with boat-shaped–elongate pollen to more advanced taxa with pollen which is boat-shaped–oblong or –elliptic. Thus it appears that primitive angiosperms had boat-shaped–elongate pollen grains. Since pollen polarity and symmetry are almost entirely determined by aperture type and pollen shape, these primitive boat-shaped–elongate, anasulcate pollen grains were also heteropolar and had bilateral symmetry.

A further examination of ranalean families considered extremely primitive on the basis of other, nonpalynological characters reveals that the pollen of many possesses a remarkable degree of psilateness (see particularly figs. 4: 3, 5; 14: 3, 4; also figs. 5: 4, 5; 9: 5; 11: 3). Also, in some ranalean families the essentially psilate pollen of the more primitive members develops into pollen with more specialized sculpturing in the more advanced taxa (compare figs. 14: 3 and 5: 1; figs. 4: 1 and 4: 2). This observation combined with the fact that most

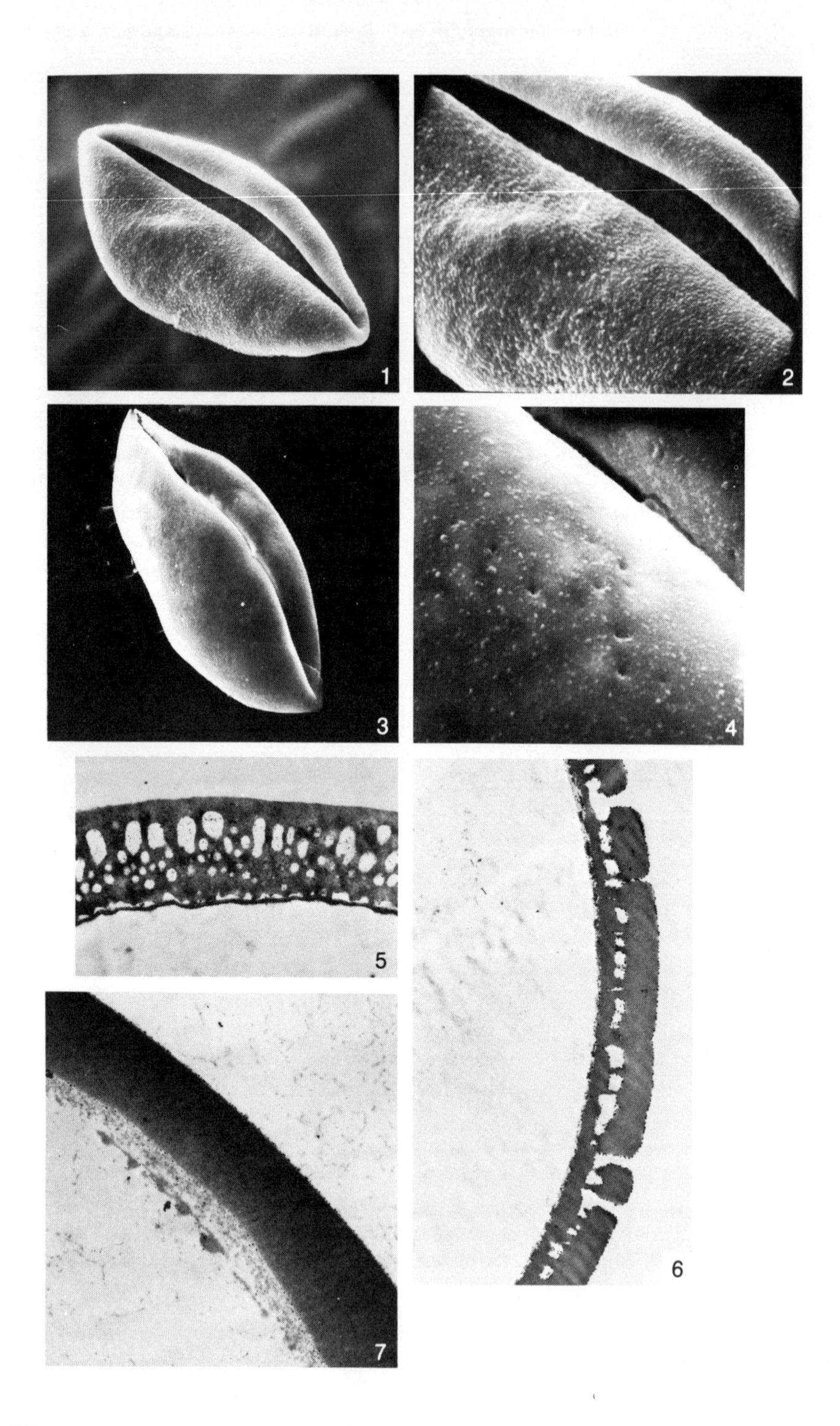
1
2
3
4
5
6
7

gymnosperm pollen is little more than scabrate (fig. 14: 1, 2), would seem to imply that primitive angiosperm pollen was more or less psilate.

As indicated earlier, a trend from large pollen grains to medium-sized or even small grains is evident within the ranalean complex as a whole, in many of its orders, and in certain of its most primitive families. A study of directional character correlation of pollen size within the ranalean complex thus supports the idea that primitive angiosperm pollen was large. Finally, it scarcely needs to be pointed out that permanent tetrads and polyads occur so sporadically within the ranalean complex that they could hardly represent a primitive type of pollen-unit. Furthermore, within the ranalean family Annonaceae, the trend is decidely from monads to tetrads and polyads (Walker, 1971b). Clearly, primitive angiosperms shed their pollen in the form of monads.

The primitive condition of one phylogenetically important character has not yet been discussed, namely, the nature of the angiosperm pollen wall itself. Although some angiosperm pollen may be semitectate or even intectate, the great majority of flowering plants have pollen which is tectate, with a nexine, columellae, and a tectum (fig. 14: 6). Most gymnosperms, on the other hand, seem to have exines with an internal honeycomblike network instead of true columellae (fig. 14: 5). In fact, the presence of a middle layer of distinct, well-defined columellae (i.e., pollen that is tectate), rather than an irregular spongy layer (i.e., pollen that is alveolate), appears to be one of the main features in which angiosperm exines differ in general from those of gymnosperms (Van Campo, 1971; Van Campo and Sivak, 1972). The question that comes to mind immediately is whether or not the pollen grains of ancestral angiosperms already had a typically angiospermous tectate exine or whether they had the alveolate exine structure characteristic of so many gymnosperms. The answer seems to be that they had neither type of exine structure.

Comparative study of the pollen of the ranalean complex has revealed a remarkable, hitherto-unrecognized characteristic of primi-

---

Fig. 14. Ancestral angiosperm pollen characters, inferred from the pollen of extant gymnosperms and primitive angiosperms. (1, 2) *Cycas revoluta* Thunb. (P-969) (1, × ca. 2500; 2, × ca. 5000). (3, 4) *Anaxagorea costaricensis* R. E. Fries (P-164) (3, × ca. 1000; 4, × ca. 10,000). (5–7) Transmission electron micrographs of pollen-wall thin sections. (5) *Cycas revoluta* Thunb. (P-969) (× ca. 7000). (6) *Calycanthus floridus* L. (P-38) (× ca. 5850). (7) *Degeneria vitiensis* I. W. Bailey and A. C. Smith (P-12) (× ca. 8125).

tive angiosperm pollen, namely, its *total lack of columellae* (Walker and Skvarla, 1975). The exine of certain ranalean families which on the basis of all their characters represent some of the most primitive of living angiosperms is completely homogeneous and devoid of any trace of either a columellate or alveolate structure (fig. 14: 7). Exine of this kind occurs in the pollen of some Magnoliaceae, the Degeneriaceae, the Eupomatiaceae, certain Annonaceae, and possibly the Himantandraceae and some members of the Nymphaeaceae. Pollen with such exine has been designated *atectate*, and species with such pollen are considered to represent *primitively columellaless angiosperms*.

Pollen which occurs in the above-mentioned ranalean families is considered to be primitively columellaless, rather than reduced, for the following reasons:

1. The order Magnoliales, to which most of these columellaless taxa belong, clearly represents the most primitive order of living angiosperms when the totality of its characters is considered.
2. The actual evolution of tectate, columellate pollen from pollen which is atectate and columellaless may be observed within the ranalean families Magnoliaceae and Annonaceae. A study of character correlation within these two families, and particularly within the latter, strongly favors recognition of a trend which goes from taxa with a homogeneous exine (e.g., *Anaxagorea*) to taxa with incipient, rudimentary columellae (*Cananga, Uvariastrum* spp.) to taxa with well-developed columellae (*Uvariastrum* spp., *Asimina*), and not the reverse.
3. Columellaless genera such as *Degeneria, Eupomatia*, and *Anaxagorea* are not submerged aquatics and do not have reduced pollen due to any specialized habit or habitat. In fact, pollen of most of the ranalean genera considered to be atectate has a relatively thick exine with no indication of a general reduction in the pollen wall which might have also resulted in a loss of columellae.
4. Secondarily reduced pollen, as observed in aquatic, saprophytic, or parasitic angiosperms among others, usually has a vestigial "columelloid layer." Transmission electron micrographs of pollen of *Degeneria* and other ranalean taxa considered to be primitively columellaless show no trace of any such columelloid layer.
5. Clearly, at some time in the past, columellae had to evolve in pollen lacking them. Since columellae obviously did not evolve externally—this would have made intectate pollen primitive,

which is certainly not the case (see Walker, 1974a)—they must have evolved internally, apparently starting with a homogeneous exine that is still preserved in the pollen of some extant angiosperms. In this connection, it is important to point out that pteridophytes (Pettitt, 1966) and some gymnosperms (Van Campo and Lugardon, 1973) have essentially structureless exines that may also be described as atectate.

By use of a combination of methods outlined in the first part of this section, we arrive at the conclusion that ancestral angiosperm pollen was: anasulcate, with a long sulcus; heteropolar; bilateral; boat-shaped–elongate; atectate and primitively columellaless; more or less psilate; solitary (shed as monads); and large. When all of these criteria are considered, it becomes evident that the three ranalean families Magnoliaceae, Degeneriaceae, and Annonaceae possess the most primitive pollen found in any living angiosperms. Pollen of the annonaceous genus *Anaxagorea* (fig. 14: 3, 4) is a good example of what ancestral angiosperm pollen probably looked like.

Comparative study of the pollen of the ranalean complex and extant gymnosperms and pteridophytes thus supports this concept of ancestral angiosperm pollen. But what does the fossil record tell us? Up to now most palynologists have considered the earliest definite fossil angiosperm pollen to be represented by the genus *Clavatipollinites* (Doyle, 1969; Muller, 1970a; see also Kemp, 1968). One reason for this is the fact that its pollen grains have distinct columellae. However, the pollen of *Clavatipollinites* appears to be too specialized in other palynological characters to be close to the basal stock of angiosperms. In addition to possessing columellae, it is much too small and is not boat-shaped–elongate enough to have been ancestral to the flowering plants. Of all extant ranalean pollen, *Clavatipollinites* appears to be closest to that of the family Chloranthaceae—a family with certain primitive features but hardly with as many primitive characters as families such as the Degeneriaceae and Magnoliaceae. In fact, Kuprianova (1967) considered *Clavatipollinites* congeneric with the chloranthaceous genus *Ascarina*, to which it appears most closely related. Clearly, if paleobotany is to tell us anything about the origin of angiosperms, pollen similar to that of *Magnolia*, *Degeneria*, and *Anaxagorea*, not *Ascarina*, must be sought in the fossil record.

Study of the ranalean complex has revealed a tremendous diversity in the pollen of living primitive angiosperms. We must not forget, however, that most of this diversity is at the level of monosulcate or monosulcate-derived pollen. Although tricolpate and tricolpate-derived pollen is essentially restricted to (dicotyledonous) angiosperms, and therefore more readily identifiable in the fossil record as angiospermous rather than gymnospermous, it itself can never provide clues about the origin and early evolution of angiosperms. Pollen belonging to this critical period in the geologic history of seed plants was at the monosulcate, not the tricolpate, level of evolution. If the present study has shown anything, it is that paleopalynologists must pay more attention to pollen grains that too often have been referred to simply as "cycadophyte pollen." This study further suggests that paleobotanists interested in the origin of the flowering plants must look for fossil gymnosperms that produced large, more or less psilate, boat-shaped–elongate pollen grains with a long anasulcus and an atectate, primitively columellaless exine. For it was probably with pollen of this type that some extinct group of seed plants crossed the line from gymnosperm to angiosperm.

Acknowledgments

I thank directors and curators of the following herbaria for use of palynological material from their collections, with special thanks to individuals listed: Gray Herbarium, Harvard University; Arnold Arboretum, Harvard University; New York Botanical Garden; U.S. National Herbarium, Washington, D.C.; Field Museum of Natural History, Chicago; Royal Botanic Gardens, Kew; British Museum (Natural History), London (N. K. B. Robson); Muséum National d'Histoire Naturelle, Paris (J.-F. Leroy, M. Malplanche); Jardin Botanique National de Belgique, Brussels (L. Liben); Rijksherbarium, Leiden (J. Muller); Botanisch Museum en Herbarium van de Rijksuniversiteit, Utrecht (W. Punt); Naturhistoriska Riksmuseet, Stockholm (S. Nilsson, B. Sparre); Bishop Museum, Honolulu (P. van Royen); Herbarium, University of Malaya, Kuala Lumpur (B. C. Stone); Division of Botany, Department of Forests, Lae, New Guinea (D. B. Foreman, J. S. Womersley); Queensland Herbarium, Brisbane (J. Parham, S. L. Everist), also in Brisbane, V. K. Moriarty and L. J. Webb; and M. Glikson and L. T. Evans (Canberra). I am particularly indebted to I. K. Ferguson (Kew) for sending numerous critical pollen samples.

I gratefully acknowledge use of the following scanning electron microscopes and help from the individuals listed: Advanced Metals Research Corp. (AMR), Burlington, Mass. (AMR-900, AMR-1000), R. Turnage and K. Benoit, operators; Japan Electron Optics Laboratory Co. (JEOL), Medford, Mass. (JSM-U3, JSM-S1, JSM-2), M. Hasegawa, operator; and Kent Cambridge Scientific, Inc. (Stereoscan), Ossining, N.Y. (S4), B. Breton, operator. I thank J. J. Skvarla and

E. S. Kemp for information gained from their thin sections of pollen of a number of ranalean species. I am indebted to J. J. Skvarla for the transmission electron micrographs of the pollen of *Calycanthus* (fig. 14: 6) and *Degeneria* (fig. 14: 7).

I thank A. Baranov and S. Feuer for translation of palynological papers in Russian and French. I acknowledge stimulating discussions on various aspects of this study with J. A. Doyle. This work was supported by NSF grant GB-35475 and by Faculty Research and Faculty Growth grants from the Research Council of the University of Massachusetts at Amherst.

References

Agababian, V. Sh. 1966. Pollen morphology of some primitive angiosperms. I [in Russian, Armenian summary]. *Biol. Zh. Armenii* 19(11): 77–89.

Agababian, V. Sh. 1967a. Contribution to the pollen morphology of the family Annonaceae [in Russian, Armenian summary]. *Biol. Zh. Armenii* 20(3): 102–5.

Agababian, V. Sh. 1967b. Germination of the microspores of the genus *Annona* [in Russian, Armenian summary]. *Biol. Zh. Armenii* 20(12): 77–80.

Agababian, V. Sh. 1968a. Notes on the morphological evolution of the microspores of Magnolianae [in Russian, Armenian summary]. *Biol. Zh. Armenii* 21(3): 24–35.

Agababian, V. Sh. 1968b. Pollen morphology of some primitive angiosperms. II [in Russian, Armenian summary]. *Biol. Zh. Armenii* 21(5): 68–78.

Agababian, V. Sh. 1968c. Pollen morphology of some primitive angiosperms. III[in Russian, Armenian summary]. *Biol. Zh. Armenii* 21(12): 42–54.

Agababian, V. Sh. 1968d. Palynology of Magnolianae. I. Palynomorphology [in Russian, Armenian summary]. *Trans. Erevan State Univ., Natur. Hist., Biol.,* 1968(3, 109): 151–67.

Agababian, V. Sh. 1969a. Pollen morphology of some primitive angiosperms. IV [in Russian, Armenian summary]. *Biol. Zh. Armenii* 22(3): 45–58.

Agababian, V. Sh. 1969b. Pollen morphology of some primitive angiosperms. V [in Russian, Armenian summary]. *Biol. Zh. Armenii* 22(7): 54–66.

Agababian, V. Sh. 1970a. Pollen morphology of some primitive angiosperms. VI [in Russian, Armenian summary]. *Biol. Zh. Armenii* 23(5): 58–69.

Agababian, V. Sh. 1970b. Palynology of Magnolianae. II. Palynosystematics [in Russian, Armenian summary]. *Trans. Erevan State Univ., Natur. Hist., Biol.,* 1970(2): 65–76.

Agababian, V. Sh. 1971a. Pollen morphology of some primitive angiosperms. VII [in Russian, Armenian summary]. *Biol. Zh. Armenii* 24(1): 58–68.

Agababian, V. Sh. 1971b. Pollen morphology of some primitive angiosperms. VIII. [in Russian, Armenian summary]. *Biol. Zh. Armenii* 24(6): 22–32.

Agababian, V. Sh. 1972a. Ultrastructure of sporoderm of some primitive angiosperms [in Russian]. *Bot. Zh.* 57: 955–59.

Agababian, V. Sh. 1972b. Pollen morphology of the family Magnoliaceae. *Grana* 12: 166–76.

Bailey, I. W., and Nast, C. G. 1943. The comparative morphology of the Winteraceae. I. Pollen and stamens. *J. Arnold Arbor.* 24: 340–46.

Bailey, I. W., Nast, C. G., and Smith, A. C. 1943. The family Himantandraceae. *J. Arnold Arbor.* 24: 190–206.

Bailey, I. W., and Smith, A. C. 1942. Degeneriaceae, a new family of flowering plants from Fiji. *J. Arnold Arbor.* 23: 356–65.

Bailey, I. W., and Swamy, B. G. L. 1948. *Amborella trichopoda* Baill., a new morphological type of vesselless dicotyledon. *J. Arnold Arbor.* 29: 245–54.

Bailey, I. W., and Swamy, B. G. L. 1949. The morphology and relationships of *Austrobaileya*. *J. Arnold Arbor.* 30: 211–26.

Barth, O. M. 1962. Catálogo sistemático dos pólens das plantas arbóreas do Brasil meridional. II. Monimiaceae e Dilleniaceae. *Mem. Inst. Oswaldo Cruz, Rio de Janeiro* 60: 405–20.

Blake, S. T. 1972. *Idiospermum* (Idiospermaceae), a new genus and family for *Calycanthus australiensis*. *Contrib. Queensland Herbarium* 12: 1–37.

Burger, W. C. 1972. Evolutionary trends in the Central American species of *Piper* (Piperaceae). *Brittonia* 24: 356–62.

Canright, J. E. 1953. The comparative morphology and relationships of the Magnoliaceae. II. Significance of the pollen. *Phytomorphology* 3: 355–65.

Canright, J. E. 1963. Contributions of pollen morphology to the phylogeny of some ranalean families. *Grana Palynol.* 4: 64–72.

Carlquist, S. 1964. Morphology and relationships of Lactoridaceae. *Aliso* 5: 421–35.

Cronquist, A. 1968. *The Evolution and Classification of Flowering Plants.* Houghton, Boston.

Dahl, A. O., and Rowley, J. R. 1965. Pollen of *Degeneria vitiensis*. *J. Arnold Arbor.* 46: 308–23.

Den Hartog, C. 1970. *Ondinea*, a new genus of Nymphaeaceae. *Blumea* 18: 413–16.

Doyle, J. A. 1969. Cretaceous angiosperm pollen of the Atlantic Coastal Plain and its evolutionary significance. *J. Arnold Arbor.* 50: 1–35.

Dufau, O. 1961. Structure cytologique du pollen de quelques angiospermes. *Rev. Gén. Bot.* 68: 73–110.

Erdtman, G. 1943. *An Introduction to Pollen Analysis.* Chronica Botanica, Waltham, Mass.

Erdtman, G. 1945. Pollen morphology and plant taxonomy. III. *Morina* L. with an addition on pollen morphological terminology. *Svensk Bot. Tidskr.* 39: 187–91.

Erdtman, G. 1946. Pollen morphology and plant taxonomy. VII. Notes on various families. *Svensk Bot. Tidskr.* 40: 77–84.

Erdtman, G. 1964. Ein Beitrag zur Kenntnis der Pollenmorphologie von *Lactoris fernandeziana* und *Drimys winteri*. *Grana Palynol.* 5: 33–39.

Erdtman, G. 1966. *An Introduction to Palynology*, vol. 1: *Pollen Morphology and Plant Taxonomy: Angiosperms.* Hafner Pub. Co., New York. Corrected reprint of 1952 edition with a new addendum.

Faegri, K., and Iversen, J. 1964. *Textbook of Pollen Analysis*, 2d rev. ed. Hafner Pub. Co., New York.

Fiser, J., and Walker, D. 1967. Notes on the pollen morphology of *Drimys* Forst., section *Tasmannia* (R. Br.) F. Muell. *Pollen Spores* 9: 229–39.

Guinet, Ph., and Le Thomas, A. 1973. Interprétation de la répartition dissymétrique des couches de l'exine dans les pollens composés: Conséquences relatives à la notion d'aperture. *Compt. Rend. Acad. Sci. Paris*, ser. D, 276: 1545–48.

Hayashi, Y. 1960. On the microsporogenesis and pollen morphology in the family Magnoliaceae. *Sci. Rep. Tôhoku Univ.*, ser. IV, Biol., 26: 45–52.

Hennig, W. 1966. *Phylogenetic Systematics.* Univ. of Illinois Press, Urbana.

Heusser, C. J. 1971. *Pollen and Spores of Chile.* Univ. of Arizona Press, Tucson.

Hotchkiss, A. T. 1955. Chromosome numbers and pollen tetrad size in the Winteraceae. *Proc. Linn. Soc. New South Wales* 80: 47–53.

Hotchkiss, A. T. 1958. Pollen and pollination in the Eupomatiaceae. *Proc. Linn. Soc. New South Wales* 83: 86–91.

Howard, R. A. 1973. Notes on the Piperaceae of the Lesser Antilles. *J. Arnold Arbor.* 54: 377–411.

Huang, T.-C. 1966. Pollen grains of Formosan plants (1). *Taiwania* 12: 1–8.

Huang, T.-C. 1967. Pollen grains of Formosan plants (2). *Taiwania* 13: 15–110.

Hull, D. L. 1967. Certainty and circularity in evolutionary taxonomy. *Evolution* 21: 174–89.

Hutchinson, J. 1964. *The Genera of Flowering Plants*, vol. 1: *Dicotyledones.* Oxford Univ. Press, London.

Ikuse, M. 1956. *Pollen Grains of Japan.* Hirokawa Pub. Co., Tokyo.

Jalan, S., and Kapil, R. N. 1964. Pollen grains of *Schisandra* Michaux. *Grana Palynol.* 5: 216–21.

Joshi, A. C. 1946. A note on the development of pollen of *Myristica fragrans* van Houtten and the affinities of the family Myristicaceae. *J. Indian Bot. Soc.* 25: 139–43.

Katz, N. I., and Katz, S. V. 1961. Pollen grains of *Euryale* in the lower Pliocene [in Russian]. *Dokl. Akad. Nauk. S.S.S.R.* 136(1): 206–8.

Kemp, E. M. 1968. Probable angiosperm pollen from British Barremian to Albian strata. *Palaeontology* 11: 421–34.

Kubitzki, K. 1969. Monographie der Hernandiaceen. *Bot. Jahrb. Syst.* 89: 78–148.

Kuprianova, L. A. 1967. Palynological data for the history of the Chloranthaceae. *Pollen Spores* 9: 95–100.

Le Thomas, A. 1972. Apport de la palynologie dans la création d'un nouveau genre d'Annonacées. *Compt. Rend. Acad. Sci. Paris*, ser. D, 274: 1652–55.

Le Thomas, A., and Lugardon, B. 1972. Sur la structure fine des tétrades de deux Annonacées (*Asteranthe asterias* et *Hexalobus monopetalus*). *Compt. Rend. Acad. Sci. Paris,* ser. D., 275: 1749–52.

Le Thomas, A., and Lugardon, B. 1974. Quelques types de structure grenue dans l'ectexine de pollens simples d'Annonacées. *Compt. Rend. Acad. Sci. Paris,* ser. D, 278: 1187–90.

Lugardon, B., and Le Thomas, A. 1974. Sur la structure feuilletée de la couche basale de l'ectexine chez diverses Annonacées. *Compt. Rend. Acad. Sci. Paris,* ser. D, 279: 255–58.

Meyer, N. R. 1964. Palynological studies on the extant and fossil Nymphaeaceae [in Russian, English summary]. *Bot. Zh.* 49: 1421–29.

Meyer, N. R. 1966a. On the development of pollen grains of Helobiae and on their relation to Nymphaeaceae [in Russian]. *Bot. Zh.* 51: 1736–40.

Meyer, N. R. 1966b. Investigation of the morphology of pollen grains of Nymphaeaceae and Helobiae for its classification and phylogeny [in Russian, English summary]. In *The Importance of Palynological Analysis for Stratigraphic and Paleofloristic Investigations,* pp. 30–34. Acad. Sci. U.S.S.R., Moscow (For the 2d Internat. Palynol. Conf., Utrecht, 1966.)

Mitroiu, N. 1963. Contributions to the palynological study of certain families of Polycarpicae (Ranales) [in Romanian, with Russian and French summaries]. *Stud. Cercle Biol.,* ser. Biol. Veg., 15: 239–50.

Mitroiu, N. 1966. A palynological study on certain representatives of some polycarpic families [in Romanian]. *Stud. Cercle Biol.,* ser. Bot., 18: 25–33.

Mitroiu, N. 1970. Études morphopolliniques et des aspects embryologiques sur les "Polycarpicae" et Helobiae, avec des considérations phylogénétiques. *Acta Bot. Horti Bucurestiensis* 1969.

Money, L. L., Bailey, I. W., and Swamy, B. G. L. 1950. The morphology and relationships of the Monimiaceae. *J. Arnold Arbor.* 31: 372–404.

Muller, J. 1970a. Palynological evidence on early differentiation of angiosperms. *Biol. Rev. Cambridge Phil. Soc.* 45: 417–50.

Muller, J. 1970b. Description of pollen grains of *Ondinea purpurea* den Hartog. *Blumea* 18: 416–17.

Nair, P. K. K. 1965. Pollen morphology of some families of Monochlamydeae. *Bot. Notiser* 118: 281–88.

Nair, P. K. K. 1967. Pollen morphology with reference to the taxonomy and phylogeny of the Monochlamydeae. *Rev. Palaeobot. Palynol.* 3: 81–91.

Nair, P. K. K. 1968. A concept of pollen evolution in the "primitive" angiosperms. *J. Palynol. (Lucknow)* 4(1): 15–20.

Nair, P. K. K. 1970. *Pollen Morphology of Angiosperms: A Historical and Phylogenetic Study.* Lucknow Pub. House, Lucknow, India.

Nast, C. G., and Bailey, I. W. 1945. Morphology and relationships of *Trochodendron* and *Tetracentron.* II. Inflorescence, flower, and fruit. *J. Arnold Arbor.* 26: 267–76.

Nast, C. G., and Bailey, I. W. 1946. Morphology of *Euptelea* and comparison with *Trochodendron.* *J. Arnold Arbor.* 27: 186–92.

Pettitt, J. M. 1966. Exine structures in some fossil and Recent spores and pollen as revealed by light and electron microscopy. *Bull. Brit. Mus. Natur. Hist.*, Geol., 13: 223–57.

Praglowski, J. 1974. Magnoliaceae Juss. Taxonomy by J. E. Dandy. *World Pollen and Spore Flora* 3: 1–45. Stockholm.

Praglowski, J., and Punt, W. 1973. An elucidation of the microreticulate structure of the exine. *Grana* 13: 45–50.

Rao, A. N., and Lee, Y. K. 1970. Studies on Singapore pollen. *Pacific Sci.* 24: 255–68.

Roland, F. 1965. Précisions sur la structure et l'ultrastructure d'une tétrade calymmée. *Pollen Spores* 7: 5–8.

Roland, F. 1968. Étude de l'ultrastructure des apertures. II. Pollens a sillons. *Pollen Spores* 10: 479–519.

Roland, F. 1969. Étude de l'ultrastructure des apertures. III. Compléments fournis par le microscope électronique à balayage. *Pollen Spores* 11: 475–98.

Roland, F. 1971a. The detailed structure and ultrastructure of an acalymmate tetrad. *Grana* 11: 41–44.

Roland, F. 1971b. Données évolutives résultant de l'étude ultrastructurale des apertures de pollens appartenant au groupe Ranales Centrospermales. *Rev. Gén. Bot.* 78: 329–38.

Rowley, J. R. 1967. Fibrils, microtubules and lamellae in pollen grains. *Rev. Palaeobot. Palynol.* 3: 213–26.

Rowley, J. R., and Flynn, J. J. 1968. Tubular fibrils and the ontogeny of the yellow water lily pollen grain. *J. Cell Biol.* 39: 159a.

Sampson, F. B. 1974. A new pollen type in the Winteraceae. *Grana* 14: 11–15.

Schodde, R. 1970. Two new suprageneric taxa in the Monimiaceae alliance (Laurales). *Taxon* 19: 324–28.

Selling, O. H. 1947. Studies in Hawaiian pollen statistics. Part II. The pollens of the Hawaiian phanerogams. *Bishop Mus. Spec. Publ.* 38: 1–430.

Shaw, G. 1971. The chemistry of sporopollenin. In *Sporopollenin*, ed. J. Brooks, P. R. Grant, M. Muir, P. van Gijzel, and G. Shaw, pp. 305–48. Academic Press, London.

Smith, A. C. 1971. An appraisal of the orders and families of primitive extant angiosperms. *J. Indian Bot. Soc.*, Golden Jubilee vol. 50A: 215–26.

Snegirevskaja, N. S. 1955. Contribution to the morphology of the pollen of Nymphaeales [in Russian]. *Bot. Zh.* 40: 108–15.

Sporne, K. R. 1949. A new approach to the problem of the primitive flower. *New Phytol.* 48: 259–76.

Sporne, K. R. 1972. Some observations on the evolution of pollen types in dicotyledons. *New Phytol.* 71: 181–85.

Straka, H. 1963. Über die mögliche phylogenetische Bedeutung der Pollenmorphologie der madagassischen *Bubbia perrieri* R. Cap. (Winteraceae). *Grana Palynol.* 4: 355–60.

Straka, H. 1966. Palynologia Madagassica et Mascarenica. *Pollen Spores* 8: 241–64.

Swamy, B. G. L. 1949. Further contributions to the morphology of the Degeneriaceae. *J. Arnold Arbor.* 30: 10–38.
Swamy, B. G. L. 1953. The morphology and relationships of the Chloranthaceae. *J. Arnold Arbor.* 34: 375–408.
Swamy, B. G. L., and Bailey, I. W. 1949. The morphology and relationships of *Cercidiphyllum*. *J. Arnold Arbor.* 30: 187–210.
Swamy, B. G. L., and Bailey, I. W. 1950. *Sarcandra*, a vesselless genus of the Chloranthaceae. *J. Arnold Arbor.* 31: 117–29.
Takhtajan, A. 1969. *Flowering Plants: Origin and Dispersal.* Smithsonian Inst. Press, Washington, D.C.
Thorne, R. F. 1968. Synopsis of a putatively phylogenetic classification of the flowering plants. *Aliso* 6(4): 57–66.
Tschudy, R. H. 1970. *Two New Pollen Genera (late Cretaceous and Paleocene) with Possible Affinity to the Illiciaceae.* U.S. Geological Survey Professional Papers 643F.
Ueno, J. 1962. On the fine structure of the pollen walls of Angiospermae. II. *Victoria. J. Biol. Osaka City Univ.* 13: 99–104.
Ueno, J. 1963. The stratigraphical structure of the pollen walls of Dicotyledoneae. I. Ranales and Amentiferae [in Japanese, English summary]. *Acta Phytotaxon. Geobot.* 19: 137–41.
Ueno, J. 1966. On the fine structure of the pollen walls of Angiospermae. IV. *Kadsura. Rep. Fac. Sci. Shizuoka Univ.* 1: 91–100.
Ueno, J., and Kitaguchi, S. 1961. On the fine structure of the pollen walls of Angiospermae. I. Nymphaeaceae. *J. Biol. Osaka City Univ.* 12: 83–89.
Van Campo, M. 1971. Précisions nouvelles sur les structures comparées des pollens de Gymnospermes et d'Angiospermes. *Compt. Rend. Acad. Sci. Paris,* ser. D, 272: 2071–74.
Van Campo, M., and Lugardon, B. 1973. Structure grenue infractectale de l'ectexine des pollens de quelques Gymnospermes et Angiospermes. *Pollen Spores* 15: 171–87.
Van Campo, M., and Sivak, J. 1972. Structure alvéolaire de l'ectexine des pollens à ballonnets des Abietacées. *Pollen Spores* 14: 115–41.
Veloso, H. P., and Barth, O. M. 1962. Catálogo sistemático dos pólens das plantas arbóreas do Brasil meridional. I. Magnoliaceae, Annonaceae, Lauraceae, e Myristicaceae. *Mem. Inst. Oswaldo Cruz, Rio de Janeiro* 60: 59–90.
Walker, J. W. 1971a. Unique type of angiosperm pollen from the family Annonaceae. *Science* 172: 565–67.
Walker, J. W. 1971b. Pollen morphology, phytogeography, and phylogeny of the Annonaceae. *Contrib. Gray Herbarium* 202: 1–132.
Walker, J. W. 1971c. Contributions to the pollen morphology and phylogeny of the Annonaceae. I. *Grana* 11: 45–54.
Walker, J. W. 1971d. Elucidation of exine structure and sculpturing in the Annonaceae through combined use of light and scanning electron microscope. *Pollen Spores* 13: 187–98.
Walker, J. W. 1972a. Chromosome numbers, phylogeny, phytogeography of

the Annonaceae and their bearing on the (original) basic chromosome number of angiosperms. *Taxon* 21: 57–65.

Walker, J. W. 1972b. Contributions to the pollen morphology and phylogeny of the Annonaceae. II. *J. Linn. Soc., Bot.* 65: 173–78.

Walker, J. W. 1974a. Evolution of exine structure in the pollen of primitive angiosperms. *Amer. J. Bot.* 61: 891–902.

Walker, J. W. 1974b. Aperture evolution in the pollen of primitive angiosperms. *Amer. J. Bot.* 61: 1112–37.

Walker, J. W., and Kemp, E. S. 1972. Preliminary studies of exine stratification in the pollen of primitive angiosperms. *Brittonia* 24: 129–30. (Abstr.)

Walker, J. W., and Skvarla, J. J. 1975. Primitively columellaless pollen: a new concept in the evolutionary morphology of angiosperms. *Science* 187: 445–47.

Wang, J. L. 1969. Morphological study on the pollen grains of Laurales of Taiwan [in Chinese, English summary]. *Bull. Taiwan Forest Res. Inst., Taipei* 175: 1–36.

Wilson, T. K. 1964. Comparative morphology of the Canellaceae. III. Pollen. *Bot. Gaz.* 125: 192–97.

Wodehouse, R. P. 1935. *Pollen Grains.* McGraw, New York.

Wodehouse, R. P. 1938. In: Smith, A. C., The American species of Myristicaceae. *Brittonia* 2: 393–510.

Seeds, Seedlings, and the Origin of Angiosperms

G. LEDYARD STEBBINS, ***Department of Genetics University of California, Davis***

THE IMPORTANCE of the ovule and seed in the origin and early evolution of the angiosperms is already implied by the name of the class. The reasons for this importance become even more evident if we regard the origin of angiosperms as a major advance in the adaptive efficiency of plants to their environment. The strong selective pressures which were undoubtedly required to produce this advance must have been exerted chiefly upon the stages of seed development and germination, and early growth of the seedling. A little reflection should enable any botanist to realize why this should be so. During the critical stages of development that elapse between flowering and the firm establishment of the seedling that will constitute the next generation, strong selective pressures of various kinds are exercised on the plant. They begin with the development of the seed from the fertilized ovule. Most seeds must contain a large amount of food material to nourish the young seedling during the germination process. Any food cache of this kind is bound to attract many kinds of predators, chiefly insects, and the developing seed is no exception. Hence its protection is of prime importance for establishing the new generation. The angiosperms have evolved a number of strategies for such protection, which have been well described by B. L. Burtt (1960), Daniel Janzen (1969), and others. Once the seed is ripe there comes the problem of dispersal. The numerous adaptive strategies that make dispersal possible are well known and need no further comment here (see Ridley, 1930; van der Pijl, 1969). Finally, the stages of seed germination and seedling establishment are critical because at these stages the developing plant is most sensitive to the rigors of the environment, particularly the effects of drought and cold.

With respect to these stages, a basic conflict exists between opposing demands of the environment. At stages preceding seed germi-

Note: The material in this contribution is part of my book, *Flowering Plants: Evolution above the Species Level,* and is reproduced by permission of the publisher, Harvard University Press.

nation, the most adaptive seed is a small one. Small seeds require shorter times for development, they can be produced in larger numbers in the same amount of space and with the same utilization of photosynthetic product as a few large ones, and they are more easily dispersed. Successful seed germination and seedling development, on the other hand, are favored by large seed size, including abundant reserves of stored food. In many species of angiosperms, a great diversity of adaptive strategies has been evolved to strike a compromise between these opposing demands. Large seeds are dispersed with the aid of various kinds of appendages or, more effectively, by means of properties which make them attractive to animal vectors. In many genera, the reduction in ovule number to one or two per flower makes possible a more flexible adjustment to the environment in a plant which produces large seeds. In an environment or during a growing season that is unfavorable, the plant can produce only a few flowers, but in a favorable season it can immediately increase its reproductive potential by producing a large number of them. The problem of protecting the development of a large seed from both the vicissitudes of weather and the attacks of predators is solved by means of various coverings. In many instances, predator attack is reduced or eliminated by bitter or toxic substances in the seed.

Obviously, the ecological relationships just described cannot be used directly to discover the course of evolution taken by the earliest angiosperms and their immediate ancestors. We must admit, however, that no direct evidence of any kind exists with respect to this problem. Consequently any kinds of relevant facts, no matter how indirect may be their application, will bring us closer to a solution of this vexing but highly important problem.

The Nature and Origin of the Distinctive Characteristics of Angiosperms

As many botanists have suggested (Bate-Smith, 1972; Kubitzki, 1972), and as I explain at length in a recent book (Stebbins, 1974), all attempts to identify any modern group of angiosperms as ancestral to the remainder of the class are futile. Even the device of adding the suffix "-like" to the name of a modern order or family is more misleading than helpful. In the absence of fossils, we simply do not know what the reproductive structures of the earliest angiosperms were like. Moreover, none of the well-known lineages of Mesozoic age can be clearly recognized as ancestral to the angiosperms. All we can say at present is that toward the end of the Paleozoic era the evolutionary line leading to the angiosperms entered a dark tunnel of ignorance, as far as we are concerned, and remained in that tunnel until the angiosperms emerged, fully differentiated, in the early part of the Cretaceous period. To discover what went on within that tunnel, we can only make guesses that are as intelligent as possible and are based upon the known processes that must have been taking place. We must cling to whatever clues we can find, however tenuous they may be. In this paper, I present my own speculations about these events.

My rationale is to concentrate on the most important characteristics of angiosperms which differentiate them from other seed plants, to consider what kinds of selective pressures or previous structural properties would have been most likely to bring them about, and then to look for fossil forms which give the best clues pointing toward the kinds of plants in which these changes might have been taking place. The most important characteristics are as follows:

1. The angiosperm carpel or ovary, which not only protects the developing ovules but also, by means of the well-developed stigma and style, enables the pollen tube to reach the ovules and effect fertilization much sooner after pollination than is possible in any group of gymnosperms.
2. The angiosperm ovule, which primitively has two integuments and is borne in the anatropous position, bent back upon its stalk or funicle. As is explained below, the probable homology of these integuments and the funicle provide instructive clues leading to fossil forms that may have been related to the ancestors of the angiosperms.
3. The reduction of the gametophyte, the presence of double fertilization and a triploid endosperm, which make possible rapid development of endosperm tissue before embryo development begins.
4. The extreme physiological flexibility of angiosperms, including both the capacity for great cellular elongation at critical stages of development, and the appearance of intercalary meristems that function for varying periods of time. This flexibility must be based upon a capacity of angiosperm tissue to produce and respond to various kinds of growth substances to a far greater degree than is possible in any other land plants.

The Significance of the Structural Characteristics of Gynoecium and Ovule

The adaptive significance of the angiosperm gynoecium is related only in part to the protective function of the ovary wall during ovule development. The protection provided by these structures is no better than that provided, for instance, by a pine cone. The difference is that the pine cone and other protective devices present in gymnosperms exist at the expense of easy and rapid pollen-tube growth. Only the angiosperm gynoecium makes possible both strong protection for the developing ovule and easy access to it by the pollen tube, thereby reducing the time required between pollination and fertilization. Most probably, therefore, the angiosperm gynoecium evolved under conditions which produced strong selective pressures favoring both protection of the ovules and a rapid cycle of flowering and seed development. Such selective pressures exist only in a climate that includes alternation between a short favorable season for reproduction and a long unfavorable one. Under such conditions, predator pressure is strong during the favorable season, developing ovules lost during this season cannot be replaced by later growth, and efficient timing of every developmental stage has a high adaptive value.

The anatropous position and double integuments of the angiosperm ovule provide clues to fossil forms which may have been related to ancestors of the angiosperms. Once the ovule is enclosed within the gynoecium, no conceivable adaptive value exists for two integuments rather than one. Consequently, the condition of double integuments must have preceded the enclosure of the ovule in a carpel or ovary. At that earlier time, the outer integument was probably the principal protective structure.

In my opinion, two characteristics of ovule integuments point toward their homology. The first is the difference between the histological structure of the two integuments in many relatively primitive angiosperms. This difference is evident from both cross sections and whole mounts in such families as Magnoliaceae (Maneval, 1914; personal observations), Dilleniaceae (Sastri, 1958), Malvaceae (van Heel, 1970) and Capparidaceae (Rao, 1938; Khan, 1950; Narayana, 1962). The second characteristic is the fact that in the families mentioned above, as well as in several others, the micropyle of the inner integument is not coincident with that of the outer integument. This condition is designated "zigzag micropyle" by Davis (1966), who records the condition in several families. An important common denominator of these families is that they are either generally recognized as close to the original angiosperm stock (Magnoliaceae, Dilleniaceae) or can be directly connected with it through evolutionary lines that do not include families lacking the character. Consequently, the primitive and original status of the zigzag micropyle is highly probable.

To explain this condition I have assumed, following Gaussen (1946), that the outer integument of the angiosperm ovule is homologous to the cupule wall of Mesozoic pteridosperms, particularly the Caytoniales and Corystospermaceae (Thomas, 1933, 1955; Harris, 1951). The resemblance of the caytonialean cupule to the anatropous ovules of angiosperms is evident from inspection of the well-known reconstructions. The difference between them is, of course, that the cupule of the Caytoniales, *s. str.*, contained several ovules, whereas the angiosperm ovule, if homologous to a cupule rather than to a gymnosperm ovule, contains only one ovule within the cupule wall or outer integument. This latter condition, however, existed in the Corystospermaceae, such as *Unkomasia,* and so could have been fixed in the line that led to the angiosperms. In this case, the evolution from a caytonialean gynoecium to one resembling *Unkomasia* and the presumable ancestors of the angiosperms would have been analogous to many examples of ovule reduction within the gynoecium of modern angiosperms, such as the trend from the follicle of more primitive Ranunculaceae to the achene of *Ranunculus, Anemone, Clematis,* and other genera.

Another bit of evidence in favor of this hypothesis exists in certain teratological forms of ovules. If the angiosperm ovule is derived from a cupule which originally contained more than one ovule, then a teratological reversion to this condition might be expected occasionally. In terms of angiosperm terminology, this would be an ovule in which the outer integument encloses two embryo sacs, each of which is surrounded by its own nucellus and inner integument. Such a condition has been found in two genera of Capparidaceae (Rao, 1938; Khan, 1950; Narayana, 1962). Although, like most other botanists, I am highly skeptical of evidence from teratology, such examples serve to strengthen somewhat a hypothesis that was developed on other grounds.

If, therefore, the cupule of the Caytoniales is homologous to the angiosperm ovule rather than to the carpel, as earlier paleontologists wrongly postulated, then what structure in the Caytoniales has been transformed into the carpel and ovary?

Two suggestions can be offered. First, the carpel could have been derived by folding a flattened rachis, which was, in turn, derived from the stemlike rachis of the Caytoniales through the initiation and activity of a marginal or intercalary meristem. The change would be analogous to that which, in the tribe Rusceae of the Lilaceae (*Asparagus, Ruscus*), transformed terete branchlets into flattened phyllodes.

Second, we might postulate that a branched, cupule-bearing rachis, similar to that found in *Unkomasia macLeanii* (Thomas, 1933), which had an axillary position on the main axis from which it emerged, became adnate to its subtending bract. The carpel would in this case be regarded as a compound structure analogous to the ovule-bearing cone scale of conifers (Florin, 1944). This possibility occurred to me because of the fossil *Lidgettonia* (Thomas, 1958), which was found in the Triassic strata of South Africa in association with leaves of the *Glossopteris* type. In this fossil the elliptical sporophyll apparently bears on its surface two rows of dehiscent cupules. Since the fossil is only an impression, and no anatomical or histological features are preserved, the way in which the cupular stalks were attached to the sporophyll cannot be determined. Such a structure could arise, however, if a dichotomously branched megasporangiophore became adnate to a bract, except for its side branches, to which cupules would have been attached. The transformation of such a structure into a primitive carpel could occur in the following manner. First, the number of ovules per cupule could become reduced to one, and the cupule could assume the anatropous orientation found in *Caytonia* and *Unkomasia.* Second, the sporophyll could become conduplicate, as has been postulated by many morphologists for the original carpel, placing the cupules or "ovules" in a submarginal or laminar position on its inner or adaxial surface.

With the evidence as scanty as it is now, I am not ready to accept unequivocally either of these hypotheses. Both suggest that the ancestors of the angiosperms were advanced seed ferns belonging to the glossopteridalean-caytonialean alliance which flourished during the Triassic and Jurassic periods, but which was probably already well evolved during the Permian. My current preference for the second hypothesis is based upon the following considerations. The modification of a *Lidgettonia*-like sporophyll in the manner described above is easier to imagine than the transformation of a secondary axis into a flattened phyllome by means of marginal growth. Also, in such primitive genera as *Degeneria* (Swamy, 1949) and *Magnolia* (Canright, 1960), the ovules are supplied by vascular strands that are separate from those that supply the carpel walls. These strands may be homologous to the ones that existed in the original megasporangiophore.

As postulated by present theory, the course of evolution in the female reproductive structures that led to the angiosperms may be summarized as follows:

1. Existence in primitive vascular plants of dichotomous sporangiophores bearing terminal sporangia.
2. Differentiation of micro- and megasporangiophores from each other, with drastic reduction of spore number in the latter.
3. Surrounding of the megasporangium with an envelope derived by modification of flattened sterile branches, as in *Lyginopteris* and many other Paleozoic seeds (Walton, 1953). The sporangium wall thus becomes the nucellus.
4. Grouping of these primitive seeds into a bowl-like cupule, as in the Carboniferous genus *Calathospermum* (Walton, 1953).
5. Reduction in size of cupules, and their arrangement on short lateral sporangiophores, as in the Caytoniales and their relatives.
6. Adnation of these sporangiophores to subtending bracts, as in several glossopteridean genera.
7. Reduction of the primary ovules to one per cupule, transforming the cupule wall into the outer integument.
8. Infolding of the bract to form the closed carpel, with the sterile tip of the bract becoming transformed into the stigma.
9. Adnation of the vascular strands of the sporangiophore to the ventral strands of the carpel wall, to give the typical three-strand carpel found in most of the apocarpous modern angiosperms.

The Significance of Cycles of Transference of Function

If the hypothesis just summarized is correct, then we can trace a succession of transfers of the function of protecting the megasporangium or embryo sac. The original protection was afforded by the megasporangial wall, which later became the nucellus. This function was then taken over by the ovule integument and later by the cupule wall, which then became the outer ovular integument. With the origin of the angiosperms, protection of the ovules became the function of the carpel or ovary wall. Within the angiosperms themselves, further transference of function took place in certain evolutionary lines, in which protection was provided, successively, by a synsepalous calyx, floral bracts, and modified leaves or bracts homologous to those which in less advanced forms subtend the inflorescence (Stebbins, 1970).

The fact must be emphasized that such cycles of transference of function took place only in certain specialized lines, both in preangiospermous plants and in the angiosperms themselves. In many other lines of evolution, the original protective structures continued to function in the same way in both unspecialized and highly specialized groups. In the eusporangiate ferns, the original sporangial wall has retained this function throughout the group. In the angiosperms, even such highly specialized flowers as those of the Balsaminaceae, Gentianaceae, and Juncaceae retain the original condition of ovular protection by the ovary wall.

We must consider, therefore, the particular environmental conditions which would favor the continuance of these cycles of transfer of function. The conclusions which I reached in a previous publication (Stebbins, 1970) can be summarized here. Cycles of transfer of function are most likely to proceed in groups that are primarily adapted to montane habitats having a variety of soil types, and in semiarid or arid climates. Their kaleidoscopic pattern of edaphic conditions and local habitats is subjected to marked changes during periods of orogeny, erosion, or overall changes in the total availability of moisture. At different stages of each cycle, selection will favor first a relatively small number of large, well-protected seeds, and then a larger number of relatively small seeds, depending upon the seasonal cycle and the type of soil to be colonized. Consequently, adaptation to changes of this kind will favor groups that are capable of undergoing this cycle. The ancestors of the angiosperms most probably belonged to such groups.

The Significance of the Angiosperm Reproductive Cycle and Flexibility of Growth

All of the characteristics in which the angiosperm reproductive cycle differs from those of other seed plants are associated with shortening of the time required from floral differentiation to seed maturity. First is the reduction of the female gametophyte to an eight-celled structure that can develop rapidly. Second is the rapid growth of the pollen tube down the stigma and style, often through tissues specialized for this purpose. Third is the fusion of the polar nuclei and double fertilization. This produces a triploid endosperm, which in an outcrossed population will be heterozygous at many loci, due to differences between the parental gametes. This endosperm is, therefore, doubly equipped for rapid growth. The triploid provides a large number of templates from which RNA molecules can be synthesized, and so makes possible more rapid protein synthesis. The added heterozygosity is likely to confer on the tissue the advantage of heterosis or hybrid vigor. The rapid growth of endosperm tissue in angiosperms is a well-known phenomenon (Maheshwari, 1950). Finally, except for the anomalous genus *Welwitschia*, angiosperms are the only seed plants in which the zygote develops directly into the embryo, without going through a coenocytic proembryo stage. All of these characteristics would have maximum adaptive value in regions with a highly seasonal climate. In the majority of angiosperms, an additional characteristic is the tricolpate pollen grain, which contributes to the rapidity of pollen-grain germination.

As many botanists have pointed out (Cronquist, 1968; Takhtajan, 1969), the great majority of angiosperm families are adapted to tropical or subtropical conditions, and adaptations to seasonal frost or freezing temperatures are secondary. Consequently, the seasonal climates that favored the evolution of angiosperms must have been those having alternating wet and dry seasons. Such climates are as widely distributed in tropical and subtropical regions as are climates that are continually moist. I believe, therefore, that the angiosperms first evolved in semiarid tropical climates or tropical climates having seasonal drought. On the basis of this hypothesis, the concentration in the moist tropics of relictual families such as the woody Ranales, having many characters which botanists regard as primitive, is a secondary phenomenon. The reasons why I believe this are explained in detail in a recent publication (Stebbins, 1974).

Finally, we must ask the question, What conditions would have provided the strongest selective pressures in favor of the flexibility of growth that characterizes the angiosperms and is responsible for much of their success? The seasonal variation already postulated could be partly responsible for such pressures. In addition, flexibility of seedling growth (particularly great cell elongation and active intercalary meristems) has a very high adaptive value in habitats that are subject to sudden shifts in the soil, which may temporarily bury the young seedling and from which it must emerge. Two kinds of habitats are particularly subject to such changes: loose talus slopes and the banks of swiftly flowing streams. I suggest that the initial success of the angiosperms may well have resulted from their ability to evolve adaptations to such habitats.

In this connection, the phylogenetic position of the angiosperms which now occupy such habitats is relevant. They belong, for the most part, to highly specialized groups. This is true, for instance of such stream-bank families as the Salicaceae, and genera such as

Vitis, Baccharis, Nerium, and *Carex.* Talus slopes are occupied by plants belonging to specialized families such as the Fumariaceae, Cruciferae, Compositae, and Gramineae. When the earliest angiosperms evolved, they were the most progressive vascular plants in existence. We might expect, therefore, that they were early colonizers of such habitats and have been rendered extinct by their more efficient descendants. Talus slopes and stream banks can support only a few species, so that when a more efficient adaptation to these habitats evolves, it is likely to drive out the less-adapted species with which it competes.

On the basis of this hypothesis, I believe that the modern angiosperms that many botanists regard as primitive are actually side lines which became secondarily adapted to more stable and permissive habitats, such as rain forests and floodplains. They are therefore archaic rather than truly primitive. The primitive ancestors of the angiosperms are, in my opinion, all extinct, and no significant fossils have been discovered that might reveal the nature of their reproductive structures. Hence we cannot tell what they were really like.

Summary

Hypotheses about the origin and early evolution of angiosperms are suggested on the basis of distinctive characters affecting the ecology of seeds and seedlings. Most of these characters represent adaptations for increasing the efficiency with which angiosperms exploit their environment. Exceptions are the presence of two ovular integuments and the anatropous position of the ovule in all angiosperms regarded as primitive or archaic. These structures suggest that the angiosperm ovule is homologous to the cupule of Mesozoic pteridosperms such as the Caytoniales and Corystospermaceae. The reduced condition of the female gametophyte, the increased efficiency of pollen-tube growth, double fertilization, rapid development of endosperm, and the elimination of a coencytic proembryo stage prior to embryo development can all be regarded as resulting from selective pressures which favor shortening the reproductive cycle. Along with other evidence, they suggest that angiosperms evolved in tropical or subtropical climates having seasonal drought. The flexibility of growth in angiosperms, which is achieved both by a greater capacity for cellular elongation and by the frequent appearance and activity of intercalary meristems, may have originated as a mechanism for colonizing talus slopes, stream banks, or both of these unstable habitats.

References

Bate-Smith, E. C. 1972. Chemistry and phylogeny of angiosperms. *Nature* 236: 353–54.

Burtt, B. L. 1960. Compositae and the study of functional evolution. *Trans. Proc. Bot. Soc. Edinburgh* 39: 216–32.

Canright, J. E. 1960. The comparative morphology and relationships of the Magnoliaceae. III. Carpels. *Amer. J. Bot.* 47: 145–55.

Cronquist, A. 1968. *The Evolution and Classification of Flowering Plants.* Houghton, Boston.

Davis, G. L. 1966. *Systematic Embryology of the Angiosperms.* Wiley, New York.

Florin, R. 1944. Die Koniferen des Oberkarbons und des unteren Perms. *Paleontographica* 85: 365–654.

Gaussen, H. 1946. *Les Gymnospermes, Actuelles et Fossiles.* vol. 2, sec. 1. Travaux du Laboratoire Forestier, Toulouse.

Harris, T. M. 1951. The relationships of the Caytoniales. *Phytomorphology* 1: 29–33.

Janzen, D. H. 1969. Seed-eaters versus seed size, number, toxicity and dispersal. *Evolution* 23: 1–27.

Khan, R. 1950. A case of twin ovules in *Isomeris arborea. Current Sci (India)* 19: 326.

Kubitzky, K. 1972. Probleme der Gross-Systematik der Blütenpflanzen. *Ber. Deut. Bot. Ges.* 85: 259–77.

Maheshwari, P. 1950. *The Embryology of Angiosperms.* McGraw, New York.

Maneval, W. E. 1914. *The Development of Magnolia and Liriodendron, including a Discussion of the Primitiveness of the Magnoliaceae.* McGraw, New York.

Narayana, H. S. 1962. Studies in the Capparidaceae, I. The embryology of *Capparis decidua* (Forsk.) Pax. *Phytomorphology* 12: 167–77.

Rao, V. S. 1938. Studies on the Capparidaceae. III. Genus *Capparis. J. Indian Bot. Soc.* 17: 69–80.

Ridley, H. N. 1930. *The Dispersal of Plants throughout the World.* L. Reeve, Ashford, England.

Sastri, R. L. N. 1958. Floral morphology and embryology of some Dilleniaceae. *Bot. Notiser* 111: 495–511.

Stebbins, G. L. 1970. Transference of function as a factor in the evolution of seeds and their accessory structures. *Israeli J. Bot.* 19: 59–70.

Stebbins, G. L. 1974. *Flowering Plants: Evolution above the Species Level.* Harvard Univ. Press, Cambridge.

Swamy, B. G. L. 1949. Further contributions to the anatomy of the Degeneriaceae. *J. Arnold Arbor.* 30: 10–38.

Takhtajan, A. L. 1969. *Flowering Plants: Origin and Dispersal.* Oliver, Edinburgh.

Thomas, H. H. 1933. On some pteridospermous plants from the Mesozoic rocks of South Africa. *Phil. Trans. Roy. Soc. London* 222B: 193–265.

Thomas, H. H. 1955. Mesozoic pteridosperms. *Phytomorphology* 5: 177–85.

Thomas, H. H. 1958. *Lidgettonia,* a new type of fertile *Glossopteris. Bull. Brit. Mus. Nat. Hist.* (Geol.) 3: 179–89.

Van Heel, W. A. 1970. Distally lobed integuments in some angiosperm ovules. *Blumea* 18: 67–70.

Van der Pijl, L. 1969. *Principles of Dispersal in Higher Plants.* Springer, Berlin.

Walton, J. 1953. The evolution of the ovule in pteridosperms. *Advan. Sci.* 10: 223–30.

Character Correlations among Angiosperms and the Importance of Fossil Evidence in Assessing their Significance

KENNETH R. SPORNE, ***The Botany School University of Cambridge, England***

THIS PAPER PRESENTS a summary of a continuing saga that started almost exactly 60 years ago. The opening chapters were written by Sinnott and Bailey in a series of papers published under the general heading, "Investigations on the phylogeny of the angiosperms." Having examined the nodal anatomy of some 400 genera of dicotyledons, belonging to 164 families, Sinnott (1914) realized that the trilacunar condition "characterizes most of the members of the Piperales, Salicales, Myricales, Juglandales and Fagales—in short, of the former Amentiferae; and is present in the great majority of the Ranales and Rosales as well." He then went on to suggest that "since in all likelihood one of these great groups approaches the primitive angiosperms (or at any rate the primitive dicotyledons) in its character, we may feel reasonably sure that the trilacunar condition became fixed in the angiosperm line very far back." Accordingly, he referred to it thereafter as "the ancient type." Sinnott and Bailey (1914a,b) then showed that the majority of plants of this type have stipules, whereas almost all plants with unilacunar nodes are without them. From this correlation they concluded that a leaf with two distinct stipules is the more ancient type.

It should be noted that up to this point in their argument no reference was made to the fossil record. However, in the fifth paper in the series, Sinnott and Bailey (1915) did, indeed, refer to numerous lists of fossil dicotyledon leaves that had been found in Cretaceous and Tertiary deposits and showed that palmate simple leaves were more abundant in past ages than at the present day. This conclusion did not depend upon the correct assignment of such remains to particular genera or families, since the point at issue concerned merely the overall shape of the leaf, but Sinnott and Bailey were aware of two possible sources of error when interpreting the fossil record. One arises from the fact that leaflets from a compound leaf might not be recognized as such and might be regarded as simple leaves. The other arises from a possible bias toward the preservation of leaves from woody plants, the leaves of herbaceous plants being "generally more delicate and less apt to be preserved in the fossil state than are the tougher ones of trees and shrubs." Quite apart from this difference in texture, there is a further possible reason for such a bias, which has only recently become apparent. Bate-Smith and Metcalfe (1957) showed that tannins occur more frequently in the leaves of woody plants than in those of herbs, and we now know (Hart and Hillis, 1972) that tannins, especially ellagitannins, have fungistatic properties which tend to prevent fungal decay.

Having looked at fossil leaves of dicotyledons, Sinnott and Bailey then showed that a correlation exists between the possession of nodes with more than one lacuna and the possession of palmate simple leaves, a correlation which is much stronger among woody plants than among herbaceous ones. Furthermore, they showed that palmate leaves are more frequent in relatively primitive groups (e.g., Ranales, Rosales, and Malvales) than in more advanced ones.

Bailey and his colleagues then turned their attention to wood characters. Bailey and Tupper (1918) examined the timber of 263 genera belonging to 118 families of dicotyledons, in addition to that of many gymnosperms, and recorded maximum, minimum, and average lengths of the various wood elements. At the same time, they observed the nature of the lignification of the wood elements, and they recognized four categories of vessel types among dicotyledons, distinguished by the nature of their end-plates: I, mainly scalariform; II, intermediate between scalariform and porous; III, oblique porous; IV, transverse porous. Correlations were demonstrated between the type of end-plate, the kind of pitting on the sidewalls of the vessels, and the average length of the vessel elements. Thus, vessels with scalariform end-plates tend to have scalariform-pitted sidewalls and to be longer than other kinds of vessel elements. The decision that evolution has proceeded from type I to type IV was based on a comparison with the wood of gymnosperms, whose wood elements were shown to be much longer than those of angiosperms. Fossil wood was included in this survey, but only that of gymnosperms, such as Cordaitales and Bennettitales; no fossil angiosperm wood was examined.

Record (1919) was interested in the occurrence of a "storied" arrangement of wood elements. After examining 200 species belonging to 86 genera (but only 19 families), he observed that there seems to be a correlation between storied arrangement and vessels with transverse perforations and pitted sidewalls. In the following year, Bailey (1920) showed that the size of tracheary elements is correlated with the size of the fusiform initials in the cambium from which they develop, and then went on (1923) to show that stratified cambium (and therefore storied wood) is correlated with short cambial initials.

In a series of three papers, Frost (1930a,b, 1931) returned to the study of the vessel, quoting numerous percentages in support of correlations between scalariform vessel end-plates, long and narrow vessel elements, angularity of outline of vessels in transverse section, uniformly thickened vessel walls, highly inclined end-plates, diffuse arrangement of vessels among the other wood elements, scalariform lateral pitting, bordered pores in the vessel end-plates, and bordered pits in their sidewalls. Throughout this work, no facts were given concerning the wood of fossil angiosperms, and the only use that Frost made of the fossil record was to compare, as did Bailey and Tupper, the average lengths of the vessel elements of dicotyledons and of Gnetales with the average lengths of the tracheids of gymnosperms (including Cordaitales and Bennettitales). This comparison was vital to the argument, however, since it allowed Frost to apply the "method of association," according to which, if it is agreed that vessel elements evolved from tracheids (which preceded them in the fossil record), then those vessel elements which are most like tracheids are primitive. This provided the clue as to the correct application of the "method of sequences," since it indicated the direction in which one should read the morphological series of vessel types.

Very similar thought processes were used by Kribs (1935) in his study of wood rays; having demonstrated correlations between heterogeneous rays and long vessel elements with oblique scalariform end-plates, he concluded that such wood rays are more primitive than homogeneous rays. Two years later, Kribs (1937) applied similar arguments in relation to the distribution of wood parenchyma. He suggested that diffuse parenchyma is the most primitive type because it is correlated with primitive vessel types, whereas vasicentric parenchyma is the most advanced.

Up to this point, none of the correlations that had been claimed had been subjected to strict statistical tests of significance. The work of Chalk (1937) marked an important change in attitude in two respects: not only did he use χ^2 tests of significance, but he also made use of fossil angiosperm wood. His primary object was to discover the extent to which the various systematic classifications of dicotyledons (by Bentham and Hooker, Engler, and Hutchinson) were in keeping with details of wood anatomy. To simplify his analysis, he chose one wood character as typical of primitive wood and one as typical of advanced wood. The primitive one was the possession of vessel elements with scalariform perforation plates. This choice was supported by a comparison of present-day species with fossil dicotyledonous woods listed by Edward (1931) (but excluding Pleistocene forms, as being too recent), for such vessel elements were shown to be more frequent in the past than they are at present. The character which Chalk chose as being advanced was the presence of storied structure. His conclusion was that Hutchinson's arrangement of the Archichlamydeae agrees more closely with the evidence obtainable from wood anatomy than does Engler's. The Metachlamydeae (Sympetalae), he found, included a mixture of primitive and advanced wood, but he was able to separate them on the basis of the type of parenchyma present: "the woods of Engler's Primulales, Plumbaginales, Contortae (part), Tubiflorae and Campanulatae have paratracheal parenchyma and constitute a very highly specialized group."

Chalk's paper had just been published when I was encouraged by Hamshaw Thomas to use similar methods in analyzing the occurrence of other characters (floral, vegetative, and biochemical), to discover whether they, too, are correlated. The first decision was whether to investigate the angiosperms as a whole or to treat the dicotyledons separately from the monocotyledons. The second course was adopted, in the hope that ultimately a comparison of the two separate groups might throw some light on their interrelationships. Another decision that had to be made was which statistical test of significance to use. I was advised to apply the simple 2×2 contingency test, using χ^2 to estimate the value of P (the probability that the observed association has occurred fortuitously) and to set the level of significance at $P = .02$. A start was made on the dicotyledons, but World War II intervened and the first results did not appear until 1949.

In my first survey, I was able to demonstrate correlations between 12 characters: woody habit, glandular leaves, alternate leaves, stipulate leaves, unisexual flowers, actinomorphic flowers, free petals, numerous stamens, numerous carpels, arillate seeds, ovules with two integuments, integuments with vascular bundles. Out of 66 possible pairings among these 12 characters, there were 37 positive correlations (in which $P \leqslant .02$) and no negative ones. It is easy to show that because primitive characters are not distributed at random among the families of the world but tend (by definition) to be "clumped" among primitive families, they should be expected to show statistical correlation. But equally, because advanced characters are clumped among advanced families (by definition), they too should be expected to show statistical correlation. Thus, although there is good reason to believe that correlated characters may be indicators of evolutionary status, there is still the question whether they are primitive or advanced. There is only one way to resolve it: to look at the fossil record. Otherwise, one is forever involved in circular arguments.

When Sinnott and Bailey made use of the fossil record, they were able to compare actual fossil leaves with modern leaves, and their correct assignment to a genus or family was relatively unimportant. For the purposes of my analysis, however, correct identification was vital, for the comparison was to be made between the flora of the world and those families known (or at any rate believed) to have been in existence by certain specified times in the past. Thus, despite the existence of many lists of fossil angiosperms based on leaf identifications, these had to be avoided because such identifications were notoriously unreliable. Instead, like Chalk, I made use of Edwards's list of pre-Pleistocene woody families, and I also used the list of families whose fruits and seeds had been identified by Reid and Chandler (1933) from the London Clay, of Eocene age.

Of the 12 correlated characters listed earlier, 8 proved to be more abundant among London Clay families than they are in the present-day world flora. Five proved to be more abundant among pre-Pleistocene woody families than among woody families of the present-day world flora. I took this to indicate that the 12 correlated characters are more likely to be primitive than advanced. However, I was well aware of the slender nature of the evidence. A statement that London Clay families tend to have a high proportion with stipules did not imply that *fossil* stipules had been observed and found to be more numerous. It referred only to *modern* families whose fruits or seeds had been identified in the London Clay. It could therefore be said, in criticism, that the involvement of the fossil record was only peripheral.

Several critics were quick to point out other weaknesses in the argument. Turrill (1950) emphasized that the two lists of fossil dicotyledons certainly did not constitute a representative sample of past floras of the world. He also questioned whether the conclusion that the correlated characters are primitive would still be valid if the dicotyledons should prove to be polyphyletic instead of monophyletic.

Stebbins (1951) criticized the interpretation of correlated characters as being either primitive or advanced. Using essentially the same statistical methods, he made a study similar to mine of character correlations, but among angiosperms as a whole instead of just the dicotyledons. His main concern was with floral characters, namely, apetaly, sympetaly, zygomorphy, reduced number of stamens, syncarpy, reduced number of ovules, parietal placentation, and epigyny. Taking these in pairs, and applying the χ^2 test of significance, he demonstrated 15 positive correlations and 2 negative ones. He further showed that, with the exception of "apetaly" and "reduced number of ovules," these characters are all negatively correlated with woodiness.

These correlations agreed fairly well with mine, but he argued that it is wrong to consider pairs of characters in this way. Instead, we ought to look at combinations ("syndromes") of several characters that occur more frequently than others. These he called "adaptive peaks." Clearly, he was correct in suggesting that some combinations of characters would be more efficient than others. Indeed, some other combinations would be ridiculous. The characters that Stebbins showed to be positively correlated could all be involved in pollination mechanisms or in seed dispersal. It does not follow, however, that efficient combinations must necessarily be advanced. It is still possible for the 12 primitive characters listed by me to have made "efficient" kinds of dicotyledons. Indeed, they do so in tropical rain forests, as we shall see later.

In due course, as further facts were assembled, more and more characters were added. "Nuclear endosperm" and "free carpels," (or, if fused, then with "axile placentation") were added in 1954. Bate-Smith and Metcalfe (1957) provided information on the occurrence of leuco-anthocyanins in leaves. The publication by Metcalfe and Chalk (1950) of their two-volume work, *Anatomy of the Dicotyledons*, gave an opportunity to check the earlier claims of Bailey and his associates concerning wood characters. Complete support was given to their conclusion that scalariform vessel end-plates, scalariform vessel sidewalls, apotracheal parenchyma, and unstoried wood are primitive characters. However, wood ray characters gave no clear results. The publication by Davis (1966) of *Systematic Embryology of the Angiosperms* provided information that allowed "binucleate pollen" and "crassinucellate ovule" to be added to the list (Sporne, 1969).

This brought the total of correlated characters to 22, conveniently referred to as "magnolioid" characters, since a large proportion of them are present in the Magnoliaceae. Of these, all except one (possession of unisexual flowers) were then used to make a rough assessment of the evolutionary status of each of Engler's 259 families. By assigning scores to each family, I calculated an "advancement index," ranging from 21% (the most primitive) to 100% (the most advanced). It was immediately clear that although the Magnoliaceae are very primitive they are no more so than the Rhizophoraceae, Flacourtiaceae, Dilleniaceae, Malvaceae, Myrtaceae, Ochnaceae, and Rosaceae. At the other extreme there are several families with a score of 100: Callitrichaceae, Hippuridaceae, Hydrostachyaceae, and Phrymaceae.

By then, several people had suggested that correlations between characters are no more than a reflection of functional interdependence between them. This might, indeed, apply to the three groups of characters enclosed in the small square boxes in figure 1. These are, respectively, wood characters, floral characters, and ovule characters. Within each of these groups there is a high degree of positive correlation. Of the 6 possible pairings between wood characters, 5 are significant; of the 14 possible pairings between floral characters, 10 are significant; and of the 10 possible pairings between ovule characters, 9 are significant. However, there can surely be no functional interdependence between such characters as the presence of stipules and the possession of nuclear endosperm. I continued to argue, therefore, that a correlation between two such characters, separated as these are both in time and space during the growth of a plant, is much more likely to arise because both characters are primitive.

Wood (1970) suggested yet another reason for the correlations. He argued that if each character, by itself, were to confer an advantage in a particular environment, then this would result in their being statistically correlated. The environment that he suggested was tropical rain forest, and he showed that the average advancement index of families recorded from areas of rain forest in Guiana, Uganda, and Borneo is lower than the average for the world flora. The difference is, indeed, highly significant, as I was able to show with the aid of a mathematical colleague (whose help was needed because the distribution of the advancement index among Engler's families is skew). This association with the tropical rain forest does not, however, prevent magnolioid characters from being primitive. But the reader must realise that, at this stage in my work, the only fossil evidence that had been invoked was that which Chalk (1937) had used. Over the intervening years, my philosophical edifice had been growing steadily, with more and more "girders" bolted into place, like those of the Eiffel Tower. However, unlike the Eiffel Tower, my edifice was in a state of unstable equilibrium, for lack of firm anchorage in the rocks.

Such a foundation was provided by Muller (1970), when he spoke at the Eleventh International Botanical Congress on "Palynological evidence on early differentiation of angiosperms." As a result of many years' work describing and identifying pollen grains from many parts of the world, he was able to provide a historical sequence from the Upper Cretaceous to the present day. The evidence that he provided can be used in two quite different ways.

The first of these arises from the fact that Muller was able to establish the order of appearance in the fossil record of different pollen types. The earliest pollen grains were sulcate and were followed by types with increasing numbers of apertures. If we look at those modern families with sulcate grains, we find that their average advancement index is 44 (compared with 56.6 for the world flora). Families with three pores have an average of 59, and those with three colpi 54. Families with polyporate grains average 62, and those with polycolpate grains 64. These figures show beyond reasonable doubt that the advancement index is indeed an indicator of evolutionary advancement and, furthermore, that the 21 correlated magnolioid characters used in its assessment are primitive.

An analysis of modern families shows that sulcate pollen is correlated with four magnolioid characters, and "pauci-aperturate" pollen (i.e., grains with fewer than four apertures) is correlated with nine characters. Accordingly, pauci-aperturate pollen is added to the list, bringing the total to 23.

The second way in which Muller's findings can be used arises from the fact that he was able to recognize many of the families whose pollen grains he had described. This result makes possible a comparison of the floral and vegetative characters of the current world flora with those of families known to have existed in early times. Nine out of the 23 magnolioid characters occurred in a significantly higher proportion of pre-Oligocene families than of present-day families.

A similar list of identifications of fossil angiosperms, compiled by Chesters, Gnauck, and Hughes (1967), gives strong support to these findings. If we compare their list of pre-Tertiary families of dicotyledons with the present world flora, we find that 11 magnolioid characters occurred in a significantly higher proportion then than now.

Only one character was less abundant in Muller's pre-Oligocene families than today: pauci-aperturate pollen. At first sight, this is rather surprising, but there is a simple explanation. Muller emphasizes the difficulty of assigning sulcate pollen grains to any particular family. It follows, therefore, that families with this kind of pollen will be underrepresented in the list of pre-Oligocene families. Equally, therefore, families with pauci-aperturate pollen will be underrepresented, since such pollen is defined so as to include sulcate pollen.

The deposits from which Muller's samples were taken were all probably tropical at the time of deposition, so perhaps they ought to be compared with present tropical floras instead of with the world flora. Such a comparison shows that eight magnolioid characters are more abundant in pre-Oligocene families than in present-day tropical floras. There is also the suspicion that rain-forest families might be overrepresented in these deposits and thus lead to a biased result. A comparison needs to be made, therefore, between pre-Oligocene families and present-day rain-forest families. The numbers with which we are now dealing are very small, but nevertheless there are two characters which are more abundant in pre-Oligocene families than in modern rain-forest families: apotracheal parenchyma and pleiomerous stamens.

Yet another possible source of error needs to be considered. Muller and the palynologists whose work he quotes were better able to identify some pollen grains than others; and this is coupled with the fact that the pollen of some families is, for various reasons, unlikely ever to find its way into geological deposits. A comparison needs to be made, therefore, between Muller's pre-Oligocene families and the complete list of families that he has been able to identify up to the present day. Such a comparison shows that three magnolioid characters are more abundant in pre-Oligocene families than in families identified in later deposits: apotracheal parenchyma, pleiomerous stamens, and ovules with two integuments.

Similar precautions need to be taken with the list of dicotyledons recorded by Chesters, Gnauck, and Hughes. Six magnolioid characters are significantly more abundant in pre-Tertiary families than in families from later deposits: woody habit, glandular leaves, presence of leuco-anthocyanins, unisexual flowers, binucleate pollen, and crassinucellate ovules.

It is now clear that, regardless of any special adaptations that magnolioid characters might confer on plants inhabiting rain forests, these characters are also primitive. Using a concept suggested by Stebbins (1972), we can describe rain forests as "museums" where archaic dicotyledons are on show for the tropical botanist to see. Without evidence from the rocks, however, we could not have been sure that they are indeed archaic.

Figure 1 lists all the correlations that have been demonstrated and includes the character "glandular tapetum," which brings the total to 24. In this figure, each symbol + or − represents a correlation whose significance (at the level of $P \leqslant .02$) has been calculated by the use of χ^2. In this respect it differs from a similar figure (Sporne, 1973) in which the bottom four rows of correlations were as-

Character Correlations in Dicotyledons	1 Woody habit	2 Scalariform end-plates	3 Scalariform side-walls	4 Apotracheal parenchyma	5 Unstoreyed wood	6 Leaves alternate	7 Stipules present	8 Secretory cells ..	9 Leuco-anthocyanins	10 Flowers unisexual	11 Flowers actinomorphic	12 Petals free	13 Stamens pleiomerous	14 Carpels pleiomerous	15 Carpels free	16 Axile placentation	17 Pollen binucleate	18 Seeds arillate	19 Two integuments	20 Integument bundles	21 Ovules crassinucellate	22 Endosperm nuclear	23 Pollen pauciaperturate	24 Tapetum glandular
1 Woody habit		+		■	■	+		+	+	+	+	+	+	+		+	+		+		+	+	+	+
2 Scalariform end-plates	+		+	+	+				+		+						+						+	
3 Scalariform side-wall		+			+					+		+				+								
4 Apotracheal parenchyma	■	+			+	+			+		+		+				+			+		−	+	
5 Unstoreyed wood	■	+	+	+						+													+	
6 Leaves alternate	+			+									+					+				+		
7 Stipules present								+	+			+	+	+		+		+	+		+	+		
8 Secretory cells ..	+						+		+		+	+	+	+					+	+	+	+		
9 Leuco-anthocyanins	+	+		+			+	+		+	+	+	+				+		+	+	+	+		
10 Flowers unisexual	+		+		+				+		+										+			
11 Flowers actinomorphic	+	+		+				+	+	+		+	+	+					+		+		+	+
12 Petals free	+		+				+	+	+		+		+	+		+		+	+		+	+	+	+
13 Stamens pleiomerous	+			+		+	+	+	+		+	+		+	+			+	+	+	+	+		
14 Carpels pleiomerous	+						+	+			+	+	+		+	+		+	+		+		+	
15 Carpels free													+	+		■			+				+	
16 Axile placentation	+		+				+					+		+	■								+	
17 Pollen binucleate	+	+		+					+															+
18 Seeds arillate						+	+					+	+	+					+	+	+	+		
19 Two integuments	+						+	+	+		+	+	+	+	+			+			+	+		
20 Integument bundles				+				+	+				+					+			+	+		
21 Ovules crassinucellate	+						+	+	+	+	+	+	+	+				+	+	+		+		
22 Endosperm nuclear	+			−		+	+	+	+			+	+					+	+	+	+			
23 Pollen pauciaperturate	+	+		+	+						+	+		+	+	+								
24 Tapetum glandular											+	+					+							
Pre-Oligocene (Muller)	+	+		+			+	+					+						+	+		+	−	
Pre-Tertiary (CGH)	+			+		+	+	+	+	+							+			+	+	+		
Rain-forest	+				−		+	+	+	+				+		+		+		+		+		

Fig. 1. Correlations among 24 characters in dicotyledons, and also between these characters and their occurrence in pre-Oligocene, pre-Tertiary and present-day rain forests. (The symbol + indicates a positive correlation at the level of significance $P \leqslant .02$, as determined by a χ^2 test, while the symbol − indicates a negative one. An empty square indicates absence of correlation. A black square indicates a meaningless correlation.)

sessed for significance by a different method and were, therefore, not strictly comparable with the rest of the figure.

The bottom line of figure 1 shows which particular characters out of the 24 are more frequent among rain-forest families than among the families of the world. The first of these (woody habit) comes as no surprise, for the rain-forest environment is created to a large extent by woody plants. Three characters, "stipules," "secretory cells" (i.e., glandular leaves), and "leuco-anthocyanins," all relate to leaves. It is not easy to imagine what special advantages the first two might confer on rain-forest plants, but there is a strong possibility that the fungistatic properties of tannins might be important in an environment so conducive to fungal growth.

Ellagitannins are probably even more important in this respect than leuco-anthocyanins, and are regarded by Bate-Smith as a primitive feature in dicotyledons. Having examined some 193 families, Bate-Smith (1962, amended in a personal communication, 1973) records ellagitannins in 57 families. These have an average advancement index of 47 (as compared with 56.6 for the world flora) and "presence of ellagitannins" shows singificant positive correlations with 5 of the 24 magnolioid characters. Bate-Smith suggests that the most primitive dicotyledons have both leuco-anthocyanins and ellagitannins in their leaves. This suggestion is supported by the fact that such families are more frequent among Muller's pre-Oligocene families than among the families of the present-day world flora, by their having an average advancement index of 43, and by their having a higher proportion with apotracheal parenchyma and secretory cells.

The possession of many carpels by a high proportion of rain-forest families could conceivably be bound up with the production of large fruits, but it is not clear why these should confer any greater advantage in rain forests than in many other parts of the world. Likewise, it is not clear why axile placentation should be particularly advantageous in rain forests. Arils, vascularized integuments, and nuclear endosperm all relate to the seed and may well be advantageous. There can be little doubt of the importance of the aril (Corner, 1964). The two other characters are associated with large size in seeds; and large seed size permits the production of large seedlings, which may be important in the rain-forest environment.

The predictive value of the advancement index (crude though it may be) has now been confirmed several times. Preliminary calculations of an advancement index showed that correlations might be expected with the two latest additions to the list of primitive characters and also with the possession of ellagitannins. Such calculations also indicated which lists of fossil dicotyledons would be worth more detailed analysis.

One of my colleagues, Lowe, carried out a similar investigation of the monocotyledons (1961). Her main results are summarized in fig-

Character Correlations in Monocotyledons	1 Perianth present	2 Perianth more than five	3 Perianth homoiochlamydeous	4 Perianth actinomorphic	5 Perianth members free	6 Flowers hermaphrodite	7 Stamens six or more	8 Carpels three or more	9 Ovary inferior	10 Ovules many	11 Endosperm helobial or cellular	12 Pollen binucleate
1 Perianth present		■	■	■	■	+		+	+	+		
2 Perianth more than five	■					+		+	+	+		
3 Perianth homoiochlamydeous	■			+	+	+	+	+				+
4 Perianth actinomorphic	■		+		+	+	+	+				+
5 Perianth members free	■		+	+		+	+	+			+	−
6 Flowers hermaphrodite	+	+	+	+	+			+	+	+		+
7 Stamens six or more			+	+	+			+				
8 Carpels three or more	+	+	+	+	+	+	+		+	+		
9 Ovary inferior	+	+				+		+		+		+
10 Ovules many	+	+				+		+	+			
11 Endosperm helobial or cellular					+							
12 Pollen binucleate			+	+	−	+			+			

Fig. 2. Correlations among 12 characters in monocotyledons. (The symbols have the same meaning as in table 1.)

ure 2; out of 62 possible pairings between 12 characters, 32 show a positive correlation (at the level of significance $P \leqq .02$) and only 1 is negative. Since the majority of these characters occur in the Amaryllidaceae, they are conveniently referred to as "amarylloid" characters.

The monocotyledons are much more difficult to analyze statistically than are the dicotyledons, for several reasons. One is that there are fewer families (and Lowe therefore used the subfamily as the statistical unit, instead of the family). Wood characters are, of course, lacking, as are any clear-cut vegetative distinctions, with the result that almost all of Lowe's characters concern the flower or the fruit. The correlations which she demonstrates could therefore be interpreted merely as a reflection of functional interdependence. However, Lowe believes that the 12 amarylloid characters are primitive, and she bases this belief on the work of Cheadle (1942, 1953), who recognized four categories of monocotyledons: (1) families with no vessels; (2) those with vessels restricted to the roots; (3) mixed families, of which some members have vessels restricted to the roots and some have vessels in other parts of the plant; (4) families with vessels throughout the plant. Cheadle showed that the vessels of plants in group 2 have chiefly scalariform perforations, while those in groups 3 and 4 have progressively higher proportions with simple perforations. From this, he concluded that the four groups represent stages in an evolutionary series. The fact that families in group 1 have a higher proportion of amarylloid characters than do those in group 4 encouraged Lowe to regard these characters as primitive rather than advanced. She then used the 12 amarylloid characters to calculate an advancement index for each family.

The fossil record was (and still is) of little help in establishing that the advancement index so calculated is really an indicator of evolutionary advance. Muller's list of pre-Oligocene angiosperms includes only three monocotyledon families. The list of pre-Tertiary angiosperms drawn up by Chesters, Gnauck, and Hughes includes nine monocotyledon families, but their average advancement index is almost exactly the same as that for the world flora of monocotyledons.

Although not strictly comparable, most of the amarylloid characters in monocotyledons correspond reasonably well to the magnolioid characters in dicotyledons. In addition, Lowe found 1 correlation supporting the idea that crassinucellate ovules are primitive in monocotyledons, as they are in dicotyledons (but she did not use this character in calculating advancement indices). However, the 1 significant correlation shown by "helobial or cellular endosperm" is at variance with the 11 positive correlations shown by "nuclear endosperm" in dicotyledons.

Some years ago (Sporne, 1956), in a discussion of the various forms that a phylogenetic classification might take, I suggested a hypothetical circular scheme, in which primitive groups would be placed near the center and more advanced ones nearer the circumference. It was only recently, however, that sufficient confidence could be placed in the advancement index of dicotyledons as a guide to evolutionary status to make worthwhile any further development of this idea.

In Figure 3, the various dicotyledon orders suggested by Cronquist (1968) have been arranged so as to reflect not only his views on closeness of affinity but also the range of advancement index of the families within each order. Thus, the six orders belonging to the Magnoliidae are grouped together within an area bounded by a bro-

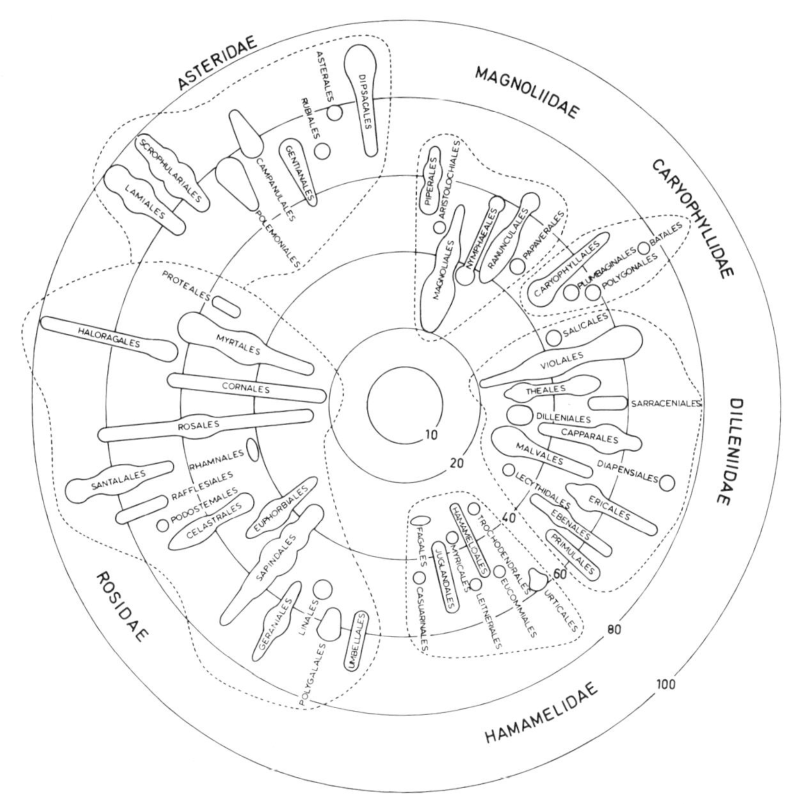

Fig. 3. Phylogenetic classification of the dicotyledons. The relationships are those suggested by Cronquist (1968). The radial extent of each order corresponds to the range of advancement indices of its constituent families.

ken line, and the Magnoliales are shown as extending from 20 (the advancement index of the Austrobaileyaceae) to 56 (the advancement index of the Hernandiaceae). Such a scheme brings out several important points. One can see at a glance that three major groups, Magnoliidae, Dilleniidae, and Rosidae, all have very primitive members. Another important point is that some orders, Violales, Sapindales, Cornales, and (especially) Rosales, cover a very wide range from primitive to relatively advanced. Finally, every member of the Asteridae has an advancement index higher than the average (56.6) for the dicotyledons as a whole.

A similar scheme has been constructed, but with less confidence, for the monocotyledons (fig. 4). Again, it becomes immediately apparent that two orders, Najadales and Restionales, cover a wide range from primitive to advanced. The most primitive monocotyle-

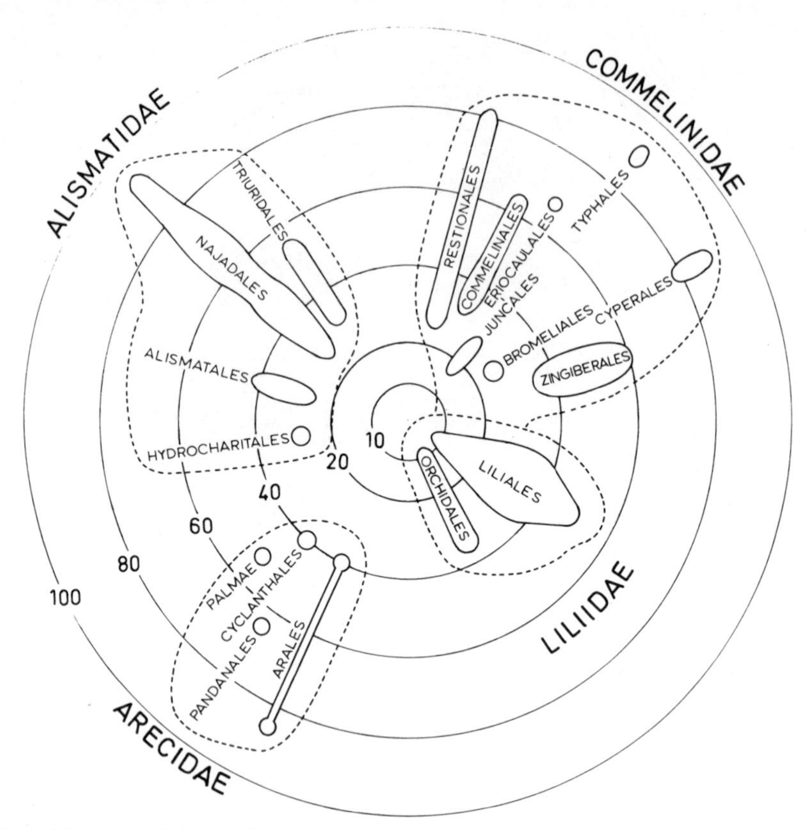

Fig. 4. Phylogenetic classification of the monocotyledons.

dons are the Amaryllidaceae and Velloziaceae (both in the Liliales), together with the Corsiaceae (in the Orchidales).

Perhaps the most important point of all is that no single order, and no single family, is near enough to the center of the dicotyledon scheme to be regarded as the common ancestor of the dicotyledons. Equally, no single order or family is near enough to the center of the monocotyledon scheme to be regarded as the common ancestor of the monocotyledons. The way in which the two circular schemes stand in relation to each other and to possible ancestors is illustrated in figure 5, where *X* and *Y* represent figures 3 and 4, respectively. Each can be thought of as forming part of a vast spherical surface, representing "now" in time. The "tree of evolution" extends back in time toward the center of the sphere and consists of organisms that lived in the past. Information concerning the past can only come from the fossil record. For the time being, details of the trunk and

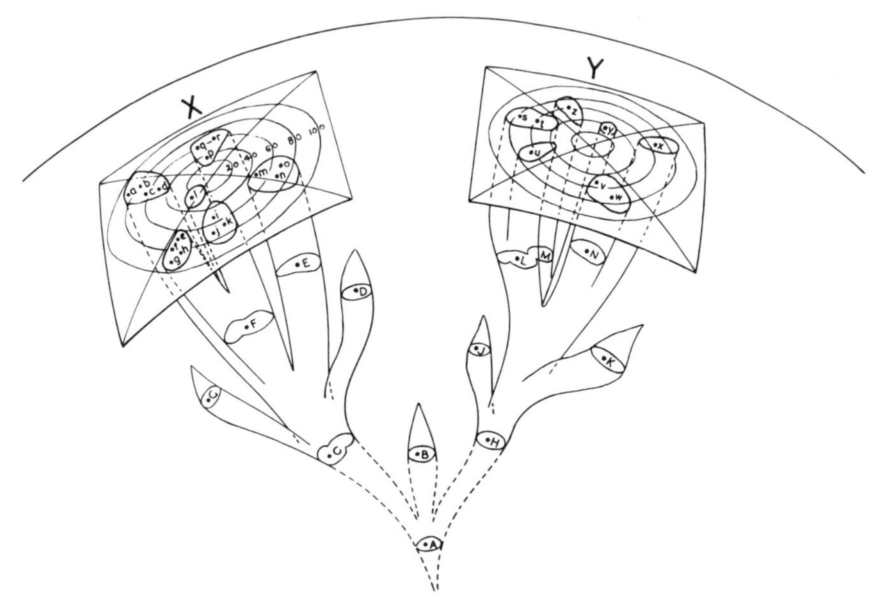

Fig. 5. Hypothetical phylogenetic classification of two related taxonomic groups *X* and *Y*, showing some of their fossil relatives. Present-day organisms are indicated by the lower-case letters *a* to *z*; fossils are indicated by the capital letters *A* to *N*. Each of the two taxonomic groups is disposed on a square which forms part of the surface of a large sphere, representing 'now' in time; concentric circles represent relative advancement (as in figs. 3 and 4). (From Sporne, 1959; reproduced by permission of the editors of *Amer. J. Bot.*)

branches of the tree of evolution of angiosperms are quite obscure, but already the fossil record (and especially that of pollen grains) has played a vital role in the understanding of those of its ultimate twigs which are displayed in the circular classification of the dicotyledons.

References

Bailey, I. W. 1920. The cambium and its derivative tissues. II. Size variations of cambial initials in gymnosperms and angiosperms. *Amer. J. Bot.* 7: 355–67.

Bailey, I. W. 1923. The cambium and its derivative tissues. IV. The increase in girth of the cambium. *Amer. J. Bot.* 10: 499–509.

Bailey, I. W., and Tupper, W. W. 1918. Size variations in tracheary cells. I. A comparison between the secondary xylems of vascular cryptogams, gymnosperms and angiosperms. *Proc. Amer. Acad. Arts Sci.* 54: 149–204.

Bate-Smith, E. C. 1962. The phenolic constituents of plants and their taxonomic significance. I. Dicotyledons. *J. Linn. Soc.*, Bot. 58: 95–173.

Bate-Smith, E. C., and Metcalfe, C. R. 1957. Leuco-anthocyanins. 3. The nature and systematic distribution of tannins in dicotyledonous plants. *J. Linn. Soc.*, Bot. 55: 669–705.

Chalk, L. 1937. The phylogenetic value of certain anatomical features of dicotyledonous woods. *Ann. Bot. (London)* 1: 409–27.

Cheadle, V. I. 1942. The occurrence and types of vessels in the various organs of the plant in the Monocotyledonae. *Amer. J. Bot.* 29: 441–50.

Cheadle, V. I. 1953. Independent origin of vessels in the monocotyledons and dicotyledons. *Phytomorphology* 3: 23–44.

Chesters, K. I. M., Gnauck, F. R., and Hughes, N. F. 1967. Angiospermae. In *The Fossil Record,* ed. W. B. Harland et al., pp. 269–88. Geological Society, London.

Corner, E. J. H. 1964. *The Life of Plants.* Weidenfeld, London.

Cronquist, A. 1968. *The Evolution and Classification of Flowering Plants.* Nelson, London.

Davis, G. L. 1966. *Systematic Embryology of the Angiosperms.* Wiley, New York.

Edwards, W. N. 1931. *Fossilium Catalogus.* II, *Plantae.* Pars 17: *Dicotyledonae (Ligna).* Berlin.

Frost, F. H. 1930a. Specialization in secondary xylem of dicotyledons. I. Origin of vessel. *Bot. Gaz.* 89: 67–94.

Frost, F. H. 1930b. Specialization in secondary xylem of dicotyledons. II. Evolution of end wall of vessel segment. *Bot. Gaz.* 90: 198–212.

Frost, F. H. 1931. Specialization in secondary xylem of dicotyledons. III. Specialization of lateral wall of vessel segment. *Bot. Gaz.* 91: 88–96.

Hart, J. H., and Hillis, W. E. 1972. Inhibition of wood-rotting fungi by ellagitannins in the heart wood of *Quercus alba. Phytopathology* 62: 620–26.

Kribs, D. A. 1935. Salient lines of structural specialization in the wood rays of dicotyledons. *Bot. Gaz.* 96: 547–57.

Kribs, D. A. 1937. Salient lines of structural specialization in the wood parenchyma of dicotyledons. *Bull. Torrey Bot. Club* 64: 177–87.

Lowe, J. 1961. The phylogeny of monocotyledons. *New Phytol.* 60: 355–87.

Metcalfe, C. R., and Chalk, L. 1950. *Anatomy of the dicotyledons.* 2 vols. Clarendon Press, Oxford.

Muller, J. 1970. Palynological evidence on early differentiation of angiosperms. *Biol. Rev. Cambridge Phil. Soc.* 45: 417–50.

Record, S. J. 1919. Storied or tier-like structure of certain dicotyledonous woods. *Bull. Torrey Bot. Club* 46: 253–73.

Reid, E. M., and Chandler, M. E. J. 1933. *The London Clay Flora.* British Museum (Natural History), London.

Sinnott, E. W. 1914. Investigations on the phylogeny of the angiosperms. 1. The anatomy of the node as an aid in the classification of angiosperms. *Amer. J. Bot.* 1: 303–22.

Sinnott, E. W., and Bailey, I. W. 1914a. Investigations on the phylogeny of the angiosperms. 3. Nodal anatomy and the morphology of stipules. *Amer. J. Bot.* 1: 441–53.

Sinnott, E. W., and Bailey, I. W. 1914b. Investigations on the phylogeny of the angiosperms. 4. The origin and dispersal of herbaceous angiosperms. *Ann. Bot. (London)* 28: 547–600.

Sinnott, E. W., and Bailey, I. W. 1915. Investigations on the phylogeny of the angiosperms. 5. Foliar evidence as to the ancestry and early climatic environment of the angiosperms. *Amer. J. Bot.* 2: 1–22.

Sporne, K. R. 1949. A new approach to the problem of the primitive flower. *New Phytol.* 48: 259–76.

Sporne, K. R. 1956. The phylogenetic classification of the angiosperms. *Biol. Rev. Cambridge Phil. Soc.* 31: 1–29.

Sporne, K. R. 1959. On the phylogenetic classification of plants. *Amer. J. Bot.* 46: 385–94.

Sporne, K. R. 1969. The ovule as an indicator of evolutionary status. *New Phytol.* 68: 555–66.

Sporne, K. R. 1973. The survival of archaic dicotyledons in tropical rainforests. *New Phytol.* 72: 1175–84.

Stebbins, G. L. 1951. Natural selection and the differentiation of angiosperm families. *Evolution* 5: 299–324.

Stebbins, G. L. 1972. Ecological distribution centers of major adaptive radiation in angiosperms. In *Taxonomy, Phytogeography and Evolution*, ed. D. H. Valentine, pp. 7–34. Academic Press, London.

Turrill, W. B. 1950. Modern trends in the classification of plants. *Advan. Sci.* 26: 238–53.

Wood, D. 1970. The tropical forest and Sporne's advancement index. *New Phytol.* 69: 113–15.

Index

Index